METHODS IN MOLECULAR BIOLOGY

For further volumes:
http://www.springer.com/series/7651

For over 35 years, biological scientists have come to rely on the research protocols and methodologies in the critically acclaimed *Methods in Molecular Biology* series. The series was the first to introduce the step-by-step protocols approach that has become the standard in all biomedical protocol publishing. Each protocol is provided in readily-reproducible step-by-step fashion, opening with an introductory overview, a list of the materials and reagents needed to complete the experiment, and followed by a detailed procedure that is supported with a helpful notes section offering tips and tricks of the trade as well as troubleshooting advice. These hallmark features were introduced by series editor Dr. John Walker and constitute the key ingredient in each and every volume of the *Methods in Molecular Biology* series. Tested and trusted, comprehensive and reliable, all protocols from the series are indexed in PubMed.

Transgenic Plants

Methods and Protocols

Edited by

Nitish Kumar

Department of Biotechnology, Central University of South Bihar, Gaya, Bihar, India

Humana Press

Editor
Nitish Kumar
Department of Biotechnology
Central University of South Bihar
Gaya, Bihar, India

ISSN 1064-3745 ISSN 1940-6029 (electronic)
Methods in Molecular Biology
ISBN 978-1-0716-4802-5 ISBN 978-1-0716-4803-2 (eBook)
https://doi.org/10.1007/978-1-0716-4803-2

This Humana imprint is published by the registered company Springer Science+Business Media, LLC, part of Springer Nature.
The registered company address is: 1 New York Plaza, New York, NY 10004, U.S.A.

Preface

To introduce natural resistance genes into plants and control various plant diseases, conventional plant breeding techniques have been widely employed. Since sexual incompatibility often prevents the introduction of one or more genes (transgene) from one plant to another, plant transformation techniques based on DNA recombinant technology can also be used to create the transformed/transgenic plant. In addition to providing the potential for genetic diversity, these recombination techniques also save time by preventing crossbreeding, enable targeted crop improvements while maintaining the desirable traits of an inbred line, and produce new kinds of resistance genes that are not found in nature.

The aim of this book is to inculcate the reader on a variety of commonly used, widely applicable, and readily reproducible methods used in the development of transgenic plants. Its detailed methodology includes a number of techniques for genetically altering plant tissues and cells as well as for obtaining entire transgenic plants from them. The creation of new vectors, the use of chloroplast engineering, and the most recent advancements in this field, such as plant genome editing, are also covered.

The development of transgenic crop plants, their application for improving agriculture, environment, ecology, and health, and their sociopolitical effects are currently significant areas of study, industry, and education. Policymakers, social activists, and funding and regulatory organizations are also interested in these topics. It is an invaluable resource for scientists, educators, students, and industrialist working in the sector, with general discussions and research papers by top international specialists.

Gaya, Bihar, India — ***Nitish Kumar***

Acknowledgments

First and foremost, I would like to praise and thank God, the almighty, who has granted countless blessing, knowledge, and opportunity to accomplish the book project work. Thanks to all the authors of the various chapters for their contributions. It had been a bit of a long process from the initial outlines to developing the full chapters and then revising them in the light of reviewer comments. I sincerely acknowledge the authors' willingness to go through this process. I also acknowledge the work and knowledge of the members of our review panels, many of which had to be done on short notice. Thanks to all the people at Springer Nature, especially Dr. John Walker and Patrick Marton with whom I corresponded for their advice and facilitation in the production of this book. I am grateful to my family members Kiran (Wife), Kartika Sharma and Laavanya Sharma (Daughters), and parents for their incredible and selfless support all the time.

Contents

Contributors

SWAPNA ALEX • *Department of Molecular Biology and Biotechnology, College of Agriculture, Thiruvananthapuram, Kerala, India*

M. T. ANDREW-PETER-LEON • *S. Thangapazham Agricultural College, (Affiliated to Tamil Nadu Agricultural University), Tenkasi, Tamil Nadu, India*

THEEYAN ANURADHA • *Department of Molecular Biology and Biotechnology, College of Agriculture, Thiruvananthapuram, Kerala, India*

SEKAR ASHOK • *ICAR-Indian Agricultural Research Institute, PUSA Campus, New Delhi, Delhi, India*

SHIVANI BHUTANI • *Department of Biotechnology, Invertis University, Bareilly, Uttar Pradesh, India*

ARJUN CHAUHAN • *Department of Biotechnology, College of Horticulture, Dr Y S Parmar University of Horticulture and Forestry, Solan, Himachal Pradesh, India; Department of Biotechnology, Institute of Applied Sciences & Humanities, GLA University, Mathura, Uttar Pradesh, India*

SELVI CHELLAMUTHU • *Department of Agricultural Entomology, Tamil Nadu Agricultural University, Coimbatore, Tamil Nadu, India*

UPPU ANITHA DEVI • *Department of Botany, Indira Priyadarshini College for Women, Hyderabad, Telangana, India*

NISHA DINKAR • *Department of Biotechnology, Khandelwal College of Management Science and Technology, Bareilly, Uttar Pradesh, India*

SUDHAKAR DURAIALAGARAJA • *Department of Plant Biotechnology, Centre for Plant Molecular Biology and Biotechnology, Tamil Nadu Agricultural University, Coimbatore, Tamil Nadu, India*

KOKILADEVI ESWARAN • *Department of Plant Biotechnology, Centre for Plant Molecular Biology and Biotechnology, Tamil Nadu Agricultural University, Coimbatore, Tamil Nadu, India*

AJITH THALIYAKKATTIL GIRIJAN • *Department of Plant Science, Central University of Kerala, Kasaragod, Kerala, India*

RAJADURAI GOTHANDARAMAN • *Department of Plant Biotechnology, Centre for Plant Molecular Biology and Biotechnology, Tamil Nadu Agricultural University, Coimbatore, Tamil Nadu, India*

DHANDAPANI GURUSAMY • *Department of Botany, Kongunadu Arts and Science College (Autonomous), Bharathiar University, Coimbatore, Tamil Nadu, India*

JULIUS JEYAPRAKASAM • *Department of Plant Sciences, School of Biological Sciences, Madurai Kamaraj University, Madurai, Tamil Nadu, India*

SHRINITHI KANNAN • *Department of Plant Sciences, School of Biological Sciences, Madurai Kamaraj University, Madurai, Tamil Nadu, India*

VASANTH KRISHNAN • *Department of Botany, Bharathiar University, Coimbatore, Tamil Nadu, India*

ABHISHEK KUMAR • *Department of Biotechnology, College of Horticulture, Dr Y S Parmar University of Horticulture and Forestry, Solan, Himachal Pradesh, India; School of Agricultural Biotechnology, Punjab Agricultural University, Ludhiana, Punjab, India*

KRISH K. KUMAR • *Department of Plant Biotechnology, Centre for Plant Molecular Biology and Biotechnology, Tamil Nadu Agricultural University, Coimbatore, Tamil Nadu, India*

SATISH KUMAR • *ICAR-National Research Centre on Plant Biotechnology, New Delhi, Delhi, India*

RAVIKUMAR SUNITHAKUMARI LEKSHMI • *Rajiv Gandhi Center for Biotechnology, Thiruvananthapuram, Kerala, India*

ARUL LOGANATHAN • *Department of Plant Biotechnology, Centre for Plant Molecular Biology and Biotechnology, Tamil Nadu Agricultural University, Coimbatore, Tamil Nadu, India*

MANJU MODGIL • *Department of Biotechnology, College of Horticulture, Dr Y S Parmar University of Horticulture and Forestry, Solan, Himachal Pradesh, India*

KANAKACHARI MOGILICHERLA • *ICAR-Indian Institute of Rice Research (IIRR), Hyderabad, Telangana, India*

RESHMA ANJUM MOHAMMED • *Sri PadmavatiMahilaVisvavidyalayam, Tirupati, Andhra Pradesh, India*

SANKARI MOHANA • *Sri PadmavatiMahilaVisvavidyalayam, Tirupati, Andhra Pradesh, India*

GOLLA NARASIMHA • *Sri Venkateshwara University, Tirupati, Andhra Pradesh, India*

JASDEEP CHATRATH PADARIA • *ICAR-National Research Centre on Plant Biotechnology, New Delhi, Delhi, India*

MAHITHA KOONJANGADAN PADINJHAREVEEDU • *Department of Plant Science, Central University of Kerala, Kasaragod, Kerala, India*

DHARAVATH PARVATHI • *Department of Botany, SRR Government Arts and Science College (A), Karimnagar, India*

M. ARUMUGAM PILLAI • *Department of Plant Breeding and Genetics, V.O.C. Agricultural College and Research Institute, Tamil Nadu Agricultural University, Killikulam, Vallanadu, Tamil Nadu, India*

VIGNESH PONNURANGAN • *Department of Plant Biotechnology, Centre for Plant Molecular Biology and Biotechnology, Tamil Nadu Agricultural University, Coimbatore, Tamil Nadu, India; Department of Plant Pathology, Centre for Plant Protection Studies, Tamil Nadu Agricultural University, Coimbatore, Tamil Nadu, India*

SUMIT PUROHIT • *Regional Centre, Uttarakhand Council for Biotechnology, Nainital, Uttarakhand, India*

SAKTHI AMBOTHI RATHNASAMY • *Department of Plant Biotechnology, Centre of Plant Molecular Biology and Biotechnology (CPMBB), Tamil Nadu Agricultural University, Coimbatore, Tamil Nadu, India*

A. SHEENA SABATINA • *ICAR-National Research Centre on Plant Biotechnology, New Delhi, Delhi, India*

LOVELESH SINGH SAINI • *Department of Life Sciences, Parul Institute of Applied Sciences, Parul University, Vadodara, Gujarat, India*

BALAJI SANTHAKUMAR • *Department of Plant Biotechnology, Centre for Plant Molecular Biology and Biotechnology, Tamil Nadu Agricultural University, Coimbatore, Tamil Nadu, India*

JASMINE M. SHAH • *Department of Plant Science, Central University of Kerala, Kasaragod, Kerala, India*

TUTIKURLA SHANKAR • *Department of Botany, Govt Degree College, Siricilla, Telangana, India*

VARANAVASIAPPAN SHANMUGAM • *Department of Plant Biotechnology, Centre for Plant Molecular Biology and Biotechnology, Tamil Nadu Agricultural University, Coimbatore, Tamil Nadu, India*

ABHISHEK SHARMA • *Department of Biotechnology and Bioengineering, Institute of Advanced Research, Gandhinagar, Gujarat, India*

SENTHILKUMAR SHRICHARAN • *ICAR-Indian Agricultural Research Institute, PUSA Campus, New Delhi, Delhi, India*

MANALI SINGH • *Department of Life Sciences, Parul Institute of Applied Sciences, Parul University, Vadodara, Gujarat, India*

JANANI SREE SIVAKUMAR • *Department of Botany, Bharathiar University, Coimbatore, Tamil Nadu, India*

KALLAMPARAMBIL BALAKRISHNAN SONI • *Department of Molecular Biology and Biotechnology, College of Agriculture, Vellayani, Thiruvananthapuram, Kerala, India*

TIRUKOVELA SRINIVAS • *Department of Botany, SRR Govt Arts & Science College, Karimnagar, Telangana, India*

RAJASEKARAN SRIPRIYA • *Department of Zoology, St. Mary's College, Thoothukudi, Tamil Nadu, India*

VELUSAMY SUMITHRA • *Department of Plant Breeding and Genetics, V.O.C. Agricultural College and Research Institute, Tamil Nadu Agricultural University, Killikulam, Vallanadu, Tamil Nadu, India*

JEBASINGH TENNYSON • *Department of Plant Sciences, School of Biological Sciences, Madurai Kamaraj University, Madurai, Tamil Nadu, India*

MASOUD TOHIDFAR • *Department of Cell & Molecular Biology, Faculty of Life Sciences & Biotechnology, Shahid Beheshti University, Tehran, Iran*

THIRUNAHARI UGANDHAR • *Department of Botany, Indira Priyadarshini Government Degree College for Women (A), Hyderabad, India*

JESHIMA KHAN YASIN • *Division of Genomic Resources, ICAR-National Bureau Plant Genetic Resources, PUSA Campus, New Delhi, India*

Chapter 1

Agrobacterium-Mediated Genetic Transformation of Indica Rice

Rajadurai Gothandaraman, Balaji Santhakumar, Vignesh Ponnurangan, Varanavasiappan Shanmugam, Krish K. Kumar, Arul Loganathan, Kokiladevi Eswaran, and Sudhakar Duraialagaraja

Abstract

Functional genomics in rice has progressed significantly due to the availability of high-quality, well-annotated genome data and the characterization of numerous genes through genetic manipulation. Over the years, various protocols for genetic transformation in rice have been developed and refined. However, different rice cultivars respond differently to these protocols, there remains potential for further improvement and optimization. In this chapter, we have provided an efficient transformation protocol with critical points for the transformation of *indica* rice with rice immature embryos.

Key words *Agrobacterium*-mediated, Indica rice, Rice transformation, Immature embryos, Protocol

1 Introduction

Rice (*Oryza sativa* L.) is among the most important crops globally, serving as the primary food source for over 2.5 billion people and the second most cultivated crop worldwide. The importance of rice as a staple food is observed because more than 50% of the global population and 80% of the population in Asia rely on it as a major component of their daily intake [1, 2]. It is grown on approximately 162 million hectares, with an annual production exceeding 756 million tonnes [3]. Numerous indica rice varieties are cultivated across India, but their production is frequently affected by various biotic and abiotic stresses. A key challenge remains to achieve a sustainable increase in yield to meet the demands of the growing population [4].

Nitish Kumar (ed.), *Transgenic Plants: Methods and Protocols*, Methods in Molecular Biology, vol. 2973, https://doi.org/10.1007/978-1-0716-4803-2_1,

Genetic engineering and genome editing of crops are a vital approach for investigating gene functions, advancing functional genomics, and facilitating crop improvement by modifying genetic pathways. *Agrobacterium tumefaciens*-mediated transformation is widely used for genetic manipulation in numerous plant species, as it enables the stable integration of single-copy, unrearranged sequences into plant genomes. In rice, an effective method for *Agrobacterium*-mediated transformation was established over a decade ago and has since become a standard protocol in many laboratories [4–8]. This method offers several advantages, including high transformation efficiency, the integration of a limited number of T-DNA copies into chromosomes, the transfer of relatively large DNA segments with defined ends, and minimal T-DNA rearrangement during transformation. Therefore, this method quickly gained more popularity in rice compared to other transformation techniques [9]. Recent studies predominately use either immature embryos or calli derived from mature seeds as explants for *Agrobacterium*-mediated transformation in rice [7–14]. While immature embryos tend to have a higher transformation efficiency, mature seeds are more convenient due to their availability in large quantities and long-term viability when stored in laboratory conditions.

In the present chapter, we have reported a more effective and reproducible *Agrobacterium*-mediated transformation protocol for indica rice using immature embryos.

2 Materials

2.1 Explant Materials

1. Prepare fields for both rice nursery and main crops (field maintained at Paddy Breeding Station, Department of Rice, Tamil Nadu Agricultural University, Coimbatore, Tamil Nadu, India).
2. Select good-quality mature rice seeds (ASD16 and CO51 varieties were used in this study), and sow in the nursery field at weekly intervals (to facilitate a continuous supply of immature embryos throughout year).
3. After 20 days of sowing, transplant the seedlings in the main crop field with 20 × 15 cm spacing and apply the recommended dose of fertilizers (120:48:48 kg N:P:K per ha).
4. Maintain the field with pest and disease-free conditions. Apply recommended chemicals to manage pest and disease infestations. If yellow-pigmented bacterial infection is caused by *Xanthomonas oryzae* pv. *oryzae* occurs, it can be controlled with a spray of streptomycin sulfate antibiotics.

2.2 Agrobacterium *Culture*

Agrobacterium tumefaciens strains LBA4404 and C58C1 were used in this study. *Agrobacterium* culture was streaked on LB Agar media with appropriate antibiotics and incubated for 48 hrs at 28 °C incubator in dark conditions (*see* **Note 1**). For our experiment, we have used 10 mg/L rifampicin and 50 mg/L kanamycin for LBA4404 and 10 mg/ L rifampicin, 50 mg/L kanamycin, 100 mg/L carbenicillin, 300 mg/L streptomycin, and 100 mg/L spectinomycin for C58C1.

2.2.1 LB Agar Media

Dissolve 10 g tryptone, 5 g yeast extract powder, and 10 g $NaCl_2$ in 900 mL of distilled water. Adjust the pH to 6.8 to 7.0, add 20 g bacteriological agar, and autoclave at 121 °C for 15 min. Cool down to a warm temperature (around 50 °C), and add appropriate antibiotics (as per the plasmids and *Agrobacterium* cultures used). Mix it well and pour 40 mL of media into each Petri plate (sterile—90 × 15 mm size) and allow for solidification for 20 min in aseptic conditions (in laminar air flow chamber). Wrap it with klin film, and store the media plates below 25 °C in dark conditions.

2.2.2 AB Minimal Media

Mix 50 mL of 20X AB salt stock with 50 mL of 20X AB buffer, and add 850 mL of distilled water. Add 14 g of Agarose Type I and autoclave at 121 °C for 15 min. Cool down to a warm temperature (around 50 °C), and add 50 mL of filter-sterilized 1.1 g/mL glucose solution, and appropriate antibiotics (as per the plasmids and *Agrobacterium* cultures used). Mix it well and pour 40 mL of media into each Petri plate (sterile—90 × 15 mm size), and allow for solidification for 20 min in aseptic conditions (in laminar air flow chamber). Wrap it with klin film, and store the media plates below 25 °C in dark conditions.

2.3 Stock Solutions

2.3.1 Major Salts

1. *N6 major salts (10X)*: Dissolve 28.3 g KNO_3, 4.63 g $(NH4)_2SO_4$, 1.66 g $CaCl_2.2H_2O$, 1.85 g $MgSO_4.7H_2O$, and 4.0 g KH_2PO_4 in 900 mL of sterile distilled water, and make up the final volume into 1000 mL. Maintain the stock in refrigerated conditions at 4 °C.
2. *MS major salts (10X)*: Dissolve 19.3 g KNO_3, 16.5 g NH_4NO_3, 4.4 g $CaCl_2.2H_2O$, 3.7 g $MgSO_4.7H_2O$, and 1.7 g KH_2PO_4 in 900 mL of sterile distilled water, and make up the final volume into 1000 mL. Maintain the stock in refrigerated conditions at 4 °C.
3. *CC major salts (10X)*: Dissolve 12.12 g KNO_3, 6.4 g NH_4NO_3, 5.88 g $CaCl_2.2H_2O$, 2.47 g $MgSO_4.7H_2O$, and 1.36 g KH_2PO_4 in 900 mL of sterile distilled water, and make up the final volume into 1000 mL. Maintain the stock in refrigerated conditions at 4 °C.
4. *AA major salts (20X)*: Dissolve 59.0 g KCl, 3.0 g $CaCl_2.2H_2O$, 10.0 g $MgSO_4.7H_2O$, and 3.0 g NaH_2PO_4.

H_2O in 900 mL of sterile distilled water, and make up the final volume into 1000 mL. Maintain the stock in refrigerated conditions at 4 °C.

5. *AB salts (20X)*: Dissolve 20 g NH_4Cl, 6.0 g $MgSO_4.7H_2O$, 3.0 g KCl, 265 mg $CaCl_2.2H_2O$, and 50 mg $FeSO_4.7H_2O$ in 900 mL of sterile distilled water, and adjust pH to 7.0. Make up the final volume into 1000 mL. Maintain the stock in refrigerated conditions at 4 °C.
6. *AB buffer (20X)*: Dissolve 20 g NaH_2PO_4 and 60 g K_2HPO_4 in 900 mL of sterile distilled water, and adjust pH to 7.0. Make up the final volume into 1000 mL. Maintain the stock in refrigerated conditions at 4 °C.

2.3.2 Minor Salts

1. *MS minor salts (100X)*: Dissolve 2.23 g $MnSO_4.4H_2O$, 1.06 g $ZnSO_4.7H_2O$, 2.5 mg $CuSO_4.5H_2O$, 25 mg $Na_2MoO_4.2H_2O$, 2.5 mg $CoCl_2.6H_2O$, 0.62 g H_3BO_3, and 83 mg KI in 900 mL of sterile distilled water, and make up the final volume into 1000 mL. Maintain the stock in refrigerated conditions at 4 °C.
2. *CC minor salts (100X)*: Dissolve 1.115 g $MnSO_4.4H_2O$, 0.576 g $ZnSO_4.7H_2O$, 2.5 mg $CuSO_4.5H_2O$, 24 mg $Na_2MoO_4.2H_2O$, 2.8 mg $CoSO_4.7H_2O$, 0.31 g H_3BO_3, and 83 mg KI in 900 mL of sterile distilled water, and make up the final volume into 1000 mL. Maintain the stock in refrigerated conditions at 4 °C.
3. *B5 minor salts (100X)*: Dissolve 1.32 g $MnSO_4.4H_2O$, 0.20 g $ZnSO_4.7H_2O$, 2.5 mg $CuSO_4.5H_2O$, 25 mg $Na_2MoO_4.2H_2O$, 2.5 mg $CoCl_2.6H_2O$, 0.30 g H_3BO_3, and 75 mg KI in 900 mL of sterile distilled water, and make up the final volume into 1000 mL. Maintain the stock in refrigerated conditions at 4 °C.
4. *FeEDTA (100X)*: Dissolve 2.78 g $FeSO_4.7H_2O$ in 900 mL of hot sterile distilled water and add 3.73 g of Na_2EDTA. Make up the final volume into 1000 mL with sterile water. Maintain the stock in refrigerated conditions at 4 °C.
5. *CuSO4 (100 mM)*: Dissolve 49.7 mg of $CuSO_4.5H_2O$ in 9 mL of sterile distilled water and make up the final volume into 10 mL. Filter-sterilize the solution with a syringe filter (0.22 μm) and store at 4 °C.

2.3.3 Vitamins

1. *B5 vitamins (100X)*: Dissolve 10 g myoinositol, 1 g thiamine hydrochloride, 100 mg pyridoxine hydrochloride, and 100 mg nicotinic acid in 900 mL of sterile distilled water, and make up the final volume into 1000 mL. Maintain the stock in refrigerated conditions at 4 °C.

2. *Modified MS vitamins (100X)*: Dissolve 10 g myoinositol, 100 mg thiamine hydrochloride, 50 mg pyridoxine hydrochloride, 50 mg nicotinic acid, and 200 mg glycine in 900 mL of sterile distilled water, and make up the final volume into 1000 mL. Maintain the stock in refrigerated conditions at 4 °C.
3. *CC vitamins (100X)*: Dissolve 9 g myoinositol, 0.85 g thiamine hydrochloride, 100 mg pyridoxine hydrochloride, 600 mg nicotinic acid, and 200 mg glycine in 900 mL of sterile distilled water, and make up the final volume into 1000 mL. Maintain the stock in refrigerated conditions at 4 °C.

2.3.4 Growth Hormones

1. *2,4-Dichlorophenoxy acetic acid (2,4-D) (100 mg/L)*: Weigh 25 mg of 2,4-D in a sterile microcentrifuge tube, and dissolve it by adding 2 mL of 100% ethanol. Drop by drop, add the dissolved solution into a 250 mL bottle containing 248 mL of double autoclaved sterile distilled water. Maintain the stock in refrigerated conditions at 4 °C.
2. *1-Naphthaleneacetic acid (NAA) (100 mg/L)*: Weigh 25 mg of NAA in a sterile microcentrifuge tube, and dissolve it by adding 2 mL 1 N NaOH. Transfer the dissolved solution into a 250 mL Duran bottle containing 248 mL of double autoclaved sterile distilled water. Maintain the stock in refrigerated conditions at 4 °C.
3. *6-Benzylaminopurine (6-BA) (100 mg/L)*: Weigh 25 mg of 6-BA in a sterile microcentrifuge tube, and dissolve it by adding 2 mL 1 N NaOH. Transfer the dissolved solution into a 250 mL Duran bottle containing 248 mL of double autoclaved sterile distilled water. Maintain the stock in refrigerated conditions at 4 °C.

2.3.5 Antibiotics

1. *Hygromycin B (50 mg/mL)*: Dissolve 1 g of hygromycin B in 20 mL of double sterilized distilled water. Filter-sterilize with 0.22 μm membrane filter and aliquot it in 1.5 mL Eppendorf tubes, and store it at −20 °C.
2. *Cefotaxime sodium (250 mg/mL)*: Dissolve 250 mg of cefotaxime sodium in 1 mL of double sterilized distilled water. Prepare cefotaxime sodium (250 mg/mL) antibiotic stocks freshly, and use it for media preparation.
3. *Kanamycin (50 mg/mL)*: Dissolve 500 mg of kanamycin acid sulfate in 10 mL of double sterilized distilled water. Filter-sterilize with 0.22 μm membrane filter and aliquot it in 1.5 mL Eppendorf tubes and store it at −20 °C.
4. *Rifampicin (20 mg/mL)*: Dissolve 200 mg of rifampicin in 10 mL of DMSO and aliquot it in 1.5 mL Eppendorf tubes, and store it at −20 °C.

5. *Spectinomycin (100 mg/mL)*: Dissolve 1 g of Spectinomycin dihydrochloride pentahydrate in 10 mL of double sterilized distilled water. Filter-sterilize with 0.22 μm membrane filter, aliquot it in 1.5 mL Eppendorf tubes, and store it at −20 °C.
6. *Streptomycin (300 mg/mL)*: Dissolve 3 g of streptomycin sulfate in 10 mL of double sterilized distilled water. Filter-sterilize with 0.22 μm membrane filter, aliquot it in 1.5 mL Eppendorf tubes, and store it at −20 °C.
7. *Carbenicillin (100 mg/mL)*: Dissolve 1 g of carbenicillin disodium salt in 10 mL of double sterilized distilled water. Filter-sterilize with 0.22 μm membrane filter, aliquot it in 1.5 mL Eppendorf tubes, and store it at −20 °C.

2.3.6 Other Stocks

1. *Amino acids (10X)*: Dissolve 8.76 g L-glutamine, 2.66 g aspartic acid, 1.74 g arginine, and 75 mg glycine in 900 mL of sterile distilled water, and make up the final volume into 1000 mL. Adjust pH to 5.8 and filter-sterilize with 0.22 μm membrane filter. Maintain the stock in refrigerated conditions at 4 °C.
2. *Acetosyringone (ACS) (100 mM)*: Dissolve 392.4 mg acetosyringone in 20 mL of 50% dimethyl sulfoxide (DMSO). Maintain the stock at below 4 °C.
3. *L-Glutamine (30 g/L)*: Dissolve 30 g L-glutamine in 900 mL of sterile distilled water, and make up the final volume into 1000 mL. Filter-sterilize with 0.22 μm membrane filter, and maintain the stock in refrigerated conditions at 4 °C.
4. *Glucose (1.1 g/mL)*: Dissolve 110 g glucose in 50 mL of sterile distilled water, and make up the final volume into 100 mL. Filter-sterilize with 0.22 μm membrane filter, and maintain the stock in refrigerated conditions at 4 °C.

2.4 Other Chemicals

1. Yeast extract powder.
2. Sodium chloride.
3. Tryptone.
4. Agar.
5. Agarose Type I.
6. Phytagel.
7. Gelrite.
8. Sucrose—molecular biology grade.
9. Glucose.
10. Proline.
11. Casein hydrolysate/vitamin assay casamino acids.
12. Maltose.

13. Mannitol.
14. Myoinositol.

2.5 Culture Media for Plant Transformation

1. *AA Infection medium*: Mix the stocks of 50 mL 20X AA major salts, 10 mL 100X FeEDTA, 10 mL 100X B5 minor salts, 10 mL 100X B5 vitamins, 100 mL 10X AA amino acids, and 1 mL 100 mM acetosyringone to 500 mL of distilled water. Add 20 g sucrose, 10 g glucose, and 0.5 g vitamin assay casamino acids in the mixture, and mix it well. Make up the volume into 1000 mL by using distilled water. Adjust pH to 5.2 and filter-sterilize with 0.22 μm membrane filter. Aliquot it in 1.5 mL microcentrifuge tube, and store them in the freezer at 4 °C. Before use, check one tube for contamination if any by incubating it at room temperature (>25 °C) for 3 days.
2. *Cocultivation medium (NB-As)*: Mix the stock solutions of 100 mL 10X N6 major salts, 10 mL 100X Fe EDTA, 10 mL 100X B5 vitamins, 20 mL of 100 mg/L 2,4-D, 10 mL of 100 mg/L NAA, and 10 mL of 100 mg/L 6-BA to 500 mL distilled water. To this mixture add and dissolve 20 g sucrose, 10 g glucose, 0.5 g proline, and 0.5 g vitamin assay casamino acids, and make up the volume into 1000 mL using distilled water. Adjust pH to 5.2 and add 8 g of agarose Type I. Autoclave at 121 °C for 15 min, and allow to cool down to warm temperature (50 °C). Add 1 mL of 100 mM acetosyringone, mix it well, and pour 40 mL into each sterile Petri plate (90 × 15 mm). Allow for solidification in the laminar air flow chamber for 40 min, and wrap it with klin film. Store the media plate at below 25 °C in dark conditions.
3. *Resting medium (CCMC)*: Mix the stock solutions of 100 mL 10X CC major salts, 10 mL 100X Fe EDTA, 10 mL 100X CC minor salts, 10 mL 100X CC vitamins, 20 mL of 100 mg/L 2,4-D, 10 mL of 100 mg/L NAA, and 2 mL of 100 mg/L 6-BA to 500 mL distilled water. To this mixture add and dissolve 20 g maltose, 36 g mannitol, 0.5 g proline, and 0.5 g vitamin assay casamino acids, and make up the volume into 1000 mL using distilled water. Adjust pH to 5.8 and add 7.5 g of gelrite. Autoclave at 121 °C for 15 min, and allow to cool down to warm temperature (50 °C). Add 1 mL of 250 mg/mL cefotaxime, mix it well, and pour 40 mL into each sterile Petri plate (90 × 15 mm). Allow for solidification in the laminar air flow chamber for 20 min, and wrap it with klin film. Store the media plate at below 25 °C in dark conditions.
4. *Selection medium (CCMCH50)*: Prepare CCMC media and adjust pH to 5.8. Add 8 g of agarose Type I and autoclave at 121 °C for 15 min. Allow to cool down to warm temperature (50 °C), add 1 mL of 250 mg/mL cefotaxime and 1 mL of

50 mg/mL hygromycin B. Mix it well, and pour 40 mL into each sterile Petri plate (90 × 15 mm). Allow for solidification in the laminar air flow chamber for 20 min, and wrap it with klin film. Store the media plate at below 25 °C in dark conditions.

5. *Pre-regeneration medium (NBPRCH40)*: Mix the stock solutions of 100 mL 10X N6 major salts, 10.12 mL (10,125 μL) 100X FeEDTA, 10 mL 100X B5 minor salts, 10 mL 100X B5 vitamins, 20 mL of 100 mg/L 2,4-D, 5 mL of 100 mg/L NAA, and 15 mL of 100 mg/L 6-BA to 500 mL distilled water. To this mixture add and dissolve 30 g maltose, 0.5 g proline, and 0.5 g casein hydrolysate, and make up the volume into 1000 mL using distilled water. Adjust pH to 5.8 and add 8 g of agarose Type I. Autoclave at 121 °C for 15 min, and allow to cool down to warm temperature (50 °C). Add 10 mL of 30 g/L L-glutamine and 0.8 mL of 50 mg/mL hygromycin B. Mix it well, and pour 40 mL into each sterile Petri plate (90 × 15 mm). Allow for solidification in the laminar air flow chamber for 20 min, and wrap it with klin film. Store the media plate at below 25 °C in dark conditions (*see* **Note 2**).
6. *Regeneration medium (RNMH30)*: Mix the stock solutions of 100 mL 10X N6 major salts, 10.15 mL (10,150 μL) 100X FeEDTA, 10 mL 100X B5 minor salts, 10.08 mL (10,080 μL) 100X B5 vitamins, and 45 mL of 100 mg/L 6-BA to 500 mL distilled water. To this mixture add and dissolve 30 g maltose, 0.5 g proline, and 0.5 g vitamin assay casamino acids, and make up the volume into 1000 mL using distilled water. Adjust pH to 5.8 and add 8 g of agarose Type I. Autoclave at 121 °C for 15 min, and allow to cool down to warm temperature (50 °C). Add 10 mL of 30 g/ L L-glutamine and 0.6 mL of 50 mg/mL hygromycin B. Mix it well, and pour 40 mL into each sterile Petri plate (90 × 15 mm). Allow for solidification in the laminar air flow chamber for 20 min, and wrap it with klin film. Store the media plate at below 25 °C in dark conditions (*see* **Note 2**).
7. *Rooting medium (RMH30)*: Add 2.04 g of Murashige and Skoog medium (without $CaCl_2$, sucrose, and agar), 220 mg $CaCl_2.2H_2O$, 10 g sucrose, 150 mg vitamin assay casamino acids, 50 mg myoinositol, and 250 mg proline to 700 mL distilled water. Make up the volume into 1000 mL using distilled water, and adjust pH to 5.8. Add 4 g of agarose Type I and 2 g of phytogel. Autoclave at 121 °C for 15 min, and allow to cool down to warm temperature (50 °C). Add 0.6 mL of 50 mg/mL hygromycin B. Mix it well, and pour 40 mL into each sterile Petri plate (90 × 15 mm). (Pour 60 mL into each sterilized clear jam bottle (400 mL capacity), and allow 40 min for solidification). Wrap it with klin film, and store the media plate/bottle at below 25 °C in dark conditions.

8. *Agar medium (0.8%)*: Add 8 g of agar into 1000 mL of distilled water. Autoclave at 121 °C for 15 min, and allow to cool down to warm temperature (50 °C). Pour 40 mL into each sterile Petri plate (90 × 15 mm), and allow 15 min for solidification. Wrap it with klin film, and store the media plate/bottle at below 25 °C in dark conditions.

2.6 Equipment

1. Zero-tip sterile dissecting forceps.
2. Stereomicroscope.
3. 90 × 15 mm sterile disposable Petri plates.
4. Sterile microfuge tubes (1.5 mL).
5. Falcon tubes (50 mL).
6. Sterile water.
7. Extra-long premium stainless-steel forceps/tweezers.
8. Tissue culture bottle with lid (sterile)—400 mL capacity.
9. Water bath.
10. Centrifuge.
11. Incubator (28 °C).
12. Plant growth chamber (26 °C).
13. Plant growth chamber (30 °C) with illumination (5000 lux).
14. Micro pipettes (variable volumes).
15. Klin film—3 inch size (plain or without color shade).
16. Syringe.
17. 0.22 μm membrane filter.
18. Jam bottles with lid.

3 Methods

3.1 Agrobacterium Culture

Streak *Agrobacterium* culture (LBA4404 or C58C1) harboring binary plasmids on LB Agar medium with respective antibiotics. Incubate it at 28 °C for 2 days in dark conditions (*see* **Note 1**).

3.2 Infection Medium Preparation

Pick 2 days of grown bacterial culture using sterile inoculation loop (very small quantity, i.e., < 1/20 of the loop), and dissolve in 1 mL of AA infection medium. The dissolved AA infection medium should be transparent without any visible bacterial culture in it. Prepare AA infection medium during the centrifugation step of the pretreatment stage. The inoculum should be freshly prepared. It is highly recommended to prepare the inoculum during the centrifugation step of the pretreatment stage.

3.3 Explants Preparation

1. Collect the fully developed immature seeds between 12 and 14 days after pollination (DAP). Collect entire panicles in a polythene cover, rinse the panicles with sufficient quantity of water, drain the excess water, and tightly close the polythene cover to prevent dehydration (*see* **Note 3**).
2. Dehusk the immature seeds with forceps without damaging them, maintain them in a Petri plate containing distilled water, then gently rinse twice with sterile water, and finally drain the water.
3. Transfer the dehusked immature seeds to a fresh sterile Petri plate (100 seeds at a time) under laminar air-flow chamber.
4. Surface sterilize the seeds by applying ~10 mL of 70% ethanol for 30 sec. Remove the ethanol using 1 mL micropipette and add ~10 mL of 2.0% sodium hypochlorite (prepare 2.0% sodium hypochlorite by using sterile distilled water). Ensure the seeds are fully immersed. Incubate the seeds for 3 min. Alternatively, 0.1% mercury chloride Tween 20 can be used.
5. Remove sodium hypochlorite solution, and rinse it three times with autoclaved sterile distilled water.
6. Completely remove the water by using the 1 mL micropipette, and place the seeds under the microscope for embryo isolation.
7. Isolate the immature embryos using sterile forceps without causing physical damage, and transfer them to a 0.8% agar medium plate.

3.4 Pretreatment

1. Transfer the healthy immature embryos in 1 mL of sterile distilled water in 1.5 mL microcentrifuge tubes. Tap the tube with a finger to allow the embryos to settle in the bottom of the tube. The embryo can be washed once with sterile distilled water to remove any adhering material.
2. Incubate the tube containing immature embryos at 43 °C for 30 min in a water bath (*see* **Note 4**). Make sure the water bath temperature reaches 43 °C before incubating the tubes. It is recommended to use a thermometer to confirm the temperature.
3. Immediately cool the microcentrifuge tube for 1 min by placing on ice.
4. Centrifuge the tube at 1100 rpm for 10 min at room temperature (25–30 °C) (*see* **Note 5**). After centrifugation, *Agrobacterium* infection should be completed within 20 min for enhanced transformation efficiency.

3.5 Infection and Cocultivation

1. Transfer (pour) the pretreated immature embryos along with water from the microcentrifuge tube to a sterile Petri plate, and transfer the leftover embryos with the help of sterile forceps.

2. Remove the water completely using the 1 mL micropipette, and transfer the immature embryos with scutellum face up to NB-As medium (30 embryos can be placed in each plate).
3. Inoculate 2–3 μL of *Agrobacterium* infection medium on each embryo. For enhanced transformation efficiency, *Agrobacterium* inoculation on pretreated embryos should be completed within 20 min after the centrifugation step.
4. Wrap the plate with klin film, and cover it with the brown sheet/aluminum foil to facilitate dark conditions. Incubate the plate at 22–25 °C for 15 min to allow the infection medium to infiltrate the NB-As medium.
5. Transfer the infected embryos to the fresh NB-As media plate, and wrap them using klin film (*see* **Note 6**).
6. Cover the plate with the brown sheet and incubate the plate at 22–25 °C for 7 days in dark conditions (*see* **Note 6**). After 7 days of incubation, very light or no visible *Agrobacterium* growth will ensure the better health of the explant.

3.6 Resting and Selection

1. After callus formation with the shoot (5–7 days after cocultivation), gently remove the elongated shoots using sterile forceps without causing injury to callus and place the calli on the CCMC medium with the scutellum face up. Wrap the plate with klin film.
2. Incubate at 30 °C for 5 days in a plant growth chamber with continuous illumination (Resting I stage).
3. Split the grown calli into small pieces (~3 to 4) (*see* Fig. 1) with sterile forceps, and transfer them to fresh CCMC medium plate with the scutellum face up (calli is made into small pieces by giving gentle press to the main calli so that it goes into small pieces and use Petri plate lid for breaking the calli).
4. Incubate at 30 °C for 10 days in a plant growth chamber with continuous illumination (Resting II stage).
5. Split the grown calli into small pieces (~4 to 6) using sterile forceps (*see* Fig. 1) and transfer them to selection (CCMCH50) medium plate added with 50 mg/L hygromycin B antibiotic and wrap the plate with klin film.
6. Incubate at 30 °C for 10 days in a plant growth chamber with continuous illumination (Selection I stage).
7. From the Selection I stage, transfer survived and proliferated calli into fresh selection medium for the second round of selection. It is essential to remove the dead or black-colored portions from the proliferating calli using forceps to avoid false selection, i.e., development of non-transformed cells. Wrap the plate using klin film. The pool of calli recovered from the single embryo is practically treated as a single transgenic event.

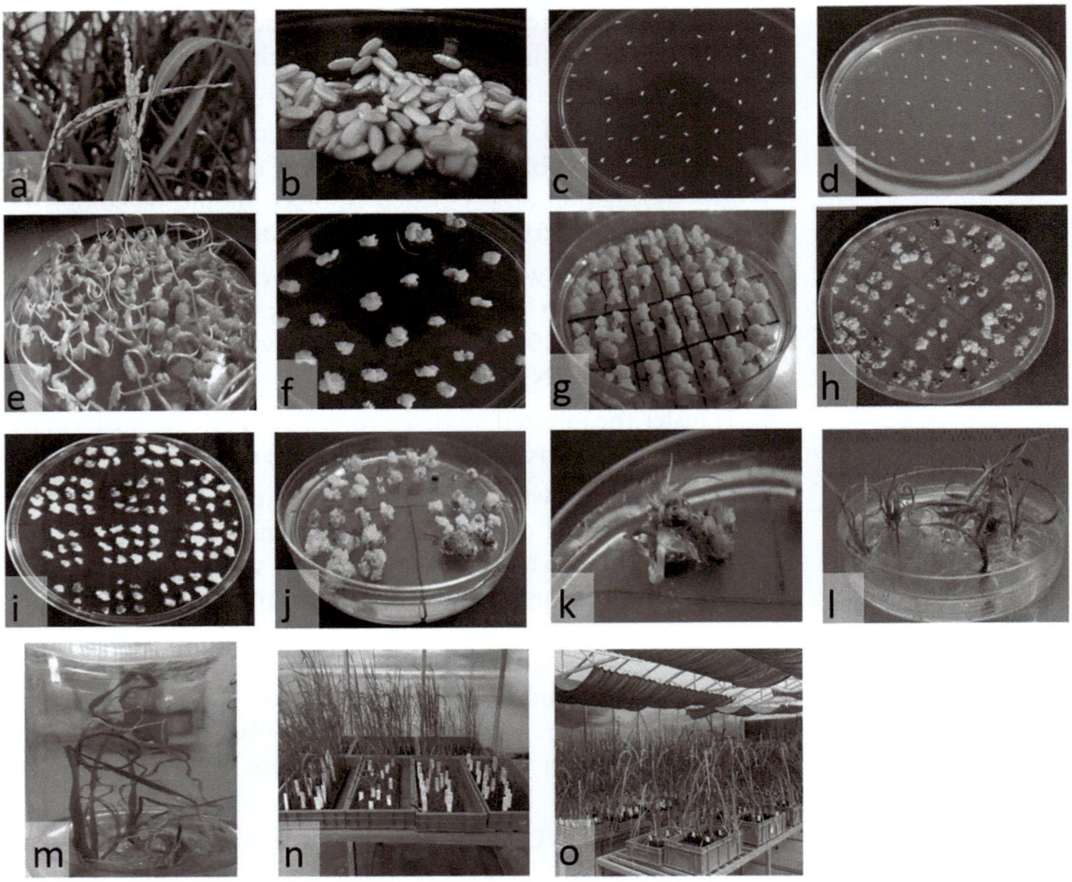

Fig. 1 *Agrobacterium*-mediated transformation of rice using immature embryos. (**a**) Immature rice panicles, (**b**) dehusked immature rice seeds, (**c**) isolation of immature embryos, (**d**) pre-cultured rice immature embryos infected with *Agrobacterium* on NB-As medium, (**e**) calli developed from immature embryos 7 days after cocultivation on NB-As medium, (**f**) subcultured calli on resting medium (CCMC) (Resting I), (**g**) subcultured calli on resting medium (CCMC) (resting II), (**h**) subcultured calli on selection medium (CCMCH50) (Selection I), (**i**) subcultured calli on selection medium (CCMCH50) (Selection II), (**j**) embryogenic calli on pre-regeneration medium (NBPRH40), (**k**) shoot development on regeneration medium (RNMH30), (**l**) root development on rooting medium (half-strength MS medium) (plate), (**m**) shoot and root elongation on rooting medium (half-strength MS medium) (jam bottle), (**n**) hardening of putative transgenic plants in soil at transgenic greenhouse, (**o**) planting putative transgenic plants in pots maintained at transgenic greenhouse

8. Incubate at 30 °C for 7 days in a plant growth chamber with continuous illumination (Selection II stage). In case of very small size of calli obtained in the specified duration, it is better to incubate in fresh plate for another 5–7 days to ensure optimum calli growth.

3.7 Regeneration of Transgenic Plants

1. Transfer well-proliferated hygromycin-resistant calli (yellowish white color) that survived from the Selection II stage into pre-regeneration medium (NBPRCH40) using sterile forceps. Avoid transferring any soft watery calli to pre-regeneration

medium. Wrap the plate with klin film. About five to ten calli can be placed in each plate (90 × 15 cm).

2. Incubate at 30 °C for 7 days in a plant growth chamber with continuous illumination (pre-regeneration stage).
3. Transfer well-proliferated calli (yellowish white or greenish calli) to regeneration medium (RNMH30) with sterile forceps (*see* Fig. 1). In a single medium plate, five to seven events can be placed. Wrap the plate with klin film (*see* **Note** 7).
4. Incubate at 30 °C for 14 days in a plant growth chamber with continuous illumination (regeneration stage).
5. Transfer the regenerated shoots (*see* Fig. 1) to the rooting medium (RMH30) using sterile forceps. Wrap the plate with klin film.
6. Incubate at 30 °C for 5–10 days in a plant growth chamber with continuous illumination (root initiation stage).
7. Transfer well-developed plantlets to rooting medium in jam bottles (to facilitate the growth and development of both shoots and roots). About four to five plantlets can be kept in a single jam bottle.
8. Incubate at 30 °C for 10 days in a plant growth chamber with continuous illumination (shoot and root development stage).

3.8 Hardening, Transplanting, and Acclimation

1. Transfer well-developed (both shoot and root) putative transformants into the portray filled with potting mixture, and place them in a tray containing water. Keep the tray in a transgenic greenhouse for 10–15 days (Hardening stage) (*see* Fig. 1).
2. Transplant the well-developed transgenic plants into pots filled with potting mix. Place the pots in a tray of water and maintain them in the transgenic greenhouse, or transplant them into a hydroponic system, providing the necessary nutrients to the transgenic plants.
3. Grow the plants in the transgenic greenhouse until maturity. Apply the required level of nutrients to grow healthy plants and to get more seeds. Harvest the seeds carefully after maturity and use them for subsequent generations.
4. Complete timeline for rice transformation protocol using immature embryos is presented in Fig. 2.

4 Notes

1. For high transformation efficiency, *Agrobacterium* cultures should be freshly streaked on LB agar media with appropriate antibiotics weekly and grown for 2 days at 28 °C. The *Agrobacterium* strains LBA4404, AGL1, and C58C1 are suitable for

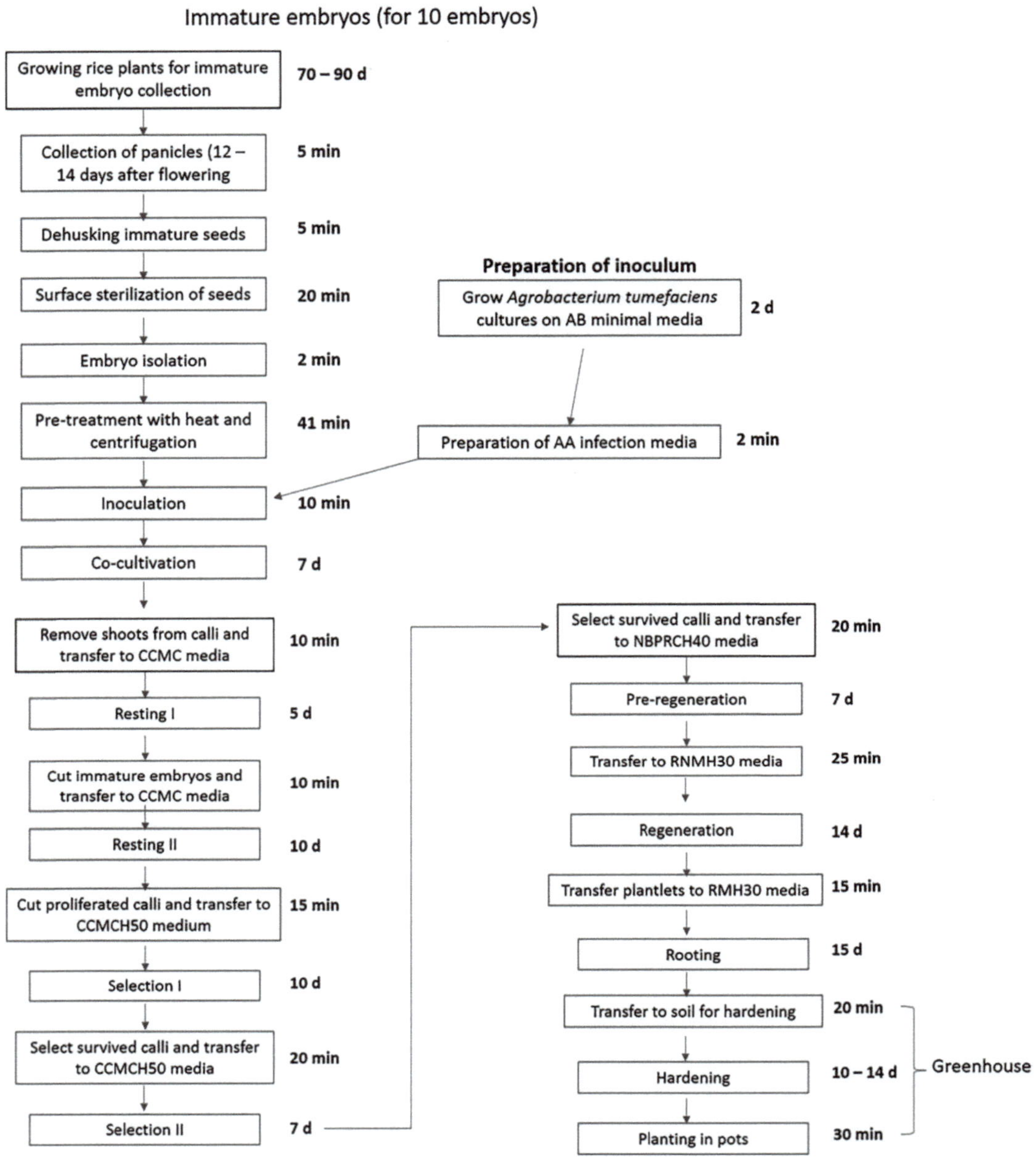

Fig. 2 Timeline for rice transformation protocol using immature embryos. By assuming ten immature embryos are used in this protocol and subsequently handled by a skilled technician in the following steps

rice transformation in the *indica* variety. Frequent PCR-based confirmation of the presence of the gene of interest, selectable marker gene, and virulence gene is highly recommended. Primers for confirming *vir* gene (~430 bp) are VF2: TCGATGT CGTGGTTCTTGAT (Forward) and VR2: ATAAACCTCCTCGTCGCGTA (Reverse). Primers for confirming *hpt* gene (~686 bp) are HptF: GACGTCTGTCGA-GAAGTT (Forward) and HptR: GCCTCCAGAAGAAGATG

(Reverse). If the PCR amplifications are found to be negative, it may indicate possible contamination with other bacteria. In this case, it is recommended to streak and grow the *Agrobacterium* cultures on AB minimal media (check Subheading 2.5 of this chapter) with appropriate antibiotics. Since AB minimal media contains only mineral salts, it allows only the *Agrobacterium* cultures to grow.

2. Add 250 mg/mL cefotaxime sodium antibiotics in pre-regeneration and regeneration medium if bacterial growth is observed alongside the calli. If no bacterial growth is observed after the second selection, cefotaxime is not needed in the pre-regeneration and regeneration stage.

3. Healthy immature embryos should be used for effective transformation. The selection of immature embryos at optimal developmental stages is crucial, and embryo size serves as a reliable indicator of this stage. The day after pollination (DAP) is necessary for embryos to attain this optimal stage vary depending on genotype (between 12 and 14 days for ASD16 and CO51 varieties) and seasonal conditions. A practical assessment method involves gently pressing the fully developed immature seeds with a thumb; if the immature seed offers slight resistance (releasing semisolid milky white color while pressing), the embryo is generally of the suitable stage. Seeds that offer no resistance or firm indicate that the embryo is either too early or too mature for use.

4. The pretreatment of embryos with heat is critical for achieving efficient transformation. The heat treatment at 43 °C in the water bath for 30 min was effective across various rice genotypes [9]. However, the optimal duration for heat treatment can vary from 25 to 45 min, depending on the condition of the immature embryos and the health of the mother plants. Thus, regular optimization and frequent verification of heat treatment duration are highly recommended.

5. Centrifugation in the pretreatment step enhances callus induction from the scutellum. However, higher centrifugation forces can lower gene transformation frequency. Optimal centrifugation conditions vary by genotype. For indica varieties, both 1100 and 2300 rpm for 10 min should be tested as starting conditions. Further optimization may be necessary for specific cultivars.

6. Transferring the infected embryos into new place in the media plate will reduce the *Agrobacterium* growth during cocultivation stage. Overgrowth of *Agrobacterium* along with calli may lead to reduce the transformation efficiency.

7. Select and transfer only the embryogenic calli to the regeneration medium. The embryogenic calli are typically small,

compact clusters of cells, appearing light yellow or greenish. In contrast, non-embryogenic calli are usually larger, softer, and semitransparent and consist of loosely held clusters of yellow or gray cells. Ensure sufficient space and media are provided for each calli for effective regeneration.

Acknowledgments

We thank the Department of Plant Biotechnology, Centre for Plant Molecular Biology and Biotechnology, Tamil Nadu Agricultural University, Coimbatore, for providing facilities to carry out experiments.

References

1. Zhao M, Lin Y, Chen H (2020) Improving nutritional quality of rice for human health. Theor Appl Genet 133(5):1397–1413. https://doi.org/10.1007/s00122-019-03530-x
2. Yadav SK, Yadav P, Chinnusamy V (2023) Nutrigenomics in cereals. In: Deshmukh R, Nadaf A, Ansari WA, Singh K, Sonah H (eds) Biofortification in cereals. Springer, Singapore. https://doi.org/10.1007/978-981-19-4308-9_12
3. FAO (2022) World food and agriculture—statistical yearbook 2022. FAO, Rome, p 380. https://doi.org/10.4060/cc2211en
4. Jain N, Khurana P, Khurana JP (2022) A rapid and efficient protocol for genotype-independent, *Agrobacterium*-mediated transformation of indica and japonica rice using mature seed-derived embryogenic calli. Plant Cell Tissue Organ Cult 151:59–73. https://doi.org/10.1007/s11240-022-02331-3
5. Nishimura A, Aichi I, Matsuoka M (2007) A protocol for *agrobacterium*-mediated transformation in rice. Nat Protoc 1(6):2796–2802. https://doi.org/10.1038/nprot.2006.469
6. Rajadurai G, Kalaivani A, Varanavasiappan S, Balakrishnan N, Udayasuriyan V, Sudhakar D, Natarajan N (2018a) Generation of insect resistant marker-free transgenic rice with a novel cry2AX1 gene. Electron J Plant Breed 9(2):723–732. https://doi.org/10.5958/0975-928X.2018.00086.8
7. Xiang Z, Chen Y, Chen Y, Zhang L, Liu M, Mao D, Chen L (2022) *Agrobacterium*-mediated high-efficiency genetic transformation and genome editing of Chaling common wild Rice (*Oryza rufipogon* Griff.) using scutellum tissue of embryos in mature seeds. Front. Plant Sci 13:849666. https://doi.org/10.3389/fpls.2022.849666
8. Chen K, Ye C, Guo J, Chen D, Guo T, Liu J, Liu C, Zhou X (2023) *Agrobacterium*-mediated transformation efficiency and grain phenotypes in six indica and japonica rice cultivars. Seed Biol 2:4. https://doi.org/10.48130/SeedBio-2023-0004
9. Hiei Y, Komari T (2008) *Agrobacterium*-mediated transformation of rice using immature embryos or calli induced from mature seed. Nat Protoc 3:824–834. https://doi.org/10.1038/nprot.2008.46
10. Rajadurai G, Varanavasiyappan S, Balakrishnan N, Udayasuriyan V, Sudhakar D, Natarajan N (2018b) Development of transgenic indica rice with a chimeric cry2AX1gene against leaffolder, Cnaphalocrocis medinalis (Guenee). Int J Curr Microbiol Appl Sci 7(07):3242–3253. https://doi.org/10.20546/ijcmas.2018.707.378
11. Luu VT, Stiebner M, Maldonado PE, Valdés S, Marín D, Delgado G, Laluz V, Wu L, Chavarriaga P, Tohme J, Slamet-Loedin IH, Frommer WB (2020) Efficient *Agrobacterium*-mediated Transformation of the Elite–*Indica* Rice Variety Komboka. Bio Protoc 10(17):e3739. https://doi.org/10.21769/bioprotoc.3739
12. Sundararajan S, Rajendran V, Nayeem S, Ramalingam S (2020) Physicochemical factors modulate regeneration and agrobacterium-mediated genetic transformation of recalcitrant indica rice cultivars – ASD16 and IR64. Biocatal Agric Biotechnol 24:101519. https://doi.org/10.1016/j.bcab.2020.101519
13. Rajadurai G, Kalaivani A, Sudhakar D, Varanavasiappan S, Kumar KK,

Balakrishnan N, Udayasuriyan V, Natarajan N (2024) Marker-free transgenic *Bt* rice engineered for resistance to yellow stem borer and leaffolder. Cereal Res Commun. https://doi.org/10.1007/s42976-024-00604-5

14. Hiei Y, Ishida Y, Komari T (2015) Rice, indica (*Oryza sativa* L.). Methods Mol Biol 1223: 155–167. https://doi.org/10.1007/978-1-4939-1695-5_12

Chapter 2

Agrobacterium-Mediated Genetic Transformation in Banana

Theeyan Anuradha, Ravikumar Sunithakumari Lekshmi, Kallamparambil Balakrishnan Soni, and Swapna Alex

Abstract

Banana (*Musa* spp.) is vital global fruit crop that is prone to significant biotic and abiotic stresses. Modern biotechnological tools, such as genome editing and genetic transformation, offer promising strategies to enhance banana resilience. Recent advancements have enabled the transformation of banana plants with desirable traits. Among these methods, *Agrobacterium*-mediated transformation of embryogenic calli has emerged as the most feasible and cost-effective technique. Optimizing this method for specific cultivars is crucial to improve transformation efficiency. This chapter outlines a successful genetic transformation protocol for the *Grand Naine* cultivar of banana, a commercially important dessert variety. Emphasizing *Agrobacterium*-mediated transformation of embryogenic calli, we highlight its economic viability and efficiency.

Key words *Agrobacterium tumefaciens*, Somatic embryogenesis, Plant transformation, *Grand Naine*, Immature male inflorescence

1 Introduction

Grand Naine is a widely favored variety for commercial plantations due to its medium height, high fruit yield, and seedless quality. Its moderate height facilitates easy harvesting and offers some resistance to wind damage. In many tropical regions, local economies heavily rely on banana production and export. Given its significance as both a staple and cash crop, extensive botanical research has been concentrated on *Grand Naine* [1]. *Grand Naine* bananas, also known as *Grande Naine*, translates to "Large Dwarf" in French. It is a cultivar of the well-known Cavendish bananas, characterized by its AAA genotype, meaning it is a triploid variant of *Musa acuminata*. This AAA cultivar, with its 33 chromosomes, produces seedless fruits through parthenocarpy [1]. Cultivated across Central America, Africa, India, and Southeast Asia, *Grand Naine* faces challenges from viral diseases. One of the most severe is banana

Nitish Kumar (ed.), *Transgenic Plants: Methods and Protocols*, Methods in Molecular Biology, vol. 2973,
https://doi.org/10.1007/978-1-0716-4803-2_2,

bract mosaic virus (BBrMV) of the *Potyvirus* genus in the Potyviridae family, which significantly reduces yields.

Efforts to genetically improve bananas through conventional breeding have faced obstacles, particularly in dessert banana breeding, which has not produced a successful hybrid in 70 years due to high sterility in AAA cultivars [2] and the absence of complete resistance genes in the banana gene pool.

To enhance crop traits, a robust and repeatable regeneration and transformation protocol is essential. Various methods, including particle bombardment, electroporation, and *Agrobacterium*-mediated genetic transformations, have been developed since the early 1990s. Explants used include embryogenic calli and embryogenic cell suspension cultures derived from floral meristems. Immature male buds from positions 0–15 have been effective in somatic embryogenesis for cv. *Grand Naine*, *Yangamba* (AAA), *French Plantain*, *Mysore and Silk* (AAB), and *Pelpita* (ABB) [3–5, and]. Notably, 74% of embryogenic cultures originate from immature male buds at positions 7 and 13 [6]. Banana regeneration and transformation are dependent on the specific cultivar/species. This chapter details the genetic transformation of the *Grand Naine* cultivar using embryogenic calli and *Agrobacterium*, highlighting the importance of cultivar-specific protocols for improving transformation efficiency.

2 Materials

All solutions should be prepared using distilled, sterilized water and chemicals of molecular biology or analytical grade. Store these solutions in sterile tubes or bottles, and, where required, use 0.22 μm syringe filters for sterilization.

1. 6-Benzylaminopurine (1 mg/L BAP): Dissolve 20 mg in 3 mL of 1 M NaOH. Adjust the volume to 20 mL with water and filter sterilize. Aliquot into 1.5 mL microfuge tubes and store at −20 °C.
2. Thidiazuron: Dissolve 1.962 g in 50 mL dimethyl sulfoxide (DMSO). Store aliquots in 1.5 mL microfuge tubes at −20 °C.
3. Rifampicin (25 mg/mL): Dissolve 625 mg in 25 mL of 0.1 N HCl. Filter sterilize and aliquot into 1.5 mL microfuge tubes.
4. Kanamycin monosulfate (50 mg/mL): Dissolve 2.5 g in 50 mL of water. Filter sterilize and aliquot into 1.5 mL microfuge tubes. Store at −20 °C.
5. Hygromycin B (50 mg/mL): Dissolve 1.25 g in 25 mL of water. Filter sterilize and aliquot into 1.5 mL microfuge tubes. Store at 4 °C.

6. 3′,5′-Dimethoxy-4′-hydroxyacetophenone (200 mM): Dissolve 1.962 g in 50 mL dimethyl sulfoxide (DMSO). Store aliquots in 1.5 mL microfuge tubes at −20 °C.
7. MS macronutrients (mg/L): 370 mg/L $MgSO_4.7H_2O$, 440 mg/L $CaCl_2.2H_2O$, 1900 mg/L KNO_3, 1650 mg/L NH_4NO_3, and 170 mg/L KH_2PO_4.
8. MS micronutrients (mg/L): 22.3 mg/L $MnSO_4.4H_2O$, 8.6 mg/L $ZnSO_4.7H_2O$, 0.025 mg/L $CuSO_4.5H_2O$, 0.025 mg/L $AlCl_3$, 0.83 mg/L KI, 6.2 mg/L H_3BO_3, 0.25 mg/L $Na_2MoO_4.2H_2O$.
9. MS iron sources (mg/L): 27.85 mg/L $FeSO_4.7H_2O$ and 37.25 mg/L Na_2EDTA.
10. MS Vitamins: Dissolve 0.5 mg/L nicotinic acid, 0.5 mg/L pyridoxine HCl, and 0.1 mg/L thiamine HCl.
11. Embryogenic callus induction medium (ECI): MS +8 mg/L BA +0.6 mg/L TDZ+ 30 g/L sucrose+5 g/L gelrite.
12. Embryo germination medium (EG): MS+ 2 mg/L BA 30 g/L sucrose+5 g/L gelrite.
13. Cocultivation (CC) medium: ½ MS +8 mg/L BA +0.6 mg/L TDZ+ 30 g/L sucrose+5 g/L gelrite+200 μM acetosyringone.
14. Washing medium (WM):): ½ MS +100 mg/L cefotaxime and without gelrite.
15. Selection medium (SM): MS +8 mg/L BA +0.6 mg/L TDZ+ 30 g/L sucrose+5 g/L gelrite+200 mg/L kanamycin.
16. Regeneration medium (RM): MS+ 2 mg/L BA +1 mg/L NAA.
17. *Agrobacterium* culture medium: LB Agar +25 mg/L rifampicin +100 mg/L streptomycin.

3 Methods

To avoid/minimize contamination, it is imperative that all steps are performed using sterile tools and media in a biological safety cabinet. Handle all genetically modified material according to DBT regulations, and follow good laboratory practice.

3.1 Preparation of Embryogenic Calli of Grand Naine

1. Collect 4-week-old immature male inflorescences (IMFs) of banana var. *Grand Naine* after post-bunch formation.
2. Surface sterilize the IMFs by wiping them with sterile cotton dipped in 90% ethanol inside a laminar flow hood.
3. Excise male flower clusters (positions 0–15) from the sterile male flower bud using sterile instruments and culture them on ECI medium (Fig. 1a).

Fig. 1 Procedure for producing transgenic banana plants: (**a**) immature male flower buds inoculated in CIM. (**b**) Initiation of embryogenic calli. (**c**) Globular somatic embryos. (**d**) Torpedo stages of somatic embryos. (**e**) Transformed embryogenic calli in selection medium. (**f**) Germinating embryos in selection medium. (**g**) and (**h**) Regenerated shoots in selection medium. (**i**) Hardened transgenic plants in coir pith compost

4. Incubate the cultures in dark at 25 °C with 50–70% relative humidity.
5. Perform subculturing in ECI medium at 1-month intervals to maintain the cultures.
6. Transfer the somatic embryos (Fig. 1b, c) in EG medium, and maintain under a 16-h photoperiod with a light intensity of 31.4 μmol m-2 s-1 to promote proper morphogenesis and the development of elongated monocot embryo structures (*see* **Note 1**).

3.2 Preparation of Agrobacterium tumefaciens Inoculum for Cocultivation

1. Streak *Agrobacterium* (GV 3103) carrying the binary vector pART27 (*see* **Note 2**) cassette onto LB medium with antibiotics for both bacterial selection (25 mg/L rifampicin and 100 mg/L spectinomycin).
2. Invert the plate and incubate at 28 °C in the dark for 2 days.
3. Isolate a bacterial colony with a loop and inoculate it into 5 mL of LB with appropriate antibiotics in a 15 mL tube. Culture overnight in an incubator shaker at 28 °C and 200 rpm.
4. Incubate the culture until the optical density (OD600) reaches 0.2–0.5.
5. Transfer around 1 mL of this culture to 50 mL of LB with antibiotics in a 250 mL flask. Continue growing overnight in an incubator shaker at 28 °C and 110 rpm.
6. When the OD600 is 0.2, transfer the culture to a 50 mL tube, and centrifuge at 5000× *g* at 4 °C for 5 min.
7. Discard the supernatant and gently resuspend the pellet in 1 mL of half-strength MS broth with 200 μM acetosyringone.

3.3 Cocultivation of Embryogenic Calli

1. Pre-culture embryogenic calli of banana on fresh ECI medium 15 days before cocultivation to maintain cells in active cell division stage.
2. Immerse pre-cultured calli in *Agrobacterium* suspension with inducer (200 μM) acetosyringone for 10 min for infection.
3. After infection, blot dry the explants on sterile filter paper and inoculate on solidified cocultivation (CC) medium without antibiotic.
4. Seal Petri plates and keep for cocultivation in dark for 48 h at 25 ± 2 °C.
5. After cocultivation period of 48 h, wash calli in washing medium (WM) medium containing 100 mg/L cefotaxime to kill the excess *Agrobacterium*.
6. Blot dry tissues with sterile filter paper and transfer to Petri plates containing half-strength ECI medium supplemented with 100 mg/L cefotaxime to curb the growth of *Agrobacterium* over the calli.
7. Subculture the calli to new Petri plates with semisolid ECI medium supplemented with 100 mg/L fresh cefotaxime for 1 week at 2-day intervals.

3.4 Selection of Transformed Embryogenic Callus

1. Use selection medium containing 200 mg/L kanamycin for selection of transformed calli.
2. Maintain tissues on the antibiotic media by subculturing once every 7 days.

3. Score the transformed and non-transformed tissues based on their response in the medium after four rounds of subculture.
4. Transfer embryogenic callus (Fig. 1e) that survives in the selection medium to the regeneration medium.

3.5 Regeneration of Putative Transgenic Plants

1. Transfer the germinating somatic embryos (Fig. 1f) that survive in the 2nd and 3rd selection to regeneration medium (MS medium supplemented with BA 2 mg L^{-1} + NAA 1 mg L^{-1}) with selection pressure for complete plantlet formation.
2. Maintain cultures at a light intensity of 31.4 μmol m^{-2} s^{-1} with a 14-h light/8-h dark photoperiod.
3. Transfer plantlets at the three to four leaf stage (Fig. 1g, h) to coir pith compost in portrays and harden in a transgenic mist chamber for a month.
4. Transfer plantlets to polybags with a soil and cow dung (1:1) mixture for hardening (Fig. 1i).

3.6 Molecular Confirmation of Putative Transgenic Plants

1. Isolate genomic DNA from leaf tissues of transgenic plants using the C-TAB method of DNA isolation.
2. Detect the presence of the nptII gene in the plantlet that survived in the selection medium (kanamycin 200 mg L^{-1}) by carrying out gene-specific PCR reaction using kanamycin primers (Fig. 2b).
3. Perform PCR amplifications for all the other genes present in the transformation cassette of pART27 (Fig. 2a, b) like cre intron, sense, and antisense coat protein genes.

4 Notes

1. This simple sterilization procedure was effective as the immature flower buds are perfectly and tightly packed within the bracts of the inflorescence. Endogenous bacterial contamination is very less in inflorescence meristem compared to shoot meristem.
2. The binary vector pART27 is utilized for constructing inverted hairpin constructs. The genetic transformation protocol for Grand Naine was developed with the primary goal of achieving resistance to banana bract mosaic virus by specifically targeting and silencing the coat protein gene.

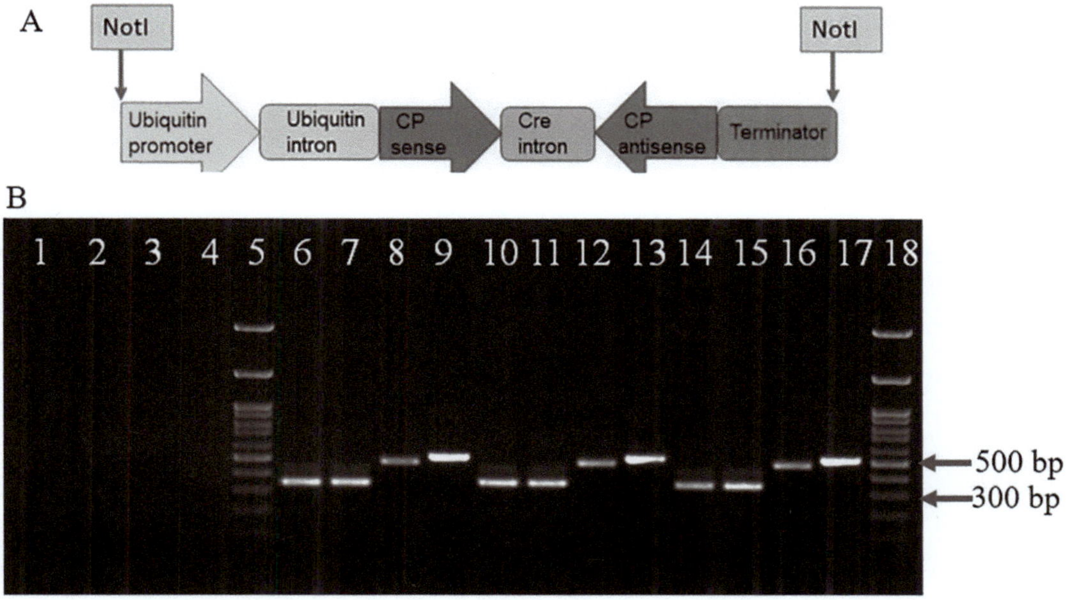

Fig. 2 (**a**) Components of gene transformation cassette, (**b**) PCR confirmation of the gene transformation cassette in transformed plants (Lanes 1 to 4: control (untransformed); underlined part represents results for one individual transformed plant (6–9, 10–13, 14–17), Lanes 6, 10, and 14: amplified sense fragment (326 bp), Lanes 7, 11, and 15: amplified antisense fragment (326 bp), Lanes 8, 12, and 16: amplified cre intron (450 bp), Lanes 9, 13, and 17: amplified *nptII* (475 bp), Lanes 5, 18: 100 bp ladder)

References

1. Ploetz RC, Kepler AK, Daniells JW, Nelson SC (2007) Banana and plantain: an overview with emphasis on Pacific Island cultivars Musaceae (banana family). In: Elevitch CR (ed) Species profiles for Pacific Island agroforestry. Permanent Agriculture Resources, Holualoa, p 27
2. Ortiz R, Vuylsteke D (1995) Factors influencing seed set in triploid Musa spp. and production of euploid hybrids. Ann Bot 75:151–155
3. Cote FX, Domergne R, Monmarson S, Schwendiman J, Teissor C, Escalant JV (1996) Embryogenic cell suspensions from male flower of Musa AAA cv. Grand Naine. Physiol Plant 97: 285–290
4. Borys C, Reyes M, Rojas L, Ocana B, Blanca P, Kosky RG, Angenon G (2012) Establishment of embryogenic cell suspension cultures and Agrobacterium-mediated transformation in banana cv. 'Dwarf Cavendish' (Musa AAA): effect of spermidine on transformation efficiency. Plant Cell Tissue Organ Cult 111:79–90
5. Vincent L, Anushma PL (2018) Micropropagation in banana using inflorescence: a review. J Cell Tissue Res 18(3):6573–6582
6. Wong WC, Jalil M, Ong-Abdullah M, Othman RY, Khalid N (2006) Enhancement of banana plant regeneration by incorporating a liquid-based embryo development medium for embryogenic cell suspension. J Hortic Sci Biotechnol 81:385–390

Chapter 3

Agrobacterium-Mediated Genetic Transformation and Genome Editing Using CRISPR-Cas9 Constructs in Rice

Andrew-Peter-Leon M. T., M. Arumugam Pillai, Krish K. Kumar, and Velusamy Sumithra

Abstract

Clustered regularly interspaced short palindromic repeats (CRISPR)-CRISPR-associated (Cas)9 has become an essential tool in every molecular plant breeding laboratory. CRISPR-Cas9 causes highly precise knock-out mutations in plants that can be exploited in crop improvement programmes. However, poor response to *Agrobacterium*-mediated genetic transformation in recalcitrant rice genotypes is a major limiting factor. This protocol describes a detailed procedure for genome editing with CRISPR-Cas9 in recalcitrant rice genotypes that otherwise show a poor response to tissue culture. With this method, high transformation efficiency can be achieved in relatively a short period.

Key words CRISPR-Cas9, *Agrobacterium*-mediated gene transfer, Recalcitrant rice

1 Introduction

Clustered regularly interspaced short palindromic repeats (CRISPR)-CRISPR-associated (Cas) is a RNA-mediated adaptive defense system found in bacteria and archaea that protect them from viruses and plasmids. The CRISPR-Cas system consists of *cas* genes which can recognize a target sequence and silence by causing double-strand breaks [1].

Among the different types of CRISPR systems, in the type II system, one of the *cas* genes is always *cas9*. The Cas9 protein has two active sites, namely, RuvC and HNH, that are responsible for the binding and cleavage of the target DNA. For this recognition and target binding, a short sequence of DNA known as "protospacer adjacent motif (PAM)" is critical [2]. Cas9 of the *Streptococcus pyogenes* requires a 5′-NGG-3′ as the protospacer adjacent motif (PAM) in the target DNA for recognition and cleavage. The PAM and Cas9 work through major and minor grove interactions. The GG dinucleotide in the noncomplementary strand is read out via

Nitish Kumar (ed.), *Transgenic Plants: Methods and Protocols*, Methods in Molecular Biology, vol. 2973,
https://doi.org/10.1007/978-1-0716-4803-2_3,

major grove interactions with arginine residues from the carboxy-terminal domain of Cas9. Another minor grove interaction between the PAM duplex and the phosphodiester group at the +1 position in the target DNA strand results in local strand separation immediately upstream of the PAM [3].

By designing the dual RNA duplex as a single-guide RNA (sgRNA) and modifying the DNA target-binding sequence, the Cas9 could be programmed to target a wide range of genome sequences [4].

This protocol elaborates on the steps in genome editing in rice using CRISPR-Cas9. The CRISPR constructs consisting of single-guide RNA and Cas9 gene is transformed in the plant cell using *Agrobacterium*. It offers many advantages such as high efficiency, integration of small numbers of copies of transfer DNA, and little rearrangement [5]. However, efficiency of transformation varies between different rice varieties. Rice varieties with low level of transformation are known as "recalcitrant" varieties. We use one such variety "Improved White Ponni" and show an efficient method to transform with high success rate. This protocol uses higher percentage of gelling agents, maltose as the carbon source, and higher concentration of 2,4-D in the tissue culture media as reported by [6] but with modifications. As such, we were able to achieve high efficiency in transformation in a standard transformation laboratory set up.

2 Materials

2.1 CRISPR-Cas9 Vector

CRISPR-Cas9 vector plasmid pRGEB32 from Yinong Yang [7] (Addgene, USA) is used here to mutate the rice genome. Key regions of the plasmid are given in Fig. 1.

2.2 Oligos for Targeted Genome Editing

The *Cas9* enzyme targets 3 bp upstream of 5′-NGG in the host as the endonuclease cutting site. The pRGEB32 plasmid can be ligated with unique 20-nucleotide-length guide sequences that target the gene regions in hosts having the above cutting site (Fig. 2). These guide sequences are complementary to the target gene sequences of the host.

1. Design these sequences as oligos using online design tools such as "CRISPR-PLANT" (Xie et al. 2014) or CRISPR-Cereal [8] that are optimized for cereals such as rice and can also show potential off-target activity.
2. Choose oligos with GC content between 40% and 80% (for increased stability) and have no off-target activity.
3. While designing, the target regions of the host is also chosen to have suitable restriction enzyme sites to facilitate detection of genome editing in plants by PCR/RE assay. For example, the

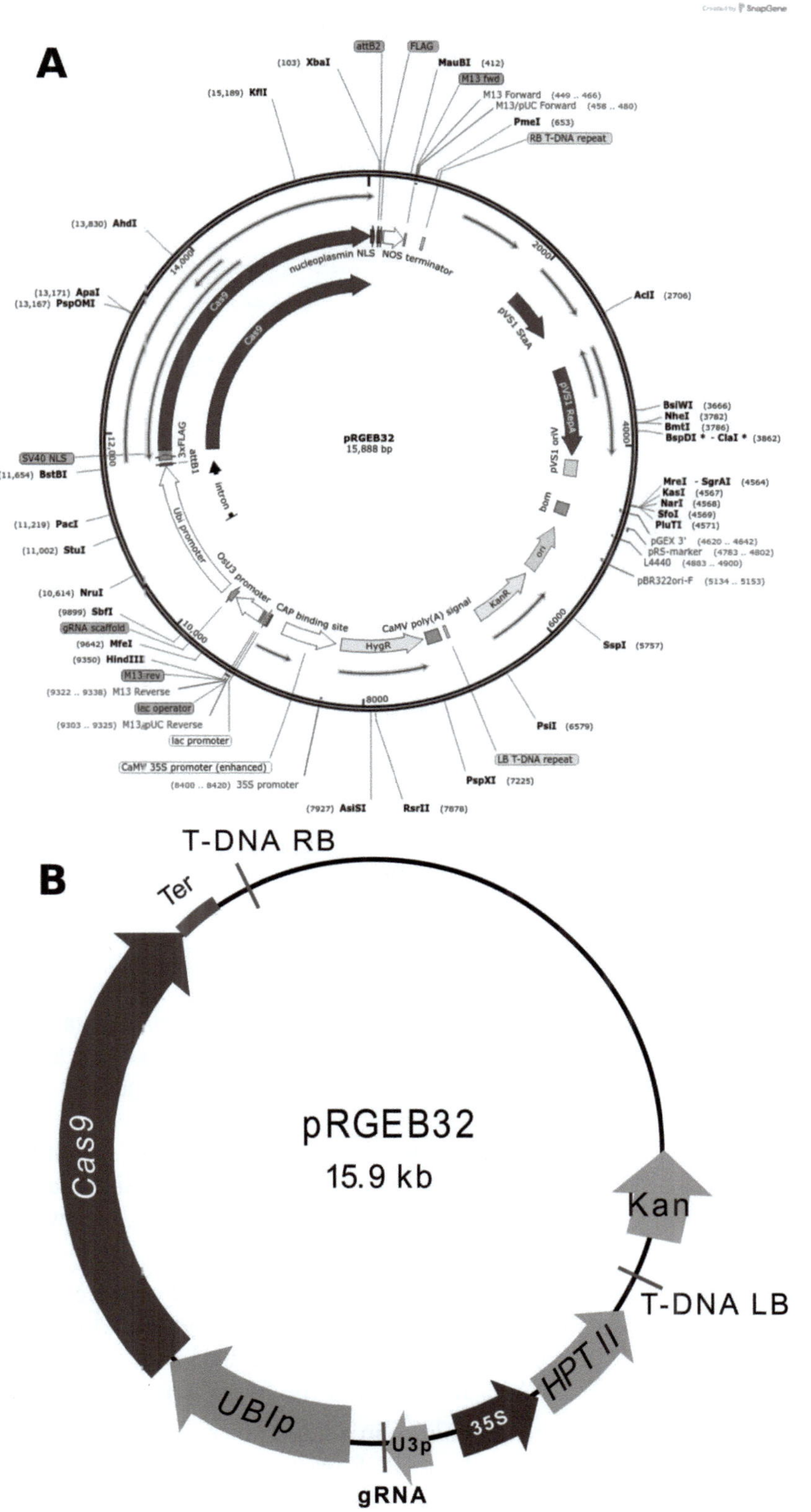

Fig. 1 Plasmid map of the pRGEB32—a CRISPR vector plasmid used in this protocol. (**a**) Detailed map of the pRGEB32 showing different sites (restriction sites, Cas9, hygromycin resistant gene, etc.); (**b**) a simple map showing key regions of the plasmid

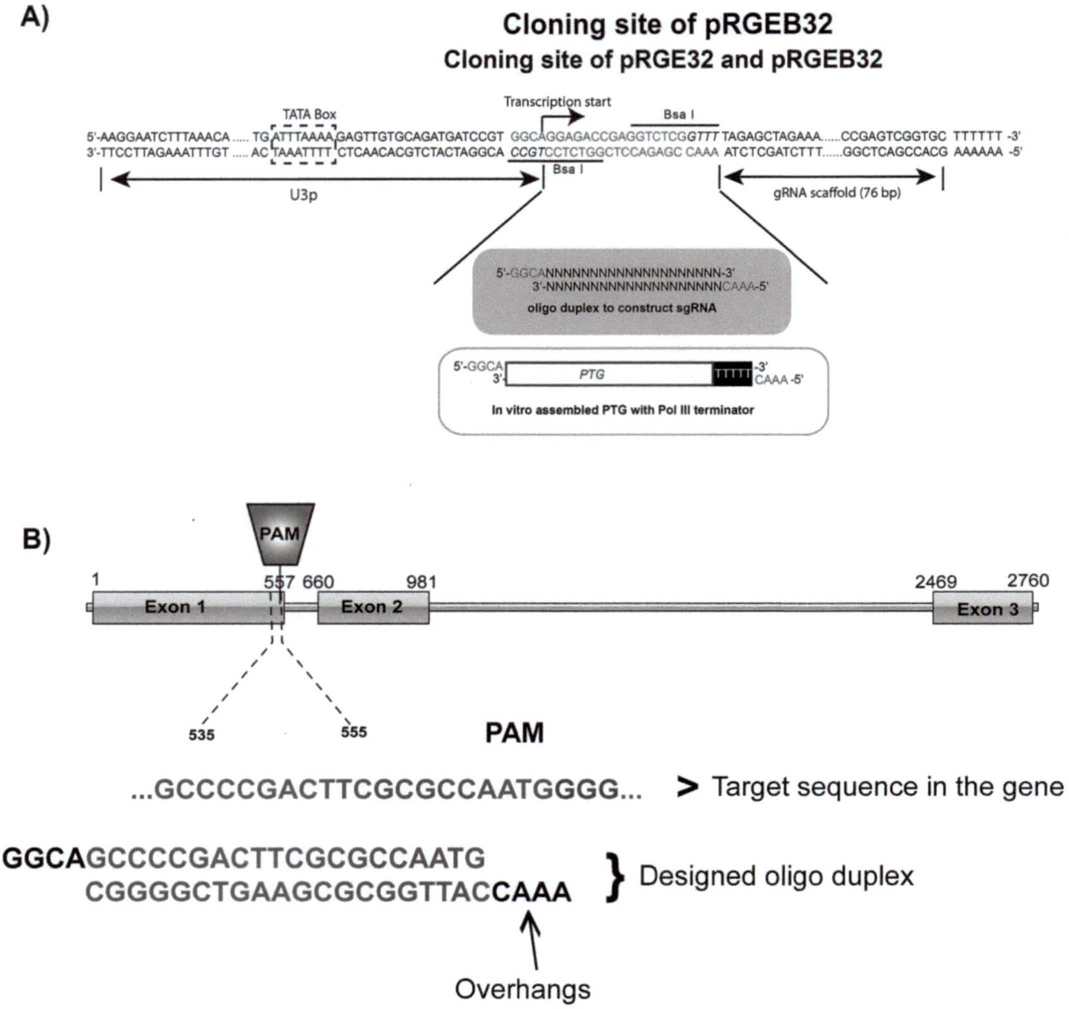

Fig. 2 Designing of sgRNA for targeted CRISPR-Cas9 editing with pRGEB32 plasmid. (**a**) The cloning site of pRGEB32, when digested with *Bsa*I enzyme leaves a four base overhang on either side. (**b**) The target region for sgRNA binding and PAM site for recognition in the SD1 gene of rice (Loc_01g66100)

SD1 region shown here has a cleavage site of *Bsa*JI enzyme. Such presence of restriction enzyme sites can be identified using the online tool NEB cutter (www.nc3.neb.com/NEBcutter).

4. pRGEB32, when digested with *Bsa*I enzyme leaves a four-base overhang on either side. Hence, before synthesis, add adapter sequences of "GGCA" at the start of the 5′→3′ direction and "CAAA" at the end of the 3′→5′ sequence (Fig. 2).
5. Synthesize oligos in a standard commercial 100 μM concentration (similar to PCR primers).

   A sample oligo pair designed to target the rice OsGA20Ox2 gene is given.

Example sd1F1— GGCAGCCCCGACTTCGCGC CAATG and sd1R1— AAACCATTGGCGC GAAGTCGGGGC.

6. Other primers: Design PCR primers to amplify the gRNA region of the plasmid. This is useful for confirmation of bacterial transformation. An example primer combination, sgs_F: A GGCGTCTTCTACTGGTGCT and sgs_R— CCTCCGTATTTGCTGACGTG.

Culture Media

2.2.1 Lysogeny Broth-Miller (for Bacterial Vultures)

1. For liquid broth weigh 25 g of LB broth, and dissolve it in 1 L of sterile water (*see* **Note 1**).
2. For solid media, add 15 g/L of agar to the liquid broth.
3. Autoclave the contents at 15 PSI for 15 min and store at 25 °C.

2.2.2 Agrobacterium Medium (AB Media)

1. AB salts (10X stock): Dissolve 10 g of NH_4Cl, 3 g of $MgSO_4.7H_2O$, 1.5 g of KCl, 100 mg of $CaCl_2.2H_2O$, and 25 mg of $FeSO_4$ in 1 L of distilled water. Sterilize by autoclaving at 15 PSI for 15 min. Store at 4 °C.
2. AB buffer (10X stock): Dissolve 10 g of NaH_2PO_4 and 30 g of K_2HPO_4 in 1 L of distilled water. Sterilize by autoclaving at 15 PSI for 15 min. Store at 4 °C.
3. Glucose solution (to be prepared fresh every time): For 1 L of AB media, dissolve 5 g of glucose in 100 mL distilled water and sterilize using 0.22 μm filter.
4. AB medium (liquid broth): Mix 100 mL each of sterile glucose solution, AB buffer, and AB salts; make up to 1000 mL using sterile distilled water under a sterile laminar air flow hood.
5. AB medium (semisolid): Add 15 g of agar in 700 mL of water, and sterilize by autoclaving. Once it gets cooled to about 60 ° C, add 100 mL each of glucose solution, AB salts and AB buffer (*see* **Note 2**).

2.2.3 Super Optimal Broth

1. Prepare super optimal broth with catabolite repression medium (SOC) [9] to recover transformed *Escherichia coli*-competent cells (*see* **Note 3**).
2. Add 0.5% yeast extract, 2% tryptone, 10 mM NaCl, 2.5 mM KCl, 10 mM $MgCl_2.6H_2O$, and 10 mM $MgSO_4.7H_2O$ in distilled water, and sterilize by autoclaving at 15 PSI for 15 min. To this, add 20 mM of glucose solution (sterilized through 0.22 μM filter). Store at 25 °C.

2.2.4 Terrific-Broth (for Overnight Culture of Transformed Bacterial Cells)

1. Modified Tartoff-Hobbs broth (terrific broth) [10] is routinely used to improve plasmid yields from recombinant *E. coli.*
2. TB medium: Mix 1.2% tryptone, 2.4% yeast extract, and 0.5% glycerol in 900 mL of distilled water. Autoclave at 15 PSI for 15 min. Store at 25 °C.
3. 10X TB salts (phosphate buffer): Mix 0.17 M of KH_2PO_4 and 0.72 M of K_2HPO_4 in distilled water. Sterilize by autoclaving at 15 PSI for 15 min and store at 25 °C.
4. Terrific broth: To 900 mL of TB medium add 100 mL of 10X TB salts under a sterile laminar air hood (*see* **Note 4**).

2.2.5 Modified AA Media (AAM)

1. Prepare and use modified AA media [11] as *Agrobacterium* infection medium.
2. AA macro (10X stock): Mix 29.5 g of KCl, 2.5 g of $MgSO_4.7H_2O$, 1.5 g of $CaCl_2.2H_2O$, and 1.6 g of $NaH_2PO_4.2H_2O$ in 1 L of distilled water. Store at 4 °C.
3. AA micro (1000X stock): Mix 7.6 g of $MnSO_4.H_2O$, 3 g of H_3BO_3, 2.0 g of $ZnSO_4.7H_2O$, 0.75 g of KI, 0.25 g of $Na_2MoO_4.2H_2O$, 0.03 g of $CuSO_4.5H_2O$, and 0.03 g of $CoCl_2.6H_2O$ in 1 L of distilled water. Store at 4 °C.
4. AA amino acids (100X stock): Mix 8.7 g of L-glutamine, 2.66 g of L-aspartic acid, 1.74 g of L-arginine, and 0.075 g of glycine in 100 mL of distilled water. Store at 4 °C.
5. MS vitamins (100X stock): Mix 10 g of myoinositol, 50 mg of nicotinic acid, 50 mg of pyridoxine-HCl, 100 mg of thiamine-HCl, and 200 mg of glycine in 1 L of distilled water. Store at 4 ° C.
6. Fe-EDTA (100X stock): Mix 2.78 g of $FeSO_4.7H_2O$ and 3.72 g of Na_2EDTA in 1 L of water. Store the solution in a brown glass bottle or an aluminum foil-covered container at 4 ° C.
7. AAM liquid medium: Mix 68.5 g of sucrose, 36 g of glucose, 500 mg of casamino acids, 100 mL of AA macro (10X), 1 mL of AA micro (1000X), 10 mL of Fe_2-EDTA (100X), 10 mL of AA amino acids (100X), and 10 mL of MS vitamins (100X), and make up the final volume to 1 L using distilled water. Adjust the pH to 5.2. Sterilize by autoclaving at 15 PSI for 15 min.

2.2.6 Antibiotic Solutions

Prepare the following antibiotics required for selecting transformed *E. coli* cells, *Agrobacterium* cells, and plant transformants, and store at −20 °C (*see* **Note 5**).

1. Dissolve 10 mg/mL of rifampicin in 100% methanol.
2. Dissolve 100 mg/mL of kanamycin in sterile molecular-grade water.

3. Dissolve 100 mg/mL of carbenicillin in sterile molecular-grade water.
4. Dissolve 250 mg/mL of cefotaxime in sterile molecular-grade water.
5. Dissolve 50 mg/mL of hygromycin B in phosphate buffered saline.

2.3 Plasmid Midi-Prep

Midi-prep of plasmid using a modified alkaline-lysis method [12] ensures good-quality plasmid for immediate use as well as for storage and future use (*see* **Note 6**).

1. Resuspension buffer: Mix 25 mM of Tris-HCl (pH 8.0), 50 mM of glucose, and 10 mM of Na_2-EDTA in sterile distilled water. Store at 4 °C.
2. Denaturing solution (prepare fresh before every use): Dissolve 0.2 N of NaOH and 1% sodium-dodecyl sulfate (SDS) in an appropriate quantity of sterile distilled water.
3. Renaturing solution: Dissolve 5 M of potassium acetate and 11.5 mL of glacial acetic acid in 100 mL of sterile distilled water. Store at 4 °C.

2.4 Plasmid and Plant Genomic DNA Mini-Prep and Gel Elution (Silica Matrix Protocol)

The silica matrix protocol [13] for purification of plasmid gives good-quality plasmids in mini-prep quantities suitable for sequencing and downstream applications (*see* **Note 7**).

1. 100 mM Tris-HCl (pH 7.5): Dissolve 0.61 g of Tris base in 40 mL of Millipore water. Adjust pH to 7.5 using conc. HCl. Make up the volume to 50 mL.
2. 100 mM EDTA (pH 7.5): Dissolve 1.86 g of Na_2_EDTA in 40 mL of Millipore water. Adjust pH to 7.5 and make up the volume to 50 mL.
3. Silica matrix: Mix 5 g of silicon dioxide with 50 mL of sterile water in a 50 mL centrifuge tube. Leave the contents undisturbed for 2 h at room temperature. Remove the supernatant (with fine silica particles). Refill the tube with sterile water and repeat the procedure once again. Make up the volume once again to 50 mL; this gives an approximate concentration of 100 mg/mL of silica. Store at room temperature.
4. Resuspension solution: Mix 50 mM Tris-HCl (pH 7.5) and 10 mM EDTA (pH 7.5) solutions and store at 4 °C. Add 100 mg/mL of RNase A before plasmid isolation.
5. Alkaline lysis solution: Mix 0.2 M of NaOH and 1% sodium dodecyl sulfate in distilled water. Store at 25 °C.
6. Neutralization solution: Dissolve 1.32 M of potassium acetate in distilled water. Adjust the pH to 4.8. Store at 4 °C.

7. Sodium iodide solution: Dissolve 6 M of NaI in distilled water. Store at 4 °C.
8. Washing solution: Mix 10 mM of Tris-HCl (pH 7.5), 100 mM NaCl, and 1 mM EDTA in 50% ethanol. Store at 25 °C.
9. SDS solution: Prepare 10% SDS solution in water and store at 25 °C.
10. P:C:I solution: Mix 25 parts of Tris-saturate phenol (pH 8.0), 24 parts of chloroform and 1 part isoamyl alcohol.

2.5 $CaCl_2$ Competent Cells

Competent cells of *E. coli* strain DH5α can be prepared by following the protocol by [14] with minor modifications. Prepare the following solutions (*see* **Note 8**).

1. 1 M $CaCl_2$: dissolve 14.7 g of $CaCl_2.2H_2O$ in 100 mL of distilled water.
2. 0.1 M $CaCl_2$: Add 10 mL of 1 M $CaCl_2$ to 90 mL of distilled water to prepare 100 mL solution.
3. 0.1 M $CaCl_2$ with 15% glycerol: Add 10 mL of 1 M $CaCl_2$ and 15 mL of 100% glycerol to 75 mL of distilled water to make 100 mL solution.
4. Sterilize the solutions by autoclaving at 15 PSI for 15 min. Store at 4 °C.

2.6 Agrobacterium

This protocol uses AGL-1 strain of *Agrobacterium*—a strain routinely used in transformation of cereals. It has a C58 chromosomal background with an insertion mutation in *recA* gene. The genome carries resistant genes for antibiotics rifampicin and carbenicillin. It contains pTiBo542DT, Ti plasmid with deleted transport (T) regions. Transformation with a binary vector containing the missing T-region will smoothly transfer the foreign DNA into host plants' genome. Maintain this strain of bacteria in YEB medium containing rifampicin 20 μg mL^{-1}.

2.7 Helper Plasmid

This protocol also uses a conjugal helper plasmid pRK2013, which helps in mobilization of non-self-transmissible plasmids of Gram-negative bacteria between different strains. It has resistance to the antibiotic kanamycin. The plasmid containing *E. coli* (DH5α) can be maintained in LB medium containing kanamycin 50 μg mL^{-1}.

2.8 Hormones for Plant Tissue Culture

Hormones for plant tissue culture are prepared in 1 mg/mL concentration and stored at 4 °C.

1. 2,4-Dichloro-phenoxyacetic acid (2,4-D): Dissolve 50 mg of 2,4-D in a few drops of absolute ethanol, and dilute to 50 mL using sterile deionized water.
2. Kinetin: Dissolve 50 mg of kinetin in a few drops of 1 N NaOH, and dilute to 50 mL with sterile deionized water.

3. 6-Benzylaminopurine (6-BA): Dissolve 50 mg of 6-BA in a few drops of 1 N NaOH, and dilute to 50 mL using sterile deionized water.
4. Alpha-naphthaleneacetic acid (NAA): Dissolve 50 mg of NAA in a few drops of 1 N NaOH and dilute to 50 mL using sterile deionized water.

2.9 Acetosyringone

Acetosyringone at 100 mM concentration enhances the transformation during co-cultivation.

Preparation: Dissolve 392 mg of 3′,5′-dimethoxy-4′-hydroxyacetophenone (Sigma, USA) in 20 mL of dimethyl sulfoxide (DMSO). Store the solution at −20 °C in 1 mL aliquots.

2.10 Media for Plant Tissue Culture

For rice plant tissue culture, this protocol uses ready-to-use basal Murashige and Skoog [15] media with minor modifications suited to recalcitrant rice genotypes (*see* **Note 9**).

2.10.1 0.7X MS (-kin) Medium for Callus Induction and 0.7X MS (+kin) for Proliferation

Use Murashige and Skoog medium [15] in reduced concentrations and with the following minor modifications for callus induction (without kinetin) and proliferation (with kinetin) (*see* **Note 10**).

1. *Callus induction media*: Dissolve 3.08 g of ready-to-use MS medium, 30 g of maltose, and 2.5 mL of 1 mg/mL solution of 2,4-D in approx. 800 mL of distilled water.
2. Adjust the pH to 5.8.
3. *Callus proliferation media*: also add 100 μL of 1 mg/mL solution of kinetin to the above.
4. Add 4 g of Gelrite and 2 g of agar, and make up the volume to 1000 mL using sterile distilled water (*see* **Note 11**).
5. Sterilize by autoclaving at 15 PSI for 15 min.
6. Cool to about 60 °C and pour into sterile Petri plates under a laminar air flow hood. Once solidified, store at dark at 25 °C (*see* **Note 12**).

2.10.2 MS Co-cultivation Medium

Agrobacterium carrying the CRISPR-Cas9 constructs and the embryogenic calli to be transformed are co-cultivated in MS co-cultivation medium.

1. Dissolve 4.4 g of ready-to-use MS media, 30 g of sucrose, 10 g of glucose, and 2.5 mL of 1 mg/mL 2,4-D solution in approx. 800 mL of distilled water and adjust the pH to 5.6.
2. Add 3 g of Gelrite, make up the volume to 1 L and sterilize by autoclaving.
3. When cooled to about 60 °C, add 1 mL of 100 mM acetosyringone, mix by swirling, and pour into sterile Petri plates.
4. Store at 25 °C and under dark.

2.10.3 MS Selection Medium for Putative Transformants

Add hygromycin 50 mg L^{-1} and cefotaxime 250 mg L^{-1} with 0.7X (+kin) to prepare selection medium. Add the antibiotics just before pouring into Petri dishes (*see* **Note 13**).

2.10.4 Regeneration Medium

MS-regeneration medium contains appropriate hormones and antibiotics to help in regeneration of the transformed calli.

1. Dissolve 4.4 g of ready-to-use MS media, 30 g of maltose, 300 mg of casamino acids (casein acid hydrolysate), 3 mL of 1 mg/mL 6-BA, and 1.5 mL of 1 mg/mL of NAA in approx. 800 mL of distilled water.
2. Adjust pH to 5.8 and add 4 g of Gelrite before making up to 1000 mL with distilled water.
3. Sterilize by autoclaving at 15 PSI for 15 min.
4. Once cooled to about 60 °C, add 1 mL of 50 mg/mL hygromycin and 1 mL of 250 mg/mL cefotaxime before pouring into Petri plates.
5. Store at 25 °C in dark.

2.10.5 Rooting Medium

Half-strength MS medium without hormones is used as the rooting medium.

1. Dissolve 2.2 g of ready-to-use MS medium and 15 g of sucrose in approx. 800 mL of distilled water. Adjust pH to 5.8.
2. Add 4 g of Gelrite before making up the volume to 1000 mL.
3. Once cooled to about 60 °C, add 600 μL of 50 mg/mL hygromycin (final concentration 30 μg/L).
4. Pour the contents into sterile culture tubes and seal with sterile cotton plugs.

3 Methods

3.1 Recovery of Plasmid (pRGEB32) Containing Cells from Stab Culture

1. If the plasmid containing cells are received as stab cultures, scrap and streak to recover the *E. coli* cells in semisolid LB media (with 50 μg/mL kanamycin).
2. After incubation at 37 °C in dark for about 12 h, single colonies appear on plates.
3. Pick a colony and grow in 50 mL of liquid LB Kan 50 medium for 8 h at 37 °C in an incubator-shaker. This culture can be used to isolate high quantity of plasmid using midi-prep procedure (Fig. 3)

3.2 Midi-Prep Procedure for Plasmid Isolation

It is better to have a high quality of concentrated original plasmids in multiple aliquots in case any failure or contamination arises

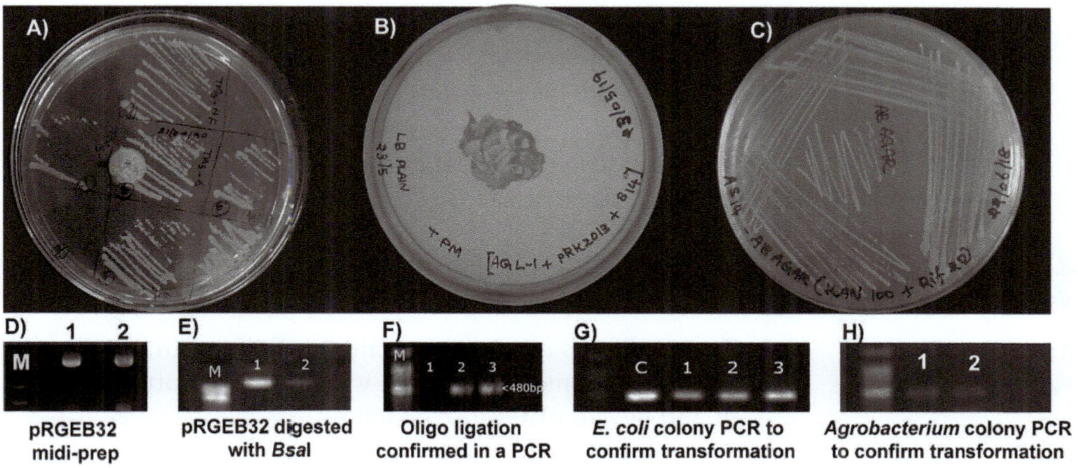

Fig. 3 Transformation of the *Agrobacterium* with CRISPR plasmid carrying the gRNA with target oligos. (**a**) Transformed *E. coli* colonies with CRISPR plasmid growing in LBKan50 medium. (**b**) Triparental mating to mobilize the plasmid to *Agrobacterium* (*E. coli* carrying pRGEB32 with custom oligo + *E. coli* carrying pRK2013 + *Agrobacterium* to be transformed). (**c**) Successfully mobilized *Agrobacterium* growing in double antibiotic media. (**d**) Plasmid isolated from colonies of *E. coli*. (**e**) *Bsa*I digestion of the pRGEB32: (1) uncut plasmid, (2) digested plasmid after gel elution. (**f**) PCR confirmation of oligo duplex ligation: (1) linearized plasmid (no ligation), (2 and 3) plasmid ligated with oligo duplex. (**g**) *E. coli* colony PCR showing amplification of target region, confirming successful transformation. (**h**) *Agrobacterium* colony PCR showing amplification of target region, confirming successful transformation

during the experiment. Midi-prep offers a better strategy to prepare high quantity of plasmids (Fig. 3d).

1. Use a refrigerated centrifuge cooled to 4 °C throughout the procedure.
2. Split the 50 mL bacterial culture into 2 × 25 mL cultures and centrifuge at 5000 rpm for 10 min.
3. Discard the supernatant and resuspend the pellet with 1.5 mL of resuspension buffer. Mix thoroughly using a wide-mouthed pipette or by vortexing until no cell clumps are visible.
4. Add 3 mL of freshly prepared denaturing solution, gently invert a few times, and incubate at room temperature for 2 min.
5. Add 2.25 mL of ice-cold renaturing solution, mix by inversion, and incubate in ice bath for 15 min. Centrifuge at 10000 rpm for 20 min.
6. Transfer the clear supernatant to a fresh 15 mL centrifuge tube, and add equal volume of phenol/chloroform (25:24) solution. Mix and centrifuge at 10000 rpm for 5 min.
7. Transfer only the top aqueous layer to a fresh centrifuge tube, and add equal volume of isopropanol. Incubate at room temperature for 5 min, and centrifuge at 10000 rpm for 15 min.

8. Discard the supernatant, wash the pellet with 1 mL of 70% ethanol, and incubate for 4 min at room temperature. Centrifuge the contents for 5 min.
9. Discard the ethanol layer, air dry the pellet at room temperature, and dissolve the contents with 1 mL of sterile molecular-grade water. Store at −20 °C

3.3 Oligo Ligation to Vector Backbone

The pRGEB32 plasmid can be linearized using *Bsa*I restriction enzyme.

3.3.1 Linearizing the Plasmid

1. Setup a 50 μL restriction reaction as follows: In a PCR tube take 41 μL of molecular-grade water, 5 μL of buffer, and 3 μL of 250 ng/μL pRGEB32 plasmid.
2. To this add 1 μL of *Bsa*I enzyme and mix by pipetting.
3. Incubate the contents in a thermocycler at 37 °C for 60 min.
4. After incubation, add 10 μL of tracking dye, and separate the fragments by electrophoresis using 1% agarose.
5. Elute the linearized plasmid with any commercial gel elution kit and dilute to a concentration of 50 ng/μL. Store at −20 °C (Fig. 3e) (*see* **Note 14**).

3.3.2 Preparing the Oligos

The forward and reverse oligos are to be annealed to double strands before ligating to the plasmid background.

1. For a 10 μL reaction, in a microcentrifuge tube, take 7 μL of molecular-grade water, 1 μL each of forward and reverse oligo (100 μM concentration), and 1 μL of ligation buffer.
2. Mix by pipetting and place the tube in a 90 °C water bath for 5 min. Then, allow the water to slowly cool to room temperature (*see* **Note 15**).

3.3.3 Vector Plasmid-Oligo Ligation

Mix the linearized vector plasmid and the double-stranded oligos together in 1:50 ratio to synthesize the CRISPR/Cas9 plasmid.

1. In a PCR tube, take 1 μL of 50 ng/μL linearized vector plasmid (1 molar ratio) and 50 molar ratio of double-stranded oligo.
2. Add 1 μL of ligase buffer, and make up the volume to 9 μL using molecular-grade water.
3. Add 1 μL of DNA ligase and mix by pipetting. Incubate the contents in a thermocycler at 16 °C for 2 h (Fig. 3f).

3.4 Preparation of Calcium Chloride Competent *E. coli* Cells

Competent cells are efficiently transformed with vector plasmids and can multiply rapidly to produce several copies. A standard DH5α strain of *E. coli* is made competent by treating them in calcium chloride solutions. Throughout this procedure, all the solutions and culture should be maintained in an ice bath. All centrifugation steps should be gentler and performed at 4 °C (*see* **Note 16**).

1. Isolate single colonies of DH5α strain of *E. coli* by streaking on a LB media plate. It carries no antibiotic and hence should be maintained free from contamination.
2. Pick a single colony and grow the bacteria to saturation in 5 mL of LB liquid medium.
3. Add 1 mL from this culture to 40 mL of sterile LB medium, and grow the bacteria for about 3 h until a silky growth is observed (OD is around 0.5). Higher OD cultures shall be diluted using sterile LB media.
4. Split the culture into 1.5 mL pre-chilled microcentrifuge tubes while maintaining in the ice bath. Incubate in the ice bath for 30 min.
5. Centrifuge the tubes at 5000 rpm for 5 min and discard the supernatant.
6. Add 100 μL of 0.1 M $CaCl_2$ solution and dissolve the pellet by gently flicking. Make up the volume to 1 mL with 0.1 M $CaCl_2$. Centrifuge the tubes at 5000 rpm for 5 min, and discard the supernatant.
7. Finally, resuspend the pellet with ice-cold 0.1 M $CaCl_2$ containing 15% glycerol and store immediately at −80 °C.
8. The competency of these cells can be assessed by transforming the pRGEB32 plasmid and growing in LB Kan 50 media.

3.5 Transforming of the Plasmid to Competent DH5α Cells

1. Thaw previously frozen competent DH5α cells by keeping them in ice bath for 30 min.
2. Add 5 μL of ligated plasmid to the cells while maintaining them in ice bath.
3. Immediately after, heat-shock the cells by keeping them in 42 ° C water bath for 45 s, and transfer to ice bath and incubate for 2 min.
4. To recover the cells, add 800 μL of SOC media to the tube and grow at 37 °C with shaking (250 rpm) for 60 min.
5. From this growing culture, pipette 100 μL and plate separately on a LB Kan50 media.
6. Centrifuge the remaining culture at 8000 rpm for 3 min, discard the supernatant, and resuspend the pellet with 100 μL of SOC media.
7. Plate this remaining culture on a fresh LB Kan50 plate, invert the plates, and incubate them at 37 °C in dark for 8–12 h (Fig. 3a).

3.6 Confirmation of Plasmid Transformation into DH5α

3.6.1 Colony PCR

1. Confirm the plasmid transformation using colony PCR with "sgs" primers targeting the gRNA region of the vector plasmid pRGEB32. Successful transformation will amplify the region, whereas plain colonies do not produce any bands.
2. Prepare a 25 μL PCR cocktail with 11.8 μL of water, 0.6 μL each of forward and reverse primers, 12.5 μL of 2X PCR mastermix, and the single colonies as the template (Fig. 3g) (*see* **Note 17**).
3. For template, pick up to five single colonies per plate that are to be confirmed and streak on a fresh LB Kan 50 media using a sterile 10 μL pipette tip.
4. Then wash the tip in the PCR cocktail prepared previously. Set up a standard 35 cycle PCR reaction in a thermocycler.

3.6.2 Plasmid Mini-Prep from Colonies Confirmed Through Colony PCR

Plasmids isolated from the colonies confirmed through colony PCR can be further confirmed through Sanger sequencing. In addition, these colonies can be stored as stocks for future transformations. The silica-matrix protocol is followed here.

1. Pick a single colony from the previously streaked LB Kan50 plate and grow in 2 mL Terrific Broth with Kan50 at 37 °C for 8–12 h with 200 rpm shaking.
2. Centrifuge the culture at 10000 rpm for 30 s to pellet the cells. Resuspend the pellet with 100 μL of resuspension solution (*see* **Note 18**).
3. Add 100 μL of alkaline lysis solution and mix by inverting the tube several times.
4. Add 100 μL of neutralization solution and mix by inversion. Centrifuge the contents at 16000× *g* for 5 min. Transfer the supernatant carefully to a fresh tube.
5. To this, add 500 μL of 6 M NaI solution and mix by inversion. Add another 20 μL of silica matrix, mix well, and incubate at room temperature for 5 min.
6. Centrifuge the contents at 16000× *g* for 10 s to pellet the silica matrix. Discard the supernatant.
7. Now, wash the matrix by adding 500 μL of washing solution and brief vortexing. Remove the supernatant by pipetting. Repeat this step once again.
8. After ensuring no residual washing solution is present, resuspend the pellet with 20 μL of sterile molecular-grade water and incubate at 70 °C for 2 min.
9. Centrifuge the tubes at 16000× *g* for 2 min; transfer the supernatant to a fresh tube and store at −20 °C.

10. Before mobilizing to *Agrobacterium*, use the mini-prep plasmid as template to confirm proper ligation of CRISPR oligos through Sanger sequencing using the sgs_F and sgs_R primers.

3.7 Mobilization of Recombinant Plasmid to Agrobacterium tumefaciens Through Triparental Mating

Follow triparental mating [16] to mobilize the plasmid into the *Agrobacterium* (Fig. 3b) (*see* **Note 19**).

1. Take the actual day of triparental mating as day +1 and follow the procedure.
2. Day −3: Streak AGL-1 strain of *Agrobacterium* on an AB (Rif20) plates and incubate at 28 °C in dark to get single colonies.
3. Day −1: Streak *E. coli* containing the plasmid to be mobilized (pRGEB32 with oligos) and the helper plasmid (pRK2013) separately on LB (Kan50) plates. Incubate overnight at 37 °C.
4. Day +1: Pick three single colonies of AGL-1 and a single colony each from pRGEB32 with oligos and pRK2013, and patch closely on a plain LB plate (no antibiotics). Mix these colonies using a sterile loop, and incubate at 28 °C in dark overnight.
5. Day +2: Scrap the bacterial growth on the LB plate with a sterile loop, and suspend in 1 mL of 0.9% sterile NaCl solution. Make serial dilutions up to 10^{-6} of this culture by adding 100 μL of the previous suspension to 900 μL NaCl. Plate 100 μL of each dilution on double antibiotic AB (Kan100 + Rif20) plates. Incubate at 28 °C for 3 to 5 days.
6. Day +5: Single colonies that appear on double antibiotic plates are usually the *Agrobacterium* with mobilized plasmids. Confirm mobilization in these colonies through colony PCR with primers that amplify the gRNA region of pRGEB32 (Fig. 3h).
7. Before mobilizing to *Agrobacterium*, use the mini-prep plasmid as template to confirm proper ligation of CRISPR oligos through Sanger sequencing using the sgs_F and sgs_R primers.

3.8 Development of Transformable Plant Material

This protocol uses embryogenic calli derived from rice seeds for transformation through *Agrobacterium* (Fig. 4).

1. Choose rice seeds that are well filled, heavy, and with no discoloration of husks (*see* **Note 20**). Before every experiment dehusk the seeds without damaging the embryo (*see* **Note 21**).
2. Set up the laminar air-flow hood to ensure sterile environment for working.
3. Take dehusked rice seeds in a sterile container. First, wash the seeds with 70% ethanol for 45 s (*see* **Note 22**). Remove the solution by pipetting.
4. Then wash the seeds with 3 mL of 0.1% $HgCl_2$ for 3 min (*see* **Note 23**). Discard the residual solution by pipetting.

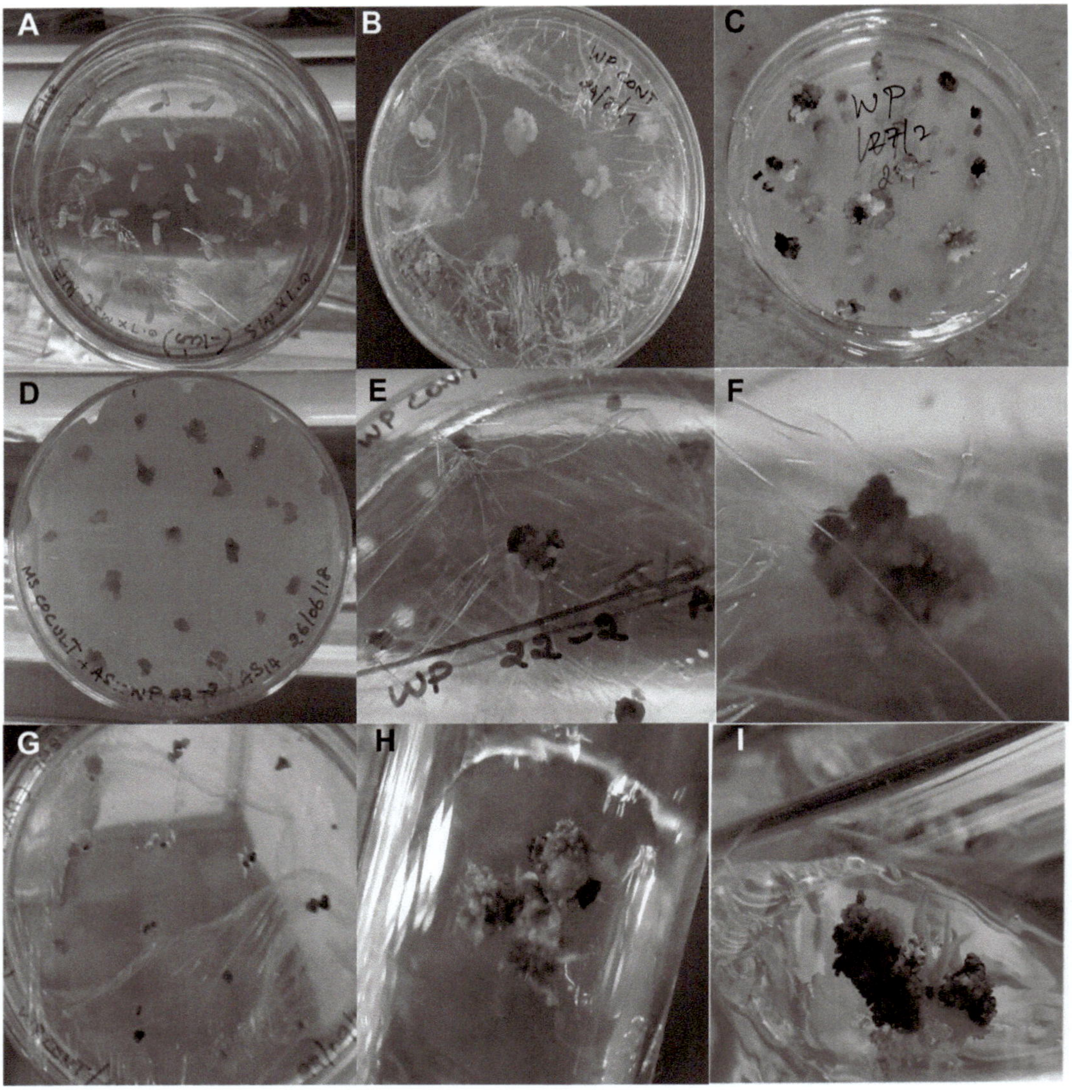

Fig. 4 Different stages in rice transformation. (**a**) Seeds inoculated in 0.7X MS(-kin) medium. (**b**) Well-proliferated calli, suitable for transformation after three subcultures. (**c**) Phenol exudation and drying of calli—these are not suitable for transformation. (**d**) Co-cultivation of calli with *Agrobacterium*. (**e** and **f**) proliferating calli in MS Selection medium indicating successful transformation. (**g**) Calli subcultured prematurely fails to develop. (H and I) Shoot regenerating from the transformed calli

5. Wash the seeds thoroughly with sterile Millipore water for at least three times.
6. For better results, soak the seeds in water for up to 6 h before transferring to callus induction media.
7. To induce callus, place about 15 sterile rice seeds on 0.7X MS (-kin) media. Seal the plates with PhytaWrap and incubate at 27 °C in dark for 7 days.

8. Observe the plates daily for contaminations, and if any, discard the plates immediately.
9. After 7 days, excise the scutellum-derived calli from the seeds, and transfer them to 0.7X MS (+kin) media for callus proliferation.
10. Once every 2 weeks, transfer the growing calli to fresh 0.7X MS (+kin) media plates. Two to three subcultures are sufficient to get transformable calli.

3.9 Co-cultivation of Agrobacterium and Rice Calli

1. Streak *Agrobacterium* culture carrying the pRGEB32 on an AB (kan100 + rif20) plate.
2. Incubate at 28 °C in dark for 3 to 5 days.
3. Meanwhile, subculture healthy rice calli (cream-colored, hard, and friable) in callus proliferation media for 5 days at 27 °C in dark.
4. After 3 to 5 days, pick about six single colonies from the *Agrobacterium* plate, and suspend in 50 mL of AB (kan100 + rif20) liquid broth; incubate at 28 °C with 250 rpm shaking to grow the culture.
5. When sufficient growth is observed (~2 days), transfer the culture to centrifuge tubes, and centrifuge at 8000 rpm for 5 min.
6. Resuspend the pellet in 5 mL of AA media and add 100 mM acetosyringone (AA-As). Adjust the optical density of the culture to 0.3, 0.5, and 1.0 at 600 nm with AA-As media.
7. Once the *Agrobacterium* culture and freshly subcultured calli are ready proceed to the callus infection stage.
8. Suspend healthy calli in different concentrations of *Agrobacterium* culture that was prepared earlier.
9. With this step we can optimize the suitable *Agrobacterium* concentration and time of infection for our choice of rice cultivar. Optimize the period of infection between 1 min and 3 min.
10. After infection, remove the calli from suspension and dry by placing them on sterile Whatman No. 1 filter paper.
11. Incubate the callus in MS-co-cultivation media plates, and overlay with a Whatman No. 1 filter paper wetted with 1 mL of AA-As medium.
12. Incubate the plates 26 °C for 3 days in dark (Fig. 4d).

3.10 Selection of Transformants

1. After co-cultivation, wash the callus with sterile water to remove excess bacterial growth and until the water is clear.
2. Give a final wash with sterile water containing 250 μg mL^{-1} of cefotaxime, and blot dry with sterile tissue papers.

3. Incubate the calli in MS selection medium at 27 °C in dark for up to 10 days.
4. After 7–10 days, transfer only the actively growing calli to a second round of selection in MS selection medium for 20 days.
5. A third round of selection may be given to reduce probable non-transformants and to enable vigorous callus growth (Fig. 4e, f).

3.11 Shoot and Root Regeneration of Transformants

1. Choose vigorously growing optimal-sized calli (about 0.5 cm dia) for regeneration. Incubate them in MS regeneration medium in dark at 27 °C for 14 days.
2. After 14 days, subculture healthy calli on fresh MS regeneration medium, and incubate them at 27 °C with light-dark cycles of 16 h and 8 h. This will produce shoots from the callus (Fig. 4h, i).
3. To induce roots, transfer calli with regenerated shoots of about 2–3 cm to rooting medium and incubate at 27 °C for 14 days (Fig. 5a).
4. Maintain light and dark cycles of 16 h and 8 h, respectively, throughout. Roots will grow sufficiently within this period.

Fig. 5 Root regeneration and hardening of CRISPR-edited plants. (**a**) Root regeneration in 0.5X MS media. (**b**) Fully regenerated plants hardened in pots

3.12 Hardening of Regenerated Plantlets

1. Transfer well-rooted plantlets to culture tubes containing sterile distilled water and seal the mouth of the tubes with aluminum foil.
2. Incubate them at 27 °C for 3 days.
3. After 3 days, remove the foil and incubate in the same tube for another 5 days. This will reduce the shock of transferring to the growing pots (Fig. 5b).
4. Transfer the plantlets to pots containing sterile soil mixture enriched with compost.
5. Maintain the potted plants under a transgenic greenhouse with appropriate day and night temperatures and light intensity.

3.13 Confirmation of Transformation

The plant genomic DNA from the putative mutants can be isolated by following the silica-matrix protocol [13].

3.13.1 Plant Genomic DNA Isolation

1. Macerate approximately 50 mg of leaf tissue sampled from the hardened plants with 400 μL of resuspension solution. Transfer the contents to a sterile microcentrifuge tube.
2. Add 30 μL of SDS solution to the tubes and mix well by inversion. Incubate at 70 °C in a water bath for 10 min.
3. Add 250 μL of P:C:I solution to the tubes, mix well by vortexing (about 30 s) and centrifuge at 16000× *g* for 10 min.
4. Carefully transfer the top aqueous layer to a fresh microcentrifuge tube, and add 500 μL of sodium iodide solution. Mix by inversion.
5. To this, add 20 μL of silica matrix and mix by inversion. Incubate the tubes at room temperature for 5 min.
6. Centrifuge the tubes at 16000× *g* for 20 s, and remove the supernatant by pipetting.
7. Wash the pelleted matrix with 700 μL of washing solution, mix by vortexing, and centrifuge at 16000× *g* for 20 s. Repeat the procedure.
8. After removing the supernatant, centrifuge the tubes once again for 20 s to collect remaining washing solution. Remove this excess solution completely using a fine tipped pipette.
9. Elute the plant genomic DNA from the matrix by adding 40 μL of sterile molecular-grade water and incubating at 70 °C. This volume can be modified as per the concentration of the DNA required.
10. Centrifuge at 16000× *g* for 2 min to collect the DNA in the supernatant.
11. Transfer the supernatant containing DNA to a fresh microcentrifuge tube without any remnant matrix using a pipette.

3.13.2 Identification of True Transformants

The hygromycin resistance gene *hptII* also gets transformed and incorporated into the rice cells. Screening the presence of this gene shows true transformants.

1. Design primers targeting the *hptII* gene of the pRGEB32 plasmid.
2. Set up a standard PCR with genomic DNA of putative mutants. Use pRGEB32 as positive control and the wild-type plant as negative control.
3. Select only the hygromycin-positive plants to confirm transformation.

3.13.3 Confirmation of Mutagenesis Through PCR/RE Assay

The *Cas9* enzyme cleaves approximately 3 bp upstream of the PAM sequence. By designing sgRNAs such that a restriction enzyme recognition site overlaps the *Cas9* cleavage site, targeted cleavage can be validated or manipulated using restriction digestion. The *OsGA20Ox2* region targeted here has a cleavage site for *Bsa*JI enzyme. Any indel mutation will affect the restriction enzyme cleavage and thus can be identified.

1. Amplify a large portion of the *Cas9* target region of the putative mutant plant using a standard PCR.
2. Use wild-type DNA as positive control.
3. Evaluate the banding pattern with a small quantity of PCR product in 2% agarose gel. If more than one band is present for a genotype, it can already be considered as a mutant and doesn't need to be digested with the restriction enzyme.
4. Clean up the remaining PCR products from the other genotypes (having a single band). Elute the DNA into approximately 50 ng/μL concentration.
5. Set up a 25 μL restriction digestion: In a PCR tube take 5 μL of template DNA, 2.5 μL of 10X restriction buffer, 0.5 μL of *Bsa*JI enzyme and nuclease-free water to 25 μL. Incubate the tubes in a thermocycler at 37 °C for 1 h.
6. Run the products in 2% agarose gel. Wild-type DNA (positive control) will break into two bands due to the enzyme.
7. Lanes with uncut bands (resistant to the enzyme) can be considered as mutagenized.
8. Choose only the plants with resistant bands and the genotypes that produced multiple bands in the PCR to evaluate the type of mutations by Sanger sequencing.

3.13.4 Phenotyping of Mutant Plants

1. Grow the mutant plants under controlled environment with 25–30 °C day and 18–20 °C night temperature with 16 h of light and 8 h of darkness.

2. Grow wild-type plants alongside the mutants as a control.
3. Evaluate the phenotype of the plants for the expression of mutant characters.

4 Notes

1. Using a commercially available LB broth in lyophilized form is much efficient to prepare microbial media. Rather than using stock solutions to prepare the LB media, this lyophilized form is much cleaner and does not get affected by the problem of stock contamination.
2. *Agrobacterium* can also be grown in other media such as YEB. But, the AB is a selective and slow-growth medium that prevents contamination by other microorganisms.
3. SOC can also be replaced with LB media for recovery of the transformed *E. coli*.
4. We observed that the terrific broth produced higher plasmid yields in relatively shorter periods of time than the LB media.
5. Prepare and store antibiotics in 1 mL aliquots in microcentrifuge tubes, and after one or two uses, discard the tubes and use from a fresh tube.
6. We observed that the mini-prep protocols may sometimes not yield better plasmids due to relatively rougher handlings. However, the midi-prep consistently produced higher quantities of good-quality plasmids.
7. The silica matrix protocol was observed to produce good quality of DNA readily useful for downstream applications. This protocol can be replaced with a commercial DNA isolation kit. However, the quantity of elution can be modified as per our requirements in the silica matrix protocol.
8. Care must be given to ensure that there is no contamination in the parental *E. coli* plates or culture.
9. Ready-to-use MS salts were preferred due to the difficulty in acquiring ammonium nitrate, a major component of the original MS media.
10. We observed that using minimal concentrations of kinetin during callus proliferation gives hard and friable calli.
11. Using of agar alone makes the media cloudy and sometimes accumulates more phenols (dependent on the rice variety). Gelrite can be replaced with agarose in the same concentration. Contaminations, if any, are also easily observed in plates with agarose/Gelrite.

12. Pour about 30 mL of media per plate. Too less quantity warrants frequent subcultures. Before storing make sure there is no condensed moisture. Wrap the plates with PhytaWrap and store upside down.
13. Some protocols may suggest the use of less concentration of hygromycin in the media. However, it is better to have less number of true transformants than having more false positives.
14. Make sure to use a high-fidelity enzyme to linearize the plasmid; ensure complete digestion and a clean elution.
15. This process can be replicated in a thermocycler by slowly reducing the temperature from 90 °C to 25 °C. But, using hot water is straightforward.
16. Use pre-cooled centrifuge tubes, and maintain them in ice throughout the procedure.
17. If colonies produce a smear (due to excess template), try making a dilution in warm water before using them as the template in PCR.
18. Keenly observe the pelleted cells. Off-colored cells (*E. coli* cells are usually creamy yellow) may indicate contamination and may explain why the PCR failed.
19. Outmost care should be given to avoid any contamination as there are no antibiotics used to select the strains.
20. Always choose the best quality of seeds for tissue culture. It is better to harvest from a single season, dry to optimum moisture. Store a reasonable quantity of seeds in a refrigerator rather than in an open lab environment where the temperature may fluctuate resulting in quick degradation of the seeds. We always observed that the seed quality affects the quality of the callus.
21. Dehusking can be done by gently rubbing the seeds sandwiched between fine-grit sandpapers. But ensure that the scutellum (embryo region) of the seed is not damaged.
22. Ethanol sterilization should not be more than a minute, as it can damage the embryo.
23. Many protocols suggest the use of other sterilizing agents such as sodium hypochlorite. However, we found that the 0.1% $HgCl_2$ gave better results. But, care should be taken while handling and disposing off of the used chemical.

References

1. Wiedenheft B, Sternberg SH, Doudna JA (2012) RNA-guided genetic silencing systems in bacteria and archaea. Nature 482(7385): 331–338
2. Gasiunas G et al (2012) Cas9–crRNA ribonucleoprotein complex mediates specific DNA cleavage for adaptive immunity in bacteria. Proc Natl Acad Sci 109(39):E2579–E2586

3. Anders C et al (2014) Structural basis of PAM-dependent target DNA recognition by the Cas9 endonuclease. Nature 513(7519): 569–573
4. Jinek M et al (2012) A programmable dual-RNA–guided DNA endonuclease in adaptive bacterial immunity. Science 337(6096): 816–821
5. Hiei Y, Komari T (2008) Agrobacterium-mediated transformation of rice using immature embryos or calli induced from mature seed. Nat Protoc 3(5):824–834
6. Kumar KK et al (2005) An improvedAgrobacterium-mediated transformation protocol for recalcitrant elite indica rice cultivars. Plant Mol Biol Report 23(1): 67–73
7. Xie K, Minkenberg B, Yang Y (2015) Boosting CRISPR/Cas9 multiplex editing capability with the endogenous tRNA-processing system. Proc Natl Acad Sci 112(11):3570–3575
8. He C et al (2021) CRISPR-cereal: a guide RNA design tool integrating regulome and genomic variation for wheat, maize and rice. Plant Biotechnol J 19(11):2141–2143
9. Hanahan D (1983) Studies on transformation of Escherichia coli with plasmids. J Mol Biol 166(4):557–580
10. Tartoff K, Hobbs K (1987) Improved media for growing plasmid and cosmid clones. Focus 9:9–12
11. Hiei Y et al (1994) Efficient transformation of rice (Oryza sativa L.) mediated by agrobacterium and sequence analysis of the boundaries of the T-DNA. Plant J 6(2):271–282
12. Addgene (2018) Recovering plasmid DNA from bacterial culture. Protocols. Available from: https://www.addgene.org/protocols/purify-plasmid-dna/
13. Li J-F, Li L, Sheen J (2010) Protocol: a rapid and economical procedure for purification of plasmid or plant DNA with diverse applications in plant biology. Plant Methods 6(1):1
14. Chang AY et al (2017) Preparation of calcium competent Escherischia coli and heat-shock transformation. JEMI Methods 1:22–25
15. Murashige T, Skoog F (1962) A revised medium for rapid growth and bio assays with tobacco tissue cultures. Physiol Plant 15(3): 473–497
16. Ditta G et al (1980) Broad host range DNA cloning system for gram-negative bacteria: construction of a gene bank of Rhizobium meliloti. Proc Natl Acad Sci 77(12): 7347–7351

Chapter 4

Characterization of Transgenic Plants by Southern Blotting in Rice

Velusamy Sumithra, M. T. Andrew-Peter-Leon, Jeshima Khan Yasin, and M. Arumugam Pillai

Abstract

This study investigates the genomic organization of the OsSAMDC gene in rice, focusing on its presence in the transgenic *japonica* rice cultivar Yukihikari. Using Southern blotting, the research examines the number and arrangement of OsSAMDC-like genes in the transgenic rice genome. The methodology involves extracting high-molecular-weight genomic DNA, digesting it with various restriction enzymes and performing Southern hybridization analysis. Multiple hybridization probes, including a full-length cDNA and gene-specific fragments, are employed to study the genomic organization of OsSAMDC and related genes. DNA sequencing and bioinformatics analyses provide insights into the gene's structure and potential gene family members. This comprehensive approach combines molecular biology techniques to elucidate the genomic context of OsSAMDC in transgenic rice, potentially contributing to our understanding of polyamine biosynthesis genes and their role in stress tolerance.

Key words Transgenic rice genome, Southern hybridization, Stress tolerance

1 Introduction

Understanding the genomic organization of genes involved in important biological processes is crucial for elucidating their function and regulation. The OsSAMDC gene involved in polyamine biosynthesis has been identified as a potential molecular marker for cold tolerance in rice. Southern hybridization analysis is a powerful method for investigating the genomic organization and copy number of genes within a transgenic plant genome [1].

Southern blotting, developed by Edwin Southern in 1975, is a technique used to detect specific DNA sequences in DNA samples. This method involves the transfer of electrophoretically separated DNA fragments to a membrane, followed by hybridization with labeled probes. Southern blotting can provide information about the presence, absence or rearrangement of specific DNA sequence

Nitish Kumar (ed.), *Transgenic Plants: Methods and Protocols*, Methods in Molecular Biology, vol. 2973,
https://doi.org/10.1007/978-1-0716-4803-2_4,

as well as the number of copies of a particular gene in the transgenic plant genome [2].

In rice, understanding the genomic context of genes like OsSAMDC can provide insights into their evolution, regulation, and potential functional redundancy. The study utilizes Southern hybridization to investigate the presence and arrangement of OsSAMDC-like genes in the genome of transgenic japonica rice cultivar Yukihikari [3].

The rice genome has been fully sequenced, revealing the presence of multiple SAMDC-like genes [3, 4]. Southern hybridization can help elucidate the specific genomic arrangement and copy number of these genes in different rice cultivars. Understanding the genomic organization of OsSAMDC and related genes can provide insights into potential gene duplication events which may have contributed to functional diversification or redundancy in the polyamine biosynthesis pathway. The results of Southern hybridization analysis can guide future studies on the regulation and expression of OsSAMDC genes potentially leading to the development of more cold-resistant rice varieties for temperate regions and high altitudes.

2 Materials

2.1 DNA Extraction (see Note 1)

1. CTAB (cetyltrimethylammonium bromide).
   Prepare the CTAB extraction buffer by mixing the following components:
   - 2 g of CTAB.
   - 1.2 g of NaCl.
   - 100 mL of 100 mM Tris-HCl (pH 8.0).
   - 20 mL of 20 mM EDTA (pH 8.0).

   Adjust the final volume to 100 mL with distilled water.

   Add 1% β-mercaptoethanol just before use to prevent oxidation of DNA (*see* **Note 2**).
2. β-mercaptoethanol.
3. Chloroform.
4. Isopropanol.
5. 70% ethanol.
6. TE buffer (10 mM Tris-HCl, 1 mM EDTA, pH 8.0) (*see* **Note 3**).

   For 1 liter of TE buffer:
   - *Tris Base*: 1.21 g (10 mM concentration).
   - Weigh out 1.21 g of Tris base using an analytical balance and add approximately 800 mL of distilled water.

- Stir the solution until the Tris is completely dissolved.
- *EDTA*: 0.372 g (achieve 1 mM concentration) are needed.
- Weigh out 0.372 g of disodium EDTA (EDTA·2Na), add the EDTA to the Tris solution while stirring. The EDTA will also dissolve in the Tris solution.
- Adjust the pH to 8.0 by adding dilute hydrochloric acid (HCl) dropwise.
- Make up the volume to 1 liter.

2.2 Restriction Enzyme Digestion

1. Restriction enzymes: Bam HI, Bgl II, Eco RI, and Hin dIII.
2. Appropriate restriction enzyme buffers.
3. Weigh 6.06 g of Tris base and add approximately 80 mL of distilled water and stir until dissolved. Add 0.58 g of NaCl to the Tris solution and add 0.6 g of $MgCl_2$ and 0.01 g of BSA and add distilled water to bring the final volume to 100 mL.
4. Adjust to pH 7.5 using dilute HCl or NaOH as needed.
5. BSA (bovine serum albumin).
6. Sterile water.

2.3 Gel Electrophoresis

1. Agarose.
2. TAE buffer (Tris-acetate-EDTA) (*see* **Notes 4** and **5**).
   - Weigh 2.42 g of Tris base, add approximately 40 mL of distilled water, and completely dissolve.
   - Carefully add 0.57 mL of glacial acetic acid to the Tris solution.
   - Weigh 0.186 g of EDTA (disodium salt, dihydrate), add it to the solution, and adjust the pH to around 8.0 using NaOH, as EDTA is more soluble at higher pH.
   - Add distilled water to bring the final volume to 100 mL.
3. Ethidium bromide.
4. DNA loading dye.
5. DNA size markers.

2.4 Southern Transfer

1. Denaturing solution (1.5 M NaCl-0.5 M NaOH).
2. Neutralizing solution (1.5 M NaCl-0.5 M TRIS-HCl pH 8.0).
3. 20x SSC buffer (*see* **Notes 6** and **7**).
   (i) Weigh out 175.3 g of NaCl and 88.2 g of sodium citrate.
   (ii) Add 800 mL of distilled water and stir the solution until the salts are completely dissolved.

(iii) Add distilled water to bring the total volume to 1 L. Store the 20x SSC buffer at room temperature or in the refrigerator.

(iv) Dilute the 20x SSC buffer by mixing 50 mL of 20x SSC with 950 mL of distilled water (*see* **Note 8**).

4. Nylon membrane (Hybond-N+; Amersham).
5. Filter paper.
6. Paper towels.

2.5 Probe Preparation

1. Full-length OsSAMDC cDNA.
2. Gene-specific DNA fragments for OsSAMDC and SAMDC.
3. Alkphos labeling system (Amersham) and Gene-specific DNA fragments (for OsSAMDC and SAMDC). PCR reagents: PCR buffer, dNTPs, Taq DNA polymerase, primers and pCRII vector (Invitrogen) (*see* **Note 9**).

2.5.1 Probe Preparation Using Alkphos Labeling System and PCR

Probes for OsSAMDC and SAMDC were prepared using PCR. The specific primers used for amplification were:

Gene-Specific Probes

- For OsSAMDC: Forward Primer: 5′-ATGGTGAACCCTGC TCCT-3′ Reverse Primer: 5′-GCAGAAGGCAAAGGAAAT-3′.
- For SAMDC: Forward Primer: 5′-ACGGAGAAAGCTGG TAAA-3′ Reverse Primer: 5′-TAAAATGACTTGTATTAT-3′.

PCR Reaction Setup

The PCR reaction mixture included: PCR Buffer; dNTPs**:** Each deoxynucleoside triphosphate (dATP, dTTP, dCTP, dGTP) at a concentration of 400 μM; Taq DNA polymerase: 2.5 U of Taq DNA polymerase is used for amplification; specific primers.

PCR Cycling Conditions

The PCR was carried out for 35 cycles with denaturation: 96 °C for 30 s. Annealing: 52 °C for 30 s. Extension: 74 °C for 1 min.

Cloning the PCR Products

The PCR products were then cloned into the pCRII vector (Invitrogen) for further analysis.

Hybridization

1. Hybridization buffer: 5x SSC (saline-sodium citrate) constituting 0.75 M NaCl and 0.075 M sodium citrate (pH 7.0) is commonly used. For use, dilute to 1x SSC with distilled water. 50% formamide is used in the hybridization buffer.
2. Sodium dodecyl sulfate (SDS)**:** 0.1% to 0.2% Tris buffer: (tris (hydroxymethyl)aminomethane) around pH 7.0 to 8.0 and hybridization oven or water bath set to 55 °C.
3. Hybridization oven or water bath.

2.6 Washing and Detection

1. Primary wash buffer (2 M urea, 0.1% SDS, 0.5 M sodium phosphate, 150 mM NaCl, 1.0 M MgCl2, 0.2% blocking reagent).
2. Secondary wash buffer (1 M TRIS base, 2 M NaCl).
3. CDP-Star reagent (Amersham).
4. X-ray film or chemiluminescence imager.

3 Methods

3.1 DNA Extraction and Southern Hybridization Analysis

1. High-molecular-weight genomic DNA was isolated from seedlings of transgenic rice cultivar "Yukihikari" by a modified cetyltrimethylammonium bromide (CTAB) procedure (*see* **Note 10**).
2. The purified DNA was digested with Bam HI, Bgl II, Eco RI, and Hin dIII.
3. Digested samples were fractionated by electrophoresis on a 0.8% agarose gel.
4. The gel was treated with denaturing solution and neutralizing solution.
5. Three DNA fragments were transferred to nylon membranes (Hybond-N+; Amersham) in 20 × SSC by the capillary transfer method.
6. Hybridization was carried out at 55 °C overnight (Fig. 1).

3.2 Probe Preparation (see Note 11)

1. Full-length OsSAMDC cDNA probe was prepared.
2. Gene-specific probes were generated: (a). 381-bp OsSAMDC (Y07766) gene-specific probe (b). 318-bp SAMDC (AJ251899) gene-specific probe.
3. Probes were labeled using the Alkphos labeling system (Amersham).

3.3 Washing and Detection

1. After hybridization, the membrane was washed twice in primary wash buffer at 55 °C for 15 min.
2. Secondary washes were performed in wash buffer for 5 min at room temperature.
3. Chemiluminescence was detected with the CDP Star reagent (Amersham) according to the manufacturer's instructions.

3.4 DNA Sequence Analysis

1. Plasmid DNA was prepared by the alkaline lysis method and purified using a Quantum Prep Plasmid Miniprep Kit (Bio-Rad).
2. Sequencing was performed by the dideoxynucleotide chain-termination method using a T7 DNA polymerase sequencing kit (Amersham Life Sciences).

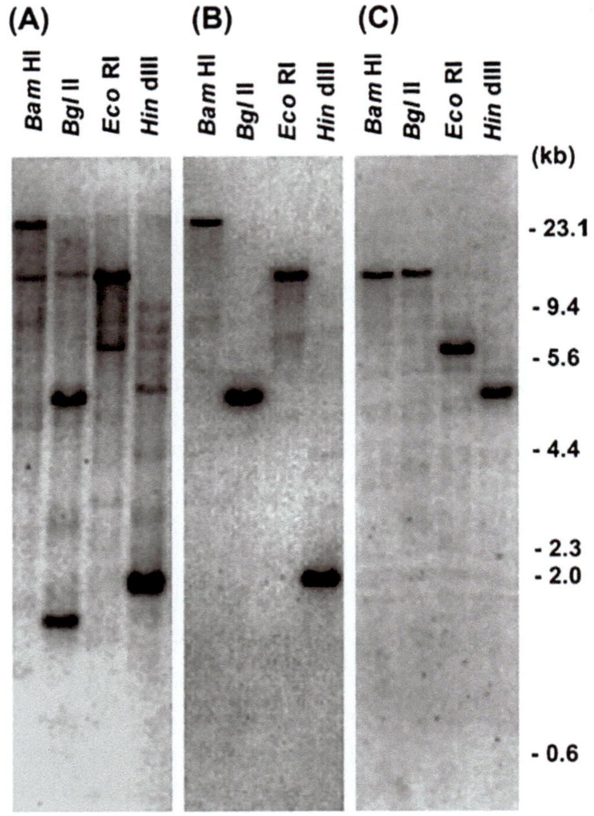

Fig. 1 Hybridization analysis of transgenic rice genomic DNA for SAMDC genes. Genomic DNA (about 10 lg per lane) isolated from transgenic cultivar "Yukihikari" seedlings was digested with either Bam HI, Bgl II, Eco RI or Hin dIII, fractionated by electrophoresis on a 0.8% agarose gel, transferred to a nylon membrane and hybridized with a full length (1.5 kb) SAMDC cDNA probe (**a**) or a 381-bp OsSAMDC (Y07766) gene-specific probe (**b**) or a 318-bp SAMDC (AJ251899) gene specific probe (**c**) obtained by PCR. Sizes based on the migration of Hin dIII fragments of kDNA are indicated on the right

3. Nucleotide sequences and deduced amino acid sequences were analyzed with the GENEWORKS software package.
4. Homology searches in databases were conducted using the BLAST program.

4 Notes

1. Always wear gloves and use proper lab safety protocols when handling chemicals and biological materials.
2. The addition of β-mercaptoethanol is crucial for preventing DNA degradation during the extraction process.

3. TE buffer can be autoclaved for sterilization if needed, but it is generally stable at room temperature.
4. TAE buffer can be used directly for gel electrophoresis at a 1X concentration (dilute the 50X stock solution spiepr A3B2 twbch ":"1:50spiepr A3B2 twbch with distilled water).
5. TAE buffer is stable for several months when stored properly.
6. For long-term storage, consider adding a small amount of sodium azide (0.02% w/v) to prevent microbial growth, but be cautious as sodium azide is toxic.
7. Always use clean equipment to avoid contamination of the buffer.
8. 20X SSC can be diluted to a working concentration of 1X by mixing 1 part of 20X SSC with 19 parts of distilled water.
9. Always include controls to ensure the specificity and efficiency of the probe.
10. The quality of the extracted DNA can be affected by the age and condition of the plant tissue, so fresh samples are recommended.
11. Follow safety guidelines when working with radioactive materials or hazardous chemicals.

References

1. Kanwar P et al (2022) Identification of gene copy number in the transgenic plants by quantitative polymerase chain reaction (qPCR). PCR Primer Des 2392:161–171
2. Southern EM (1975) Detection of specific sequences among DNA fragments separated by gel electrophoresis. J Mol Biol 98(3):503–517
3. Pillai M, Akiyama T (2004) Differential expression of an S-adenosyl-L-methionine decarboxylase gene involved in polyamine biosynthesis under low temperature stress in japonica and indica rice genotypes. Mol Gen Genomics 271:141–149
4. Chen M et al (2014) Down-regulation of S-adenosylmethionine decarboxylase genes results in reduced plant length, pollen viability, and abiotic stress tolerance. Plant Cell Tissue Organ Cult (PCTOC) 116:311–322

Chapter 5

Characterization of Transgenic Arabidopsis Plants by Western Blotting

Senthilkumar Shricharan, Sekar Ashok, M. Arumugam Pillai, and Jeshima Khan Yasin

Abstract

Western blotting is a widely used technique for identifying the expression of a particular protein. This chapter presents a comprehensive methodology for characterizing transgenic plants in *Arabidopsis thaliana* using Western blotting. Western blotting allows for the accurate evaluation of protein expression levels, which is essential for confirming the integration and functionality of transgenes. The technique consists of several stages, starting with the preparation of plant samples and protein extraction. Then extracted proteins are subjected to electrophoretic separation on polyacrylamide gels and then transferred onto a membrane. Subsequently, the membrane is examined using specialized antibodies that specifically bind to the transgenic product, enabling the identification and measurement of the produced protein. Important factors such as the specificity of antibodies, the stability of proteins, and the use of loading controls are taken into account to guarantee the precision and consistency of the outcomes. This technique provides researchers with a systematic approach to efficiently analyze transgenic plants.

Key words Antibody, Arabidopsis, Proteins, SDS-PAGE, Western blotting

1 Introduction

Western blotting is a widely used analytical technique in molecular biology, biochemistry, and cell biology to detect specific proteins in a complex mixture. The technique combines protein electrophoresis with immunodetection to provide qualitative and semiquantitative data about protein expression, posttranslational modifications, and protein-protein interactions [1, 2]).

1.1 Overview of the Western Blotting Process

The Western blotting protocol consists of several key steps, each crucial for the successful detection and analysis of target proteins [3–5].

Sample Preparation: Proteins are extracted from cells or tissues using lysis buffers that may contain various detergents and

Nitish Kumar (ed.), *Transgenic Plants: Methods and Protocols*, Methods in Molecular Biology, vol. 2973, https://doi.org/10.1007/978-1-0716-4803-2_5,

protease inhibitors to ensure protein stability and integrity. The choice of buffer can influence the efficiency of protein extraction.

SDS-PAGE (Sodium Dodecyl Sulfate-Polyacrylamide Gel Electrophoresis): Proteins are denatured and separated based on their molecular weight. SDS, an anionic detergent, binds to proteins, giving them a uniform negative charge and linearizing them. This allows proteins to migrate through a polyacrylamide gel matrix under an electric field, separating them by size.

Protein Transfer: After electrophoresis, the separated proteins are transferred from the gel onto a membrane (PVDF or nitrocellulose) using either a wet or semi-dry transfer system. This step is critical for subsequent immunodetection [6, 7].

Blocking: The membrane is incubated with a blocking buffer (typically containing nonfat dry milk or BSA) to prevent the nonspecific binding of antibodies to the membrane.

Primary Antibody Incubation: The membrane is incubated with a primary antibody specific to the target protein. This antibody binds to the protein of interest.

Secondary Antibody Incubation: A secondary antibody, conjugated to an enzyme such as horseradish peroxidase (HRP), is used to detect the primary antibody. This secondary antibody amplifies the signal.

Detection: The membrane is incubated with a chemiluminescent substrate that reacts with the enzyme on the secondary antibody, producing light. This light is detected using an imaging system or X-ray film, allowing visualization of the protein bands.

2 Materials

2.1 Materials for SDS PAGE

Lysis Buffer (suitable for plant tissues): 50 mM Tris-HCl (pH 7.4), 150 mM NaCl, 1% Triton X-100, 1% sodium deoxycholate, 0.1% SDS, and protease and phosphatase inhibitors (e.g., 1 mM PMSF, 1 μg/mL aprotinin, 1 μg/mL leupeptin, and 1 mM Na_3VO_4, 1 mM NaF) (*see* **Note 1**).

1. Grinding media (e.g., liquid nitrogen and mortar and pestle).
2. Protein Assay Reagents (e.g., BCA Protein Assay Kit).
3. Laemmli Sample Buffer (2×): 4% SDS, 20% glycerol, 10% 2-mercaptoethanol, 0.004% bromophenol blue, and 0.125 M Tris-HCl (pH 6.8).

4. Running Buffer (1×): 25 mM Tris, 190 mM glycine, and 0.1% SDS.
5. Transfer Buffer: 25 mM Tris, 190 mM glycine, and 20% methanol.
6. Gel Solutions.
   - Separating gel buffer (1.5 M Tris, pH 8.8): 1.5 M Tris-HCl. Adjust pH to 8.8 with HCl.
   - Stacking gel buffer (0.5 M Tris, pH 6.8): 0.5 M Tris-HCl. Adjust pH to 6.8 with HCl.
   - 10% SDS solution.
   - Acrylamide/bis-acrylamide (30% acrylamide:0.8% bis-acrylamide).
   - Ammonium persulfate (APS) solution (10% w/v).
   - TEMED (N,N,N′,N′-Tetra methyl ethylenediamine).
7. Protein Ladder/Markers.
8. Protein Samples.
9. Deionized Water.
10. *RIPA Buffer*: 50 mM Tris-HCl (pH 7.4), 150 mM NaCl, and 1% NP-40 (or 1% Triton X-100) (*see* **Note 2**).

    0.5% Sodium deoxycholate, 0.1% SDS and protease and phosphatase inhibitors (e.g., 1 mM PMSF, 1 μg/mL aprotinin, 1 μg/mL leupeptin, 1 mM Na3VO4, 1 mM NaF).

2.2 Materials for Western Blot

1. Running Buffer (1×): 25 mM Tris, 190 mM glycine and 0.1% SDS.
2. Transfer Buffer: 25 mM Tris, 190 mM glycine, and 20% methanol.
3. Blocking Buffer: 5% Nonfat dry milk or 5% BSA in TBS-T (Tris-buffered saline with 0.1% Tween 20).
4. Primary Antibody: Diluted in blocking buffer (typically 1: 1000, but depends on antibody).
5. Secondary Antibody: Diluted in blocking buffer (typically 1: 5000 to 1:10000 but depends on antibody).
6. Chemiluminescent substrate (e.g., ECL).
7. TBS-T Buffer (1×): 20 mM Tris-HCl (pH 7.6), 150 mM NaCl, and 0.1% Tween 20.

2.3 Materials for Protein Transfer to PVDF or Nitrocellulose Membrane

1. Transfer Buffer (Wet Transfer): 25 mM Tris, 190 mM glycine, and 20% methanol. Optionally, add 0.1% SDS to improve transfer efficiency for high-molecular weight proteins.
2. Transfer Buffer (Semi-Dry Transfer): 48 mM Tris, 39 mM glycine, and 20% methanol.

3. PVDF Membrane: Requires activation in methanol before use.
4. Nitrocellulose Membrane: Does not require activation (*see* **Note 3**).
5. Filter Paper: Pre-cut to the size of the gel and membrane.
6. Sponges (for Wet Transfer): Pre-cut to the size of the gel and membrane.
7. Electrophoresis Transfer Apparatus: Wet transfer tank or semi-dry transfer apparatus.
8. Ice Pack (for Wet Transfer): To maintain a low temperature during transfer.

2.4 Materials for Transfer to Membrane

1. *Prepare transfer buffer*: 25 mM Tris, 190 mM glycine, and 20% methanol.
2. *Transfer proteins*
   - Follow the standard protocol for transferring proteins to a PVDF or nitrocellulose membrane using a wet or semi-dry transfer system.

3 Methods

3.1 Sample Preparation

3.1.1 Tissue Collection

1. Harvest Arabidopsis tissues at the desired developmental stage or treatment condition. Immediately flash freeze the tissue in liquid nitrogen and store at −80 °C until use

3.2 Protein Extraction

3.2.1 Cell Lysis Using RIPA Buffer

- Wash cells twice with cold PBS.
- Add cold RIPA buffer containing protease and phosphatase inhibitors to the cells. Typically, use 200–500 μL of RIPA buffer per 10^6 cells.
- Scrape or detach cells from the culture dish.
- Incubate on ice for 30 minutes with occasional vortexing.
- Centrifuge the lysate at 14,000× g for 10 min at 4 °C.
- Collect the supernatant containing the proteins.

3.2.2 Cell lysis with Phosphate Buffer

1. Grind the frozen tissue to a fine powder using liquid nitrogen and mortar and pestle or by shaking with chilled ball bearings in a pre-cooled tube.
2. Add ice-cold lysis buffer to the homogenized tissue (typically 100–200 μL per mg of tissue).
3. Incubate on ice for 30 minutes with occasional vortexing.

4. Centrifuge the lysate at high speed (14,000× *g*) for 10 min at 4 °C.
5. Collect the supernatant containing the extracted proteins.

3.2.3 Sample Preparation for SDS-PAGE

(a) Mix the protein sample with 2× Laemmli sample buffer at a 1: 1 ratio.
(b) Boil the mixture at 95 °C for 5 min.
(c) Cool the sample on ice.

3.2.4 Protein Quantification

- Determine protein concentration using a protein assay kit according to the manufacturer's instructions.

3.3 SDS-PAGE Electrophoresis

1. Prepare an SDS-PAGE gel (typically 10% or 12% resolving gel).
2. Load equal amounts of protein (20–50 μg) into each well of the gel.
3. Load a protein ladder into one well for molecular weight reference.
4. Run the gel at 100–150 V until the dye front reaches the bottom of the gel.

3.3.1 Gel Electrophoresis

1. *Assemble Gel Apparatus*
   - Place the polymerized gel in the electrophoresis tank.
   - Fill the inner and outer chambers with 1× running buffer.
2. *Load Samples*
   - Carefully remove the comb from the stacking gel.
   - Rinse the wells with running buffer to remove unpolymerized acrylamide.
   - Load equal amounts of protein samples (typically 20–50 μg) into the wells.
   - Load a protein ladder/marker into one well.
3. *Run the Gel*
   - Attach the gel apparatus to a power supply.
   - Run the gel at 80 V through the stacking gel (until the dye front enters the separating gel).
   - Increase the voltage to 100–120 V and continue running until the dye front reaches the bottom of the gel.
4. *Gel Staining (Optional)*

   We recommend running two gels simultaneously with the same set of samples. Stain and view one gel for immediate analysis, while using the other gel for blotting. This approach ensures thorough validation of results and facilitates the capture of high-quality images suitable for publication.

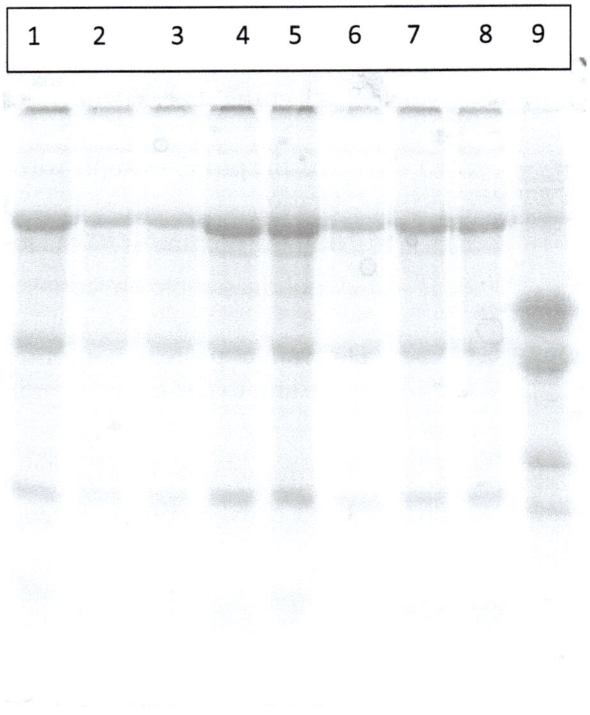

Fig. 1 SDS page profile of Arabidopsis transgenic lines (lanes 2–8, transgenic Arabidopsis lines; 1, col.; 9, protein ladder)

Coomassie Blue Staining

- After electrophoresis, remove the gel from the glass plates.
- Stain with Coomassie Brilliant Blue R-250 solution (0.1% Coomassie Blue R-250, 40% methanol, 10% acetic acid) for 1 h.
- Destain with a solution of 40% methanol and 10% acetic acid until clear protein bands are visible (Fig. 1).

3.4 Protein Transfer

Electrophoretic transfer of proteins from polyacrylamide gels to nitrocellulose sheet using protocol Towbin et al. [6].

1. Soak the gel and PVDF or nitrocellulose membrane in a transfer buffer.
2. Assemble the transfer sandwich (sponge, filter paper, gel, membrane, filter paper, sponge).
3. Transfer proteins from the gel to the membrane using a wet or semi-dry transfer system at 100 V for 1–2 h or 30 V overnight at 4 °C.

3.4.1 Preparing the Gel and Membrane

1. *After Electrophoresis*
   - Turn off the power supply and remove the gel from the electrophoresis apparatus.

- Carefully open the gel cassette and trim any excess gel if necessary.

2. *Prepare the PVDF Membrane*
   - Activate the PVDF membrane by soaking it in 100% methanol for 30 s.
   - Rinse the membrane in distilled water for a few seconds.
   - Equilibrate the membrane in a transfer buffer for at least 10 min.
3. *Prepare the nitrocellulose membrane*
   - Equilibrate the nitrocellulose membrane in a transfer buffer for at least 10 min.

3.4.2 Wet Transfer Method

1. *Set up the transfer sandwich*
   - In a transfer tank filled with transfer buffer, assemble the sandwich as follows (from the bottom to top):
     1. Sponge.
     2. Filter paper.
     3. Gel.
     4. PVDF or nitrocellulose membrane.
     5. Filter paper.
   - Ensure there are no air bubbles between the layers by gently rolling a pipette or a roller over the surface.
2. *Transfer conditions*
   - Insert the cassette into the transfer tank and add ice packs to keep the buffer cool.
   - Attach the electrodes (black to the cathode and red to the anode).
   - Run the transfer at 100 V for 1–2 h, or 30 V overnight at 4 °C.

3.4.3 Semi-dry Transfer Method

1. *Set up the transfer sandwich*
   - In the semi-dry transfer apparatus, assemble the sandwich as follows (from the bottom to top):
     1. Filter paper soaked in semi-dry transfer buffer.
     2. Gel.
     3. PVDF or nitrocellulose membrane.
     4. Filter paper soaked in semi-dry transfer buffer.
   - Ensure there are no air bubbles between the layers by gently rolling a pipette or a roller over the surface.

2. *Transfer conditions*
   - Place the assembled sandwich into the semi-dry transfer apparatus.
   - Close the apparatus and attach the electrodes.
   - Run the transfer at 15 V for 30–60 minutes.

3.4.4 Post-transfer

1. *Stain the membrane (Optional)*
   - Stain the membrane with Ponceau S solution (0.1% Ponceau S in 5% acetic acid) for 5 minutes to verify protein transfer (*see* **Note 4**).
   - Rinse with water until protein bands are visible.
2. *Blocking the membrane*
   - Incubate the membrane in blocking buffer (5% nonfat dry milk or 5% BSA in TBS-T) for 1 h at room temperature or overnight at 4 °C with gentle shaking.

3.4.5 Primary Antibody Incubation

1. *Primary Antibody*
   - Incubate the membrane with the primary antibody diluted in blocking buffer (typically 1:1000, but varies) for 1–2 h at room temperature or overnight at 4 °C with gentle shaking.

3.4.6 Washing

- Wash the membrane three times for 5 minutes each with TBS-T buffer.

3.4.7 Secondary Antibody Incubation

Secondary Antibody
- Incubate the membrane with HRP-conjugated secondary antibody diluted in blocking buffer (typically 1:5000 to 1:10000, but varies) for 1 h at room temperature with gentle shaking.

3.4.8 Final Washing

- Wash the membrane three times for 5 minutes each with TBS-T buffer.

3.4.9 Detection by Chemiluminescence Detection Protocol

(A) *Incubation with Chemiluminescent Substrate*

1. *Prepare Substrate*
   - Mix equal volumes of the luminol/enhancer solution and the peroxide solution (or follow the manufacturer's instructions if different).
   - Typically, 1–2 mL of substrate mixture is enough to cover a standard membrane (7 × 8.5 cm).

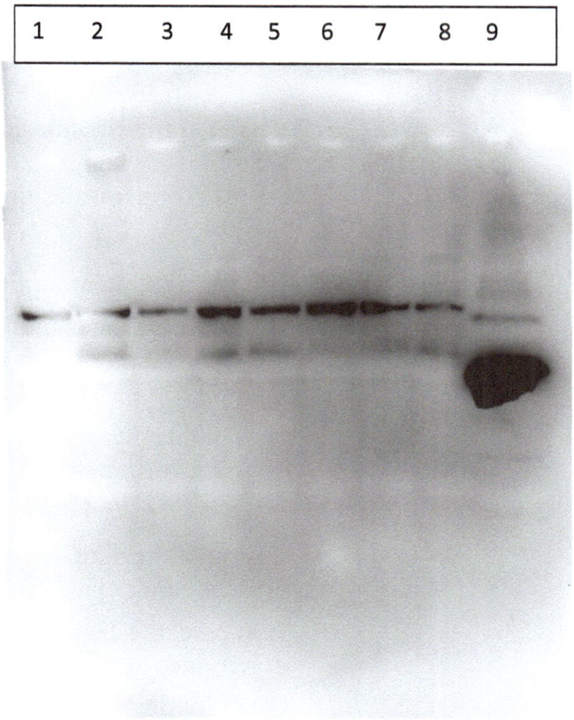

Fig. 2 Western blot of myosin over-expression lines of Arabidopsis (lanes 2–8, transgenic Arabidopsis lines; 1, col.; 9, protein ladder)

2. *Incubate the Membrane*
   - Place the membrane protein side up on a clean, flat surface (e.g., inside a clean plastic sheet or on a piece of plastic wrap).
   - Add the prepared chemiluminescent substrate mixture evenly over the membrane, ensuring complete coverage.
   - Incubate for 1–5 min at room temperature. Do not let the membrane dry out.

(B) *Visualization*

1. *Remove Excess Substrate*
   - Carefully drain any excess substrate from the membrane. It can be helpful to gently tap the edge of the membrane on a paper towel.
2. *Expose the Membrane*
   - Place the membrane inside a plastic wrap or a plastic sheet protector to prevent drying.

- If using X-ray film:
  - In a darkroom, place the membrane in close contact with an X-ray film.
  - Expose the film for 1 s to 5 min, depending on the signal strength and the substrate used. Start with a short exposure time and adjust as necessary.
  - Develop the film using an X-ray film processor.
- If using a chemiluminescent imaging system:
  - Follow the manufacturer's instructions for capturing the image.
  - Typically, the imaging system will allow for real-time adjustment of exposure times and image enhancement.

(C) *Analyze the Results*

- Examine the film or digital image for the presence of protein bands (Fig. 2).
- Compare the bands to the molecular weight marker to determine the size of the detected proteins.
- Quantify the signal intensity using appropriate software if quantitative analysis is required.

3.5 Optional: Stripping and Reprobing the Membrane

1. *Stripping the Membrane*
   - If you need to probe the membrane with another antibody, you can strip the membrane to remove the first set of antibodies.
   - Use a stripping buffer (e.g., 62.5 mM Tris-HCl, pH 6.8, 2% SDS, 0.1 M β-mercaptoethanol) and incubate the membrane at 50 °C for 30 minutes with gentle shaking.
   - Wash the membrane thoroughly with TBS-T (three times for 5 minutes each).
2. *Reblock the Membrane*
   - Reblock the membrane using the same blocking buffer as initially used (5% nonfat dry milk or 5% BSA in TBS-T) for 1 h at room temperature.
3. *Reprobe the Membrane*
   - Repeat the steps for primary and secondary antibody incubations with the new antibodies, followed by chemiluminescence detection as described above.

3.6 Applications

Western blotting is a versatile technique used for

1. *Confirming the Presence of Specific Proteins*: It allows researchers to verify whether a particular protein is present in a sample.

2. *Comparing Protein Expression Levels*: By analyzing the intensity of protein bands, researchers can compare the expression levels of proteins across different samples or conditions.
3. *Studying Protein Overexpression*: It is particularly useful for validating the overexpression of proteins. This is often achieved by comparing the protein levels in samples transfected with overexpression constructs to control samples.
4. *Analyzing Protein Modifications*: The technique can detect posttranslational modifications such as phosphorylation, ubiquitination, or glycosylation by using modification-specific antibodies.
5. *Investigating Protein-Protein Interactions*: Western blotting can be used in co-immunoprecipitation experiments to study interactions between proteins.

4 Notes

1. *Phosphate Buffer Method*: Suitable for basic cell lysis and protein extraction with minimal interference, especially when protease inhibitors are not needed.
2. *RIPA Buffer Method*: Provides more stringent lysis, particularly for cells with strong cytoskeletal components or when a more thorough extraction is needed, including the use of various detergents and protease inhibitors to maintain protein integrity.
3. *Stripping and reprobing*: We do not recommend reusing the membrane. However, if you are standardizing the protocol, you may reuse it.
4. *Post-transfer staining:* We do not recommend this step if your antibodies are highly specific. Staining the membrane can often interfere with the final result, particularly in chemiluminescence assays.

References

1. Gilda JE, Gomes AV (2013) Stain-Free total protein staining is a superior loading control to β-actin for Western blots. Anal Biochem 440(2): 186–188. https://doi.org/10.1016/j.ab.2013.05.027
2. Moritz CP (2020) Tubulin or not tubulin: heading toward total protein staining as loading control in Western blots. Proteomics 20(3–4): e1900110. https://doi.org/10.1002/pmic.201600189
3. Mahmood T, Yang PC (2012) Western blot: technique, theory, and trouble shooting. N Am J Med Sci 4(9):429. https://doi.org/10.4103/1947-2714.100998
4. Taylor SC, Posch A (2014) The design of a quantitative western blot experiment. Biomed Res Int 2014:1. https://doi.org/10.1155/2014/361590
5. Burnette WN (1981) "Western Blotting": Electrophoretic transfer of proteins from sodium dodecyl sulfate–polyacrylamide gels to unmodified nitrocellulose and radiographic detection with antibody and radioiodinated protein

A. Anal Biochem 112(2):195–203. https://doi.org/10.1016/0003-2697(81)90281-5

6. Towbin H, Staehelin T, Gordon J (1979) Electrophoretic transfer of proteins from polyacrylamide gels to nitrocellulose sheets: procedure and some applications. Proc Natl Acad Sci 76(9):4350–4354. https://doi.org/10.1073/pnas.76.9.4350

7. Kurien BT, Scofield RH (2006) Western blotting. Methods 38(4):283–293. https://doi.org/10.1016/j.ymeth.2005.11.007

Chapter 6

Agrobacterium-Mediated Genetic Transformation in *Solanum torvum* L. Using Leaf Explants

Thirunahari Ugandhar, Dharavath Parvathi, Uppu Anitha Devi, Tutikurla Shankar, and Tirukovela Srinivas

Abstract

A particularly promising method involves the use of leaf explants cultured on Murashige and Skoog (MS) medium, supplemented with 0.5 mg/L indole-3-acetic acid (IAA) and benzylaminopurine (BAP). Following co-cultivation with a disarmed strain of *Agrobacterium tumefaciens* LBA 4404 harboring a plant expression binary vector, the explants were further treated with 100 mg/L kanamycin and 200 mg/L cefotaxime to facilitate selection and prevent bacterial contamination. The co-cultivation process successfully produced transgenic adventitious shoots, with an average of 2.3 ± 0.4 shoots per explant. These transgenic plants were subsequently vernalized in a greenhouse, where they grew to maturity and displayed normal morphological characteristics. Progeny analysis of the independent transgenic lines confirmed the successful integration of the gus gene into the genome, with its inheritance following a Mendelian pattern in three independent lines.

Keywords Agrobacterium, b-Glucuronidase (GUS), Kanamycin, Cefotaxime, Solanine, Solasodine, Bronchial asthma, Hybridization, Transgenic, *Solanum torvum*, Leaf explants

1 Introduction

The *Agrobacterium tumefaciens*-mediated transformation system has emerged as one of the most efficient methods for introducing foreign genes into plants. This method is favored due to its broad host range, allowing it to be used in various economically important crops [1]. The process involves transferring genes of interest into the plant genome using *Agrobacterium*, with applications in crop improvement and resistance development. Although this technique has been successful in crops such as soybean [2], broad bean [3], sesame [4], and sunflower [5], certain crops like Brassica have shown recalcitrance to genetic transformation due to difficulties in regenerating transformed tissues [6, 7].

Nitish Kumar (ed.), *Transgenic Plants: Methods and Protocols*, Methods in Molecular Biology, vol. 2973,
https://doi.org/10.1007/978-1-0716-4803-2_6,

Despite the challenges, *Solanum torvum* offers potential as a model species for *Agrobacterium*-mediated transformation. This technique can enable the development of transgenic varieties with improved traits, including pest and disease resistance, enhanced growth rates, and higher yields. However, achieving reproducible transformation in *S. torvum* requires careful optimization of pre-culture and co-cultivation conditions [1].

Recent advances in plant biotechnology have focused on improving the efficiency and reproducibility of genetic transformation protocols. Techniques such as the use of non-oncogenic strains of *Agrobacterium* as vectors have enhanced the success rate of genetic transformation in previously recalcitrant crops [8]. Histochemical assays, such as the GUS (β-glucuronidase) assay, are commonly employed to verify the successful integration of transgenes into plant genomes [5].

While genetic transformation has seen progress in various crop species, *Solanum torvum* remains a challenging candidate due to its specific in vitro culture requirements. Developing a reliable protocol for the transformation of *S. torvum* using *A. tumefaciens* requires further research into the optimization of pre-culture and co-cultivation periods, as well as the use of efficient regeneration systems to ensure successful gene integration and plant recovery. To successfully carry out tissue culture and genetic transformation studies using *Solanum torvum*, a comprehensive setup of supplies, equipment, reagents, and media is required. Below is a detailed list of these items.

2 Materials

2.1 Instruments

1. *GA7 tissue culture boxes with lids*: Used to house cultures in a sterile environment.
2. *Laminar flow hood*: Ensures a sterile workspace for tissue culture manipulations.
3. *Forceps*: For handling explants and other materials without contamination.
4. *Scalpels*: For precise cutting of plant tissues.
5. *Sterile, disposable Petri dishes*: For culturing plant tissues and maintaining sterility.
6. *Sterile filter paper*: Used for drying or filtering in sterile conditions.
7. *Cork borers (0.7 mm)*: For preparing explant samples by cutting uniform tissue sections.
8. *Cork borer sharpener*: To maintain the sharpness of the cork borers.

9. *Bunsen burners*: Used for sterilizing tools by flame.
10. *Inoculation loops*: For transferring *Agrobacterium* or other microbial cultures.
11. *Environmental shaker incubators (25 °C)*: Provide controlled shaking and temperature conditions for bacterial cultures.
12. *Environmental growth chambers*: Maintain optimal plant growth and tissue regeneration conditions.
13. *Micropipettes (P20, P200, P1000, P5000)*: For precise measurement and transfer of liquids.
14. *1.5 mL microfuge tubes*: For small-volume reagent storage and centrifugation.
15. *15 and 50 mL capped centrifuge tubes*: For larger volume sample preparation and centrifugation.
16. *15% (v/v) Household bleach*: Used for sterilization of plant tissues before culture.
17. *70% ethanol*: For sterilizing surfaces and tools.
18. *Laboratory sealing film (Parafilm)*: For sealing culture containers and preventing contamination.
19. *Heated water bath (55 °C)*: For temperature-controlled incubation of reagents or samples.
20. *Disposable 10 mL sterile syringes*: Used for sterile transfer of liquid media or reagents.
21. *Acrodisc 0.2 μm syringe filter sterilization units*: For sterilizing solutions by filtration.

2.2 Chemicals and Reagents

1. *1/2X MSO (pH 5.8)*: Half-strength Murashige and Skoog (MS) medium solidified with 0.8% phytagar; serves as the basic medium for plant culture.
2. *Agrobacterium strains*: Disarmed (non-gall-forming) laboratory strains such as EHA 101, 105, C58, or LBA4404 containing an engineered binary transformation vector; used for genetic transformation.
3. *YEP medium (pH 7.2)*: Yeast extract-peptone medium, comprising:

   5.0 g/L Bacto-yeast extract, 10.0 g/L Bacto-peptone, 10.0 g/L NaCl, 15.0 g/L Bacto-agar.
4. *Filter-sterilized MS20IM Agrobacterium induction medium (pH 5.25)*:

   MS salts and vitamins, 2% (w/v) sucrose, 100 μM acetosyringone, 1 mM betaine phosphate or proline, 2.5 mM MES (2-(4-morpholino) ethane sulfonic acid).
5. *Co-cultivation medium (pH 5.8)*: MS medium supplemented with:

4.5 μM benzylaminopurine (BA), 0.5 μM naphthalene acetic acid (NAA), 0.8% (w/v) phytagar

6. *MSBN1.1 shoot regeneration medium (pH 5.8)*: Similar to the co-cultivation medium but with added selective agents for regenerating transgenic shoots.
7. *MSHF rooting medium (pH 5.8)*: MS medium solidified with 0.8% phytagar, with selective agents for rooting transgenic plants.

3 Methods

3.1 Plant Material and Seed Sterilization

Seeds of *Solanum torvum* were collected from mature plants grown in the research field of the Department of Botany. Mature seeds were dried and subjected to surface sterilization to ensure aseptic conditions during germination. Initially, the seeds were soaked in sterile distilled water for 24 h to enhance water uptake and soften the seed coat. Subsequently, the seeds were sterilized using 0.1% (w/v) aqueous mercuric chloride ($HgCl_2$) solution for 3–5 min, ensuring the removal of surface contaminants. To eliminate any residual $HgCl_2$, the seeds were rinsed thoroughly three times with sterile distilled water. After sterilization, the seeds were dried on sterile tissue paper under a laminar-flow hood to further prevent contamination.

For germination, 20 seeds per culture bottle were aseptically placed on Murashige and Skoog (MS) basal medium, which was supplemented with 3% (w/v) sucrose and solidified with 0.8% (w/v) agar. The culture bottles were then incubated under controlled conditions at 25 ± 1 °C with a 16-h photoperiod. Light intensity was maintained at 50–60 μmol m^{-2} s^{-1}, provided by cool white, fluorescent tubes. After 30 days of germination, leaf explants from these seedlings were harvested for transformation experiments.

3.2 Bacterial Strain and Culture Conditions

For the transformation experiments, *Agrobacterium tumefaciens* strain LBA4404, harboring the binary plasmid pBIN19, was used. This plasmid carries the *nptII* (neomycin phosphotransferase II) gene, which confers resistance to kanamycin, and the *uidA* gene encoding β-glucuronidase (GUS), which serves as a reporter gene. The *Agrobacterium* strain was cultured on Luria-Bertani (LB) medium, composed of 5.0 g/L sodium chloride (NaCl), 10.0 g/L Bacto-tryptone, and 5.0 g/L yeast extract. The medium was solidified with 7.0 g/L Bacto-agar and adjusted to a pH of 7.0 using NaOH. Kanamycin (100 mg/L) was added to the medium to maintain selection pressure for the plasmid. *Agrobacterium* cultures were incubated at 28 °C until adequate growth was achieved, after which they were used for plant transformation experiments (*see* **Note 1**).

3.3 Transformation and Plant Regeneration

The transformation of *Solanum torvum* was executed using *Agrobacterium tumefaciens* strain LBA4404, which carries the binary plasmid pBIN19. This plasmid includes the *nptII* gene (neomycin phosphotransferase II) and the *uidA* gene (GUS). The transformation protocol involved the following steps:

3.4 Co-cultivation with Agrobacterium Suspension

1. *Preparation of Agrobacterium Culture*: Two colonies from a freshly streaked *Agrobacterium* plate were inoculated into 10 mL of liquid Luria-Bertani (LB) medium. The culture was incubated overnight at 28 °C with shaking at 250 rpm. To ensure plasmid stability, 100 mg/L kanamycin (KM) was included in the medium [1] (see **Note 2**).
2. *Infection of Explants*: Hypocotyl explants from 30-day-old *S. torvum* seedlings were submerged in the *Agrobacterium* suspension for 10 min with gentle agitation to ensure even infection. After incubation, explants were blotted dry on sterile filter paper to remove excess bacterial suspension.
3. *Co-cultivation Conditions*: The explants were transferred to shoot regeneration (SR) medium containing Murashige and Skoog (MS) salts, 0.5 mg/L indole-3-acetic acid (IAA), and 3.0 mg/L benzylaminopurine (BAP). The co-cultivation was conducted under a 16-h photoperiod with light intensity of 50–60 μmol m^{-2} s^{-1} at 25 ± 2 °C for 3 days (Table 1) [9] (*see* **Note 3**).

Table 1
Successive growth media used for the transformation and selection of transgenic shoots from leaf explants of *S. torvum*

No.	Steps and components of media	MS 1 medium	MS 2 medium
1	Time for *A. tumefaciens* inoculation	8 min	–
2	Co-cultivation with *A. tumefaciens*	2 min	–
3	Proliferation of shoots	–	4–5 weeks
4	Kanamycin selection and shoot regeneration	–	6–7 weeks
5	MS salts (g/L)	4.5	4.3
6	Sucrose (g/L)	25	30
7	Hormone (mg/L) IAA + BAP	(0.5) + (3.0)	(0.5) + (3.0)
8	Cefotaxime (mg/L)	–	200
9	Kanamycin (mg/L)	–	100
10	pH	5.8	5.8
11	Agar-Agar (g/L)	8	8

3.5 Post Co-cultivation and Selection

1. *Washing and Selection*: After co-cultivation, explants were washed in MS liquid medium to remove residual *Agrobacterium*. They were then transferred to selective shoot regeneration (SR) medium (MS1) containing 200 mg/L cefotaxime to eliminate *Agrobacterium* and 100 mg/L kanamycin to select for transformed cells [10]. A control group of non-transformed explants was maintained under identical conditions (*see* **Notes 4** and **5**).
2. *Shoot Proliferation*: After 4 weeks, shoots were excised from the explants and sub-cultured onto fresh selective medium (MS2) with 100 mg/L kanamycin to promote further shoot proliferation. The shoots underwent two to three passages in selective medium, involving repeated excision and exposure to the elongation medium to ensure that only transformed shoots continued to grow [1].
3. *Rooting and Shoot Elongation: Root Induction*: Green, healthy shoots were transferred to MS medium supplemented with 0.1 mg/L indole-3-butyric acid (IBA) and 100 mg/L kanamycin for root induction. Roots developed successfully after 2–3 weeks, indicating successful transformation and regeneration of *S. torvum* (Fig. 1b–d) [9].

3.6 Histochemical GUS Assay

The histochemical GUS assay was conducted to detect β-glucuronidase (GUS) activity in the transformed tissues (Fig. 1; Table 2). The procedure followed was adapted from previous methods [11].

1. *Preparation of X-Gluc Stock and Staining Buffer*

   X-Gluc stock solution was prepared by dissolving 20 mg of X-Gluc (5-bromo-4-chloro-3-indolyl β-D-glucuronide) in 1 mL of dimethyl sulfoxide (DMSO) to make a 20 mg/mL stock solution.

   The staining buffer was prepared by mixing 0.85 μL of sterile distilled water, 100 μL of monosodium phosphate buffer (pH 7.0), 5 mL of X-Gluc stock solution, and 5 μL of Triton X-100 in an Eppendorf tube.
2. *Staining Procedure*

   Explants were placed in the staining buffer and incubated overnight at 37°C.

   After incubation, the explants were treated with a 1:3 (v/v) mixture of aceto alcohol (a combination of one-part glacial acetic acid and three parts ethanol) to remove chlorophyll. Following this, the tissues were fixed in 70% ethanol to preserve the staining results.

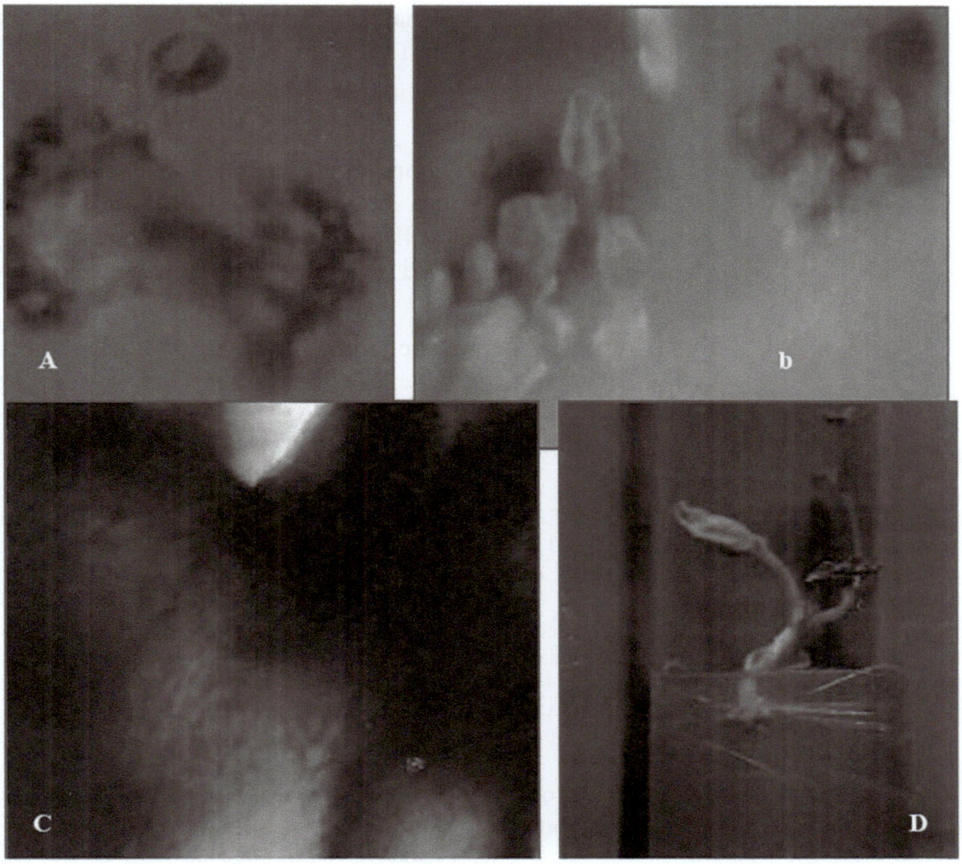

Fig. 1 Plant regeneration from leaf explants of *S. torvum* transformed with gus gene. (**a**) Blue color formation on selection medium with 100 mg/L kanamycin. (**b**) Shoots elongated from the kanamycin-resistant leaf explants after 4 weeks of culture. (**c**) GUS expression in transformed tissue (developing shoots after 6 weeks of culture). (**d**) In vitro rooting from kanamycin resistant micro shoots after 8 weeks of culture

Table 2
Successive growth media used for the transformation and selection of transgenic shoots from leaf explants of *S. torvum*

Explant	Strain	No. of explants culture	Explants bleached	Explants with callus	Shoots	Mean no. of shoots/ explants (S.E.)
Leaf	–	40	–	3.0 ± 1.0	7.3 ± 0.3	–
Leaf	LBA4404	40	15	15 ± 03	2.3 ± 2.0	–

4 Notes

1. Inoculate the strain or culture in fresh liquid culture or medium and preserve it for 2 weeks. Every 2 weeks once sub-culturing of the culture is must.

2. Usually cultures inoculate in the early hours because it takes hardly more than 4–8 h to reach the OD600 of 1.0.
3. Co-cultivation incubation should not exceed 48 h in the co-cultivation medium.
4. Antibiotic stock solutions should be filter-sterilized and stored at −20 °C. Antibiotics should be added to the medium when the temperature is less than 60 °C. Antibiotic-containing plate media should be labeled with the preparation date and stored at 4 °C.
5. The washing step in the freeze/thaw method helps to remove the antibiotics and benefit to the recovery of transformants.

References

1. Gelvin SB (2003) Agrobacterium-mediated plant transformation: the biology behind the "gene-jockeying" tool. Microbiol Mol Biol Rev 67(1):16–37
2. Hinchee MA (1988) Herbicide resistance in soybean through Agrobacterium-mediated transformation. Plant Cell Rep 7(6):427–432
3. Jelenic S et al (2000) Transformation of Vicia faba using Agrobacterium tumefaciens. Plant Cell Tissue Organ Cult 62(2):147–152
4. George EF, Hall MA, De Klerk GJ (1987) Plant tissue culture procedure: background. Springer
5. Weber E et al (2003) Regeneration and transformation of sunflower (Helianthus annuus L.). Plant Cell Tissue Organ Cult 74(1):23–30
6. Godwin ID, Ford-Lloyd BV, Newbury HJ (1991) In vitro regeneration of Brassica oleracea subsp. italica and Brassica napus subsp. oleifera. Plant Cell Rep 9(6):419–423
7. Toriyama K, Arimoto Y, Uchimiya H, Hinata K (1991) Transgenic rice plants after direct gene transfer into protoplasts. Bio/Technology 9 (10):966–968. https://doi.org/10.1038/nbt1091-966
8. Lindsey K (1992) Genetic transformation of Brassica species: progress and prospects. Plant Physiol Biochem 30(3):201–209
9. Murashige T, Skoog F (1962) A revised medium for rapid growth and bio assays with tobacco tissue cultures. Physiol Plant 15(3): 473–497
10. Hood EE, Gelvin SB, Melchers LS, Hoekema A (1993) New agrobacterium helper plasmids for gene transfer to plants. Transgenic Res 2(4):208–218
11. Jefferson RA, Kavanagh TA, Bevan MW (1987) GUS fusions: β-glucuronidase as a sensitive and versatile gene fusion marker in higher plants. EMBO J 6(13):3901–3907

Chapter 7

Development of Fungal Resistant Transgenic Plants Using RNAi Approach in Apple

Arjun Chauhan, Abhishek Kumar, and Manju Modgil

Abstract

Fungal pathogens threaten apple (*Malus* X *domestica*) production, causing economic losses and reduced fruit quality. Excessive fungicide use has led to resistant fungal strains, environmental pollution, health risks, biodiversity loss, and soil degradation emphasizing the need for sustainable pest management practices. The protocol mentioned here outlines RNA interference (RNAi) via *Agrobacterium*-mediated gene transfer to induce host-induced gene silencing (HIGS) in apple plants, boosting fungal resistance.

Key words Apple, Transgenic, RNA interference, Fungal pathogen, HIGS

1 Introduction

The domestic apple (*Malus* X *domestica*) holds significant economic importance globally due to its flavor, health benefits, and versatile use. Fungal diseases, caused by various pathogens, are one of the major contributors to this decline [1, 2]. This presents a substantial threat to crop yield and quality, often leading to significant economic losses. In HP, the State Department of Horticulture releases annual spray schedules in collaboration with Dr. Y.S. Parmar University of Horticulture and Forestry, focusing on fungicidal management [3]. However, the overuse of fungicides has led to resistance in pathogens. This resistance highlights the need for more sustainable approaches to disease management, especially given the rising consumer demand for organic produce. RNA interference (RNAi) offers a promising strategy for fungal disease management. RNAi is an RNA-dependent gene-silencing mechanism that can be triggered by double-stranded RNA (dsRNA) from various sources, including transgenes in genetically modified organisms [4]. This method has been successfully used to target specific fungal genes, offering a precise approach to control [4]. Host-induced gene silencing (HIGS) is an emerging transgenic

Nitish Kumar (ed.), *Transgenic Plants: Methods and Protocols*, Methods in Molecular Biology, vol. 2973,
https://doi.org/10.1007/978-1-07-6-4803-2_7,

technology where plants produce siRNA molecules that silence target fungal genes during pathogen invasion [5]. In this chapter, we will elaborate on the protocol for developing fungal-resistant transgenic apple plants using the RNAi approach. Specifically, we will discuss the steps involved in *Agrobacterium*-mediated gene transfer in apple and screening of the transgenic plants for applying HIGS technology to silence pathogenic fungal genes, offering a sustainable solution to fungal disease management in apple cultivation.

2 Materials

2.1 Establishment of In Vitro Apple Mother Culture

1. Explant material: Fresh cuttings of apple cv. Red Chief. were obtained from the Department of Fruit Science, Dr. Y.S. Parmar University of Horticulture and Forestry, Nauni, and were used as source of explants (*see* **Note 1**).
2. M S basal medium: Murashige and Skoog (MS) medium stock solutions (MS stocks I, II, III, IV, V, and VI) [6]. Store at 4 °C (Table 1 and *see* **Note 2**).
3. Surface sterilization: 0.1% Bavistin; 70% alcohol; 1% NaOCl (*see* **Note 3**).
4. Polyvinylpyrrolidone (PVP) (*see* **Note 4**).
5. Plant growth hormones: 200 μM stock solutions each of 6-benzylaminopurine (BA), gibberellic acid (GA_3), naphthalene acetic acid (NAA), and indole-3-butyric acid (IBA). Store at 4 °C.
6. 3% sucrose and 0.5% charcoal.
7. Shoot multiplication medium (SMM): MS medium supplemented with 1.0 mg/l BA, 0.5 mg/l GA_3, and 0.1 mg/l IBA and solidified with 8 g/l agar, pH 5.8 [7].
8. Shoot regeneration medium (SRM): MS medium supplemented with 5 mg/l benzylaminopurine (BA), 0.2 mg/L naphthalene acetic acid (NAA), and solidified with 0.8% agar, pH 5.8.

2.2 Growth of Agrobacterium *and Inoculum Preparation*

1. *Agrobacterium* strain: Strain containing RNAi-pRI 101-AN-HSP (heat shock protein) binary vector construct (Fig. 1) (*see* **Notes 5** and **6**).
2. Luria Bertani (LB) medium: 10.0 g/L of tryptone, 5.0 g/L of yeast extract, 10 g/l of NaCl, 15 g/l of agar (only for solidified medium). LB medium with rifampicin and kanamycin at 15 mg/l and 50 mg/l, respectively (*see* **Note 7**).
3. ½ Strength MS medium: For resuspension of *Agrobacterium* pellet.
4. 50 mL centrifuge tubes.

Table 1
Composition of MS medium stocks

Stock	Amount (mg/l)	Stock (g/l)
Macronutrients		
Stock I		**20X**
NH_4NO_3	1650	33
KNO_3	1900	38
$MgSO_4.7H_2O$	370	7.4
KH_2PO_4	170	3.4
Stock II		**100X**
$CaCl_2$	440	44
Micronutrients		
Stock III		**200X**
KI	0.83	0.166
$MnSO_4$	22.3	4.46
$ZnSO_4$	8.6	1.72
Na_2MoO_4	0.25	0.05
H_3BO_3	6.2	0.012
Stock IV		**10,000X**
$CuSO_4.6H_2O$	0.025	0.25
$CoCl.6H_2O$	0.025	0.25
Stock V		**200X**
$FeSO_4$	27.8	5.56
$Na_2EDTA.2H_2O$	33.6	6.72
Stock VI (vitamins)		**100X**
Inositol	100	10
Nicotinic acid	0.5	0.05
Pyridoxine HCL	0.5	0.05
Thiamine HCL	0.1	0.01
Glycine	2.0	0.2
Sucrose	3%	

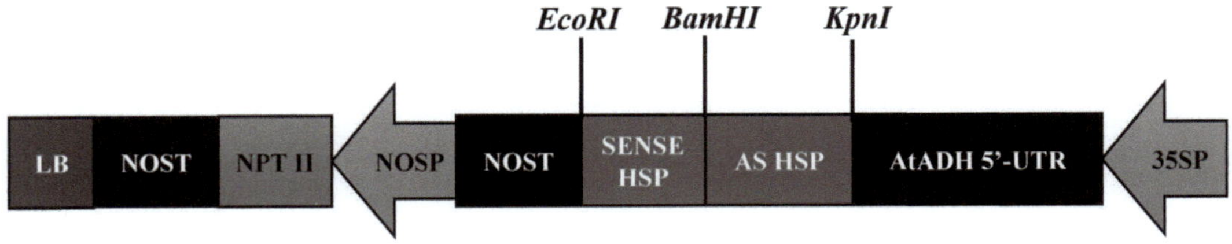

Fig. 1 T-DNA map of RNAi-pRI 101-AN–HSP binary vector (**a**) and RNA folding prediction as generated by mfold

5. Spectrophotometer: For measuring OD at 560 nm.
6. Rotary incubator shaker: Set to 28 °C and 220 rpm.
7. Centrifuge: Capable of 5000 rpm.
8. Sterile pipettes and pipette tips.

2.3 Preparation and Infection of Leaf Explants

1. In vitro multiplied plant culture: Four-week-old in vitro cultures with fully expanded leaves.
2. 100 × 15 mm sterile Petri dishes.
3. Sterile scalpel blades.
4. Sterile Whatman filter paper: For blotting excess medium from leaf sections before transfer to the control medium.
5. Shoot regeneration medium.
6. *Agrobacterium* cell suspension containing RNAi—pRI 101-AN—HSP.
7. Sterile forceps.

2.4 Co-cultivation

1. *Agrobacterium* cell suspension containing the desired vector construct.
2. 100 × 15 mm Petri dishes for holding leaf sections during co-cultivation.
3. Co-cultivation medium (CM): Shoot regeneration medium without antibiotics, pH 5.8.
4. Sterile filter paper for blotting excess *Agrobacterium* suspension from leaf sections.
5. Parafilm for sealing Petri dishes.
6. Incubator set to 25 ± 2 °C, with dark conditions for incubation.

2.5 Selection and Regeneration of Transgenic Apple Shoots

1. Selection medium: Shoot regeneration medium as mentioned above, augmented with 8 mg/L kanamycin and 500 mg/L cefotaxime for selection of *Agrobacterium*-infected leaf sections and suppression of *Agrobacterium* overgrowth, respectively (*see* **Notes 8** and **9**).

2. Shoot regeneration medium.
3. Shoot multiplication medium.

2.6 Rooting and Acclimatization of the Putative Transgenic Shoots

1. 50 mL test tubes
2. Rooting medium: ½ MS liquid medium supplemented with 3 mg/l IBA and selective agents (*see* **Note 10**).
3. 250 mL conical flasks with solidified MS medium: Contains 0.6% agar, used for further rooting in the dark at 25 ± 2 °C.
4. Cocopeat, vermiculite, and vermicompost mixture (3:1:1): For initial acclimatization of rooted shoots in a growth chamber.
5. Soil and cocopeat mixture (1:1).

2.7 Molecular Analysis of Putative Transgenic Shoots

2.7.1 DNA Extraction

1. CTAB Extraction Buffer: Composed of 10 mL 1 M Tris–HCl (pH 8), 28 mL 5MNaCl, 4 mL 0.5 M EDTA, 2 g CTAB, 0.8 g polyvinylpyrrolidone (PVP), and 1 mL β-mercaptoethanol. Dissolve properly, make the final volume to 100 mL with molecular-grade water, and store them at room temperature (*see* **Note 11**).
2. Chloroform/Isoamyl Alcohol Solution: Mix chloroform and isoamyl alcohol in a 24:1 ratio and keep at 4 °C.
3. Isopropanol with Sodium Acetate: Store isopropanol at 4 °C with one-tenth volume of 3 M sodium acetate.
4. 70% Ethanol: Maintain 70% ethanol at 4 °C.
5. TE Buffer: Prepare 1 M Tris-HCl (pH 8.0) with 0.5 M EDTA (see **Note 12**).

2.7.2 Polymerase Chain Reaction

1. Master Mix (2X): Taq DNA polymerase, buffer components, magnesium chloride, and dNTPs; store at 20 °C.
2. HSP Gene-Specific Primers: Forward Primer - GAGCGAC GATTCAGATATTG and Reverse Primer -CACTAAAGCAGC CATCATCT.
3. Primer stock solution: 100 pmol and store at 20 °C (*see* **Note 13**).

2.7.3 Agarose Gel Electrophoresis

1. 1.3% Agarose gel containing 0.1 μg/mL ethidium bromide (final concentration) in Tris borate EDTA buffer (pH 8.8) (*see* **Note 14**).
2. 2X loading dye (Genei, India); gel documentation system (Biovis, England); BioPhotometer D30 (Eppendorf, Germany).

3 Methods

3.1 Establishment of In Vitro Apple Mother Culture

1. Collect the bud explant during the months of June and July, and surface sterilize using 0.1% Bavistin (15 min), 70% alcohol (50 sec), and 1% NaOCl (20 min) one after another. Wash the buds with autoclaved distilled water following each treatment.
2. Transfer buds to ½ strength MS medium supplemented with 1 g/l polyvinylpyrrolidone (PVP), and place on Rotary incubator shaker at room temperature, 250 rpm for 5 h to rinse the phenols, and then blot excess medium using the Whatman filter paper.
3. Inoculate buds on MS medium supplemented with 1 mg/l BA, 0.5 mg/l GA_3, 0.1 mg/l IBA, and activated charcoal (AC) (100 mg/l) for initial 15 days at 25 ± 2 °C with a 16-h lights photoperiod, and then transfer to the same medium except activated charcoal for shoot induction (Fig. 2) (*see* **Note 15**).
4. Once the sprouted bud attains 3–4 cm height, then cut the shoots from base, and transfer to the 50 mL multiplication medium in 150 mL conical flask (Fig. 2).
5. Allow the shoots to multiply for 4–6 weeks by incubating culture flask under fluorescent lights at 25 ± 2 °C with a 16-h photoperiod (Fig. 2) (*see* **Note 15**).
6. Excise leaf explants from the multiplied shoots, and inoculate them on shoot regeneration medium to standardize the regeneration protocol, and cut the remaining stem between nodes, and inoculate them into 150 mL flask containing 50 mL multiplication medium.

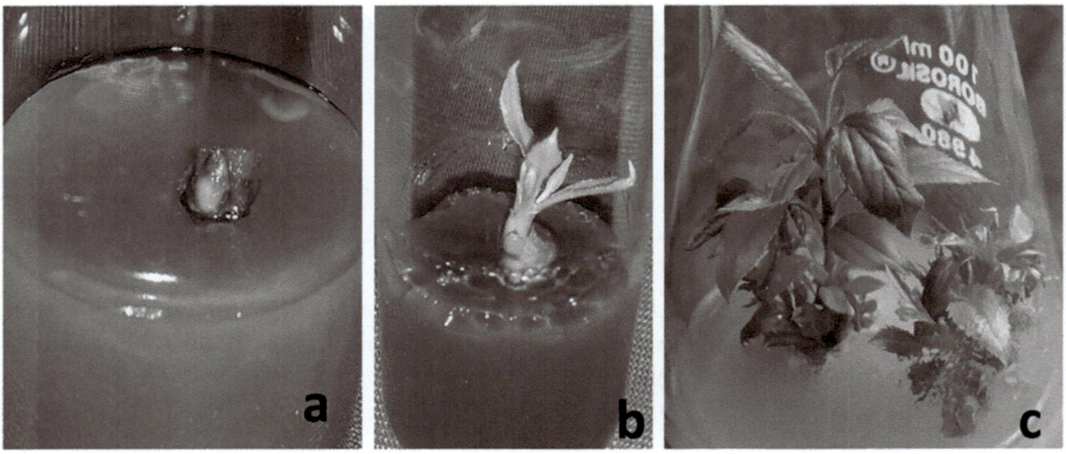

Fig. 2 Establishment of the mother culture. (**a**) Bud explant used for the establishment of culture, (**b**) sprouted bud explant, (**c**) multiplication of established culture

Fig. 3 (**a**) Regeneration of the shoots on shoot regeneration medium, (**b** and **c**) regeneration of the shoots on the selection media

7. Similarly, inoculate the leaf explants on selection media prior to transformation in order to evaluate the regeneration frequency with and without selection pressure in control (untransformed).
8. Regeneration will appear after 3–4 weeks (Fig. 3).
9. The multiple shoot cultures were maintained by sub-culturing the shoots after 4 weeks on the multiplication medium consisting of MS salts and vitamins supplemented with 30 g/l sucrose, 0.8 mg/l BA, 0.5 mg/l GA_3, and 0.1 mg/l IBA and solidified with 8 g/l agar. pH of the medium was adjusted to 5.7–5.8 before adding agar (Fig. 2) .

3.2 Growth of Agrobacterium *and Inoculum Preparation*

1. Inoculate a loopful of actively growing culture (*Agrobacterium* strain containing RNAi-pRI 101-AN-HSP binary vector construct) in 50 mL centrifuge tube containing 25 mL of LB media containing rifampicin and kanamycin at 15 mg/L and 50 mg/L, respectively (*see* **Note 16**).
2. Incubate the culture tube at 28 °C in incubator Rotary incubator shaker at 220 rpm overnight.
3. Take the OD of the culture at 560 nm in spectrophotometer. Then dilute the culture to the OD of 0.6 to final volume of 25 mL.
4. Centrifuge the 25 mL of culture at 5000 rpm for 5 min in a 50 mL centrifuge tube.
5. Discard the supernatant and dissolve the pellet in 20 mL of ½ strength MS medium.
6. Use the dissolved bacterial culture for the transformation of the apple explants.

3.3 Preparation and Infection of Leaf Explants

1. Remove fully expanded leaves from the 4-week-old in vitro multiplied culture and place them in 100 mm Petri dishes containing sterile ½ strength MS liquid medium.

2. Cut the small section of 0.5 cm from the leaves using the sterile scalpel blade, and dip them in the ½ strength MS medium. Prepare leaf sections in bulk of approximately 25/plate (*see* **Note 16**).
3. Place one set of leaf sections approx. 25 for control experiment on shoot regeneration medium after gently blotting away excess medium using sterile Whatman filter paper.
4. The remaining leaf sections were transferred to Petri dish containing *Agrobacterium* cell suspension, as prepared previously. Incubate at room temperature (~25 °C) for 5–15 min with occasional mixing (*see* **Note 17**).

3.4 Co-cultivation

1. Remove each leaf section individually from the *Agrobacterium* suspension, and blot excess culture using sterile filter paper without damaging the explants.
2. Place these infected sections in the 100 × 15 mm Petri dishes having co-cultivation media. Place approximately 25 leaf section in each plate (Fig. 4).
3. To place large number of events, we usually cut approximately 300 leaf explants on routine basis, and each Petri dish was inoculated with approximately 25 leaf section per plate (*see* **Note 18**).
4. Seal each Petri dish with Parafilm.
5. Incubate cultures at 25 ± 2 °C in the dark condition for 2 days (*see* **Note 19**).

3.5 Selection and Regeneration of Transgenic Apple Shoots

1. Subculture the co-cultivated leaf sections to selective medium having appropriate antibiotics, which are infected with *A. tumefaciens* (Fig. 4).
2. In addition to that, the control leaf explants (those which are not infected with *A. tumefaciens*) should be transferred to shoot regeneration medium having the appropriate selective agents in Petri dishes.

Fig. 4 (**a**) Infection of the *Agrobacterium* to the leaf explants, (**b**) regeneration of the putative shoots on the selection media, (**c**) multiplication of the regenerated shoots

3. Incubate the cultures at 25 ± 2 °C in the dark conditions for initial 15 days, and then transfer to the light condition with 16 h light and 8 h dark.
4. Check regularly for any contamination.
5. Subculture leaf explants every 4 weeks to the fresh selection medium, and keep the expanded leaf explant in good contact with the media.
6. The leaf explants will expand in size and develop callusing over time (Fig. 4).
7. Transformed shoots will appear within 6–8 weeks (Figs. 3c and 4b).
8. Multiply the transformed shoots on multiplication media (Fig. 4c).

3.6 Rooting and Acclimatization of the Transgenic Shoots

1. Excise the regenerated shoots from the base using forceps and scalpel blades without damaging leaves.
2. Place them in 50 mL test tube containing 10 mL rooting medium (1/2 MS liquid medium+ IBA 3 mg/l, with selective agents), for 10 days only (*see* **Note 10**).
3. Transfer the shoots to the 250 mL conical flasks containing 50 mL solidified MS medium using 0.6% Agar (only four shoots per flask), and place in the dark condition at 25 ± 2 °C.
4. Roots initiation will be visible within 10–15 days (Fig. 5).
5. The well-rooted shoots can be acclimatized by putting them in cocopeat/vermiculite/vermicompost (3:1:1) mixture and incubating them in growth chamber at 25 ± 2 °C with 16 h light and 8 h dark condition for initial 15 days (Fig. 5).

Fig. 5 (**a**) Rooting of transgenic shoots (**b**) hardening of the transgenic shoots

6. Transfer the shoots to the soil: cocopeat (1:1) mixture and place them in glasshouse conditions to acclimatize to the environmental conditions.

3.7 Molecular Analysis of Putative Transgenic Shoots

Total genomic DNA of control untransformed shoots and putative transgenic shoots developed from different events were isolated using CTAB method [8] with slight modifications [9] as follows:

3.7.1 DNA Isolation Procedure

1. Grind 200–250 mg of leaf material from each putative transgenic lines in 800 μL of pre-chilled DNA extraction buffer using a pre-chilled mortar and pestle.
2. Transfer the ground leaf material to a 2 mL microcentrifuge tube containing 800 μL of pre-warmed (65 °C) DNA extraction buffer and mix gently by inverting the tube.
3. Incubate the samples at 65 °C for 1 h.
4. Add 1.5 mL of chloroform alcohol (24:1) to the tube, mix by gentle inversion to emulsify, and centrifuge at 12,000 rpm for 20 min at room temperature.
5. Carefully transfer the upper aqueous phase to a new 1.5 mL microcentrifuge tube, avoiding the interphase.
6. Add 2/3rd volume of isopropanol to the aqueous phase, mix by gentle inversion, and incubate at −20 °C for 1 h.
7. Centrifuge at 10,000 rpm for 15 min to collect the DNA pellet.
8. Wash the DNA pellet with 200 μL of 70% ethanol and centrifuge at 8000 rpm for 5 min.
9. Resuspend the DNA pellet in sterile double-distilled water, and store at −20 °C.
10. Treat the DNA with RNase for purification.

3.7.2 Qualitative and Quantitative Analysis of Genomic DNA

1. Take 2 μL of genomic DNA sample.
2. Mix it with 3 μL of 2X loading dye.
3. Prepare a 0.8% agarose gel by dissolving agarose powder in 1X TE buffer (45 mM Tris base, 1 mM EDTA, pH 8.0).
4. Add ethidium bromide to the agarose gel for DNA visualization.
5. Load the DNA sample into the wells of the agarose gel.
6. Run the gel under an appropriate voltage setting until bands are separated.
7. Visualize the integrity of the DNA bands under a gel documentation system.
8. Measure the absorbance of the DNA sample at 260 nm and 280 nm using BioPhotometer D30.

9. Calculate the A260/A280 ratio to assess the purity of the DNA (*see* **Note 20**).

3.7.3 PCR Amplification of Antisense HSP Gene-Specific Primers

1. Prepare the PCR reaction mixture (20 μL total volume) as follows:
2. Add 1 unit Taq DNA polymerase per reaction.
3. Add 1X Taq buffer (with $MgCl_2$). Ensure the buffer contains the necessary $MgCl_2$ concentration.
4. Use a final concentration of 10 μM for forward and reverse primers.
5. Add dNTPs to a final concentration of 250 μM.
6. Add 100 ng of template DNA.
7. For control reactions, use 20 ng of untransformed shoot DNA, plasmid DNA, and nuclease-free water as the template.
8. Adjust the reaction to a final volume of 20 μL with nuclease-free water if needed.
9. Briefly vortex and spin down the reaction mixture to ensure all components are evenly mixed.
10. The amplification reaction consisted of an initial denaturation step at 95 °C for 5 min, 35 cycles comprising denaturation at 95 °C for 1 min, annealing at 60 °C for 1 min, extension at 72 ° C for 1 min, and a final extension step at 72 °C for 5 min.
11. PCR Product Visualization: Analyze PCR products on a 1.3% agarose gel, prepared by dissolving 1 g of agarose in 100 mL of electrophoresis buffer, with 1 μL of a 10 mg/mL ethidium bromide solution added. Mix samples with a gel loading dye containing both loading and tracking dyes before loading into the wells. After electrophoresis, visualize the gel using a gel documentation system.
12. Run PCR products on agarose gel electrophoresis and check for desired bands (Fig. 6) for confirming the transgenic shoots.

4 Notes

1. Different types of tissues like leaf, stem, roots, etc. can be used to regenerate shoots under the optimum conditions and plant growth regulator concentration and combination [10].
2. Prepare stock solutions of Murashige and Skoog medium and store at 4 °C (Table 1). After dissolving all the stock solutions in enough deionized water, make it up to 1 L, adjust the pH to 5.8 (add 8 g/L agar for solid medium), and autoclave for 25 min at 120 °C.

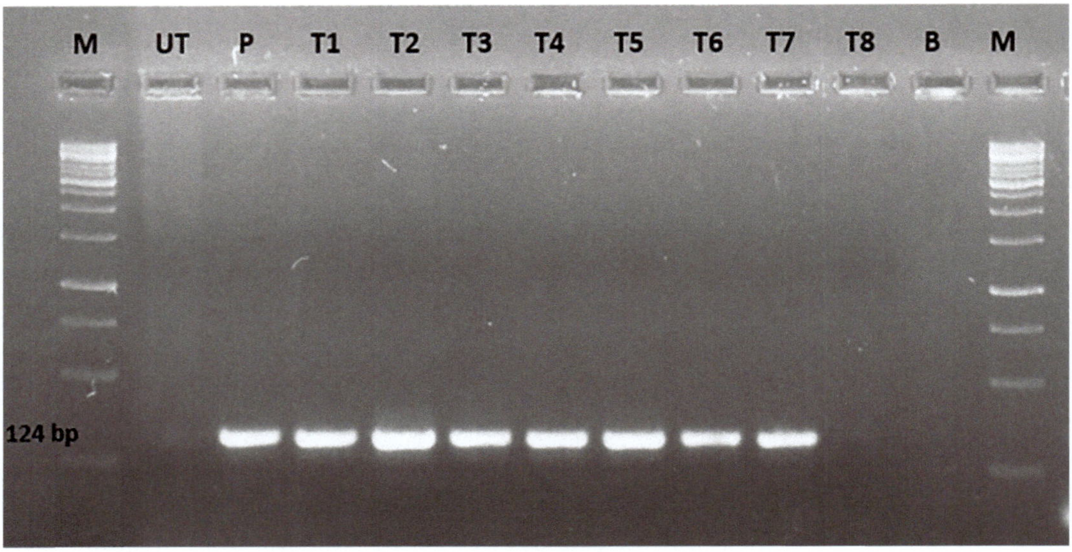

Fig. 6 Confirmation of the transgene integration in the transgenic shoots by gene-specific primer yielding 124 bp band. M, 100 bp DNA marker; UT untransformed control, P plasmid vector, T1-T8 Transgenic plants, B water control

3. Some researchers also use $HgCl_2$ which is very hard to dissolve in water and has health hazardous properties. Prepare 0.1% Bavistin, 70% alcohol, and 1% NaOCl solution by dissolving in sterilized distilled water prior to surface sterilization.
4. PVP is added to media to overcome the browning of media. Other agents can also be used, viz., ascorbic acid, activated charcoal, etc.
5. Selection of the *Agrobacterium* strain to be used for genetic transformation can be decided based on the genetic background of genotype under study. GV3101, EHA105, and LBA4404 are some of the most common *Agrobacterium* strains used for apple transformation.
6. The *Agrobacterium* strain carrying the RNAi-pRI 101-AN-HSP binary vector, with the HSP gene in both sense and antisense orientations, has been utilized. A fungal virulence or pathogenicity gene can be selected based on the specific pathogen, as the HSP gene plays a crucial role in fungal growth, development, and pathogenicity.
7. Antibiotics will depend upon the plasmid used.
8. Selective agents (antibiotics) used for this purpose are usually prepared as stock solutions that are typically 500 to 1000 fold more concentrated than their working strength in cultures. The prepared antibiotic solution must be stored as filter-sterilized solutions in a freezer (−20 °C) for up to 2 months.

9. The selection pressure of the antibiotics to select transgenic shoots is cultivar-dependent and needs to be standardized prior to the transformation experiments. Kanamycin (8 mg /L) is used for selection of transgenic plants (var. Red Chief), and cefotaxime 100–300 mg/L is used for elimination of *Agrobacterium* from the apple cultures after co-cultivation [11, 12].
10. Two types of rooting methods (one-step or two-step) are used for induction of roots in the multiplied shoots or regenerated shoots of apple genotypes. In the case of one-step method, shoots were cultured directly into the MS-solidified medium supplemented with low concentration of IBA (0.1–0.5 mg/l). While in the case of two-step method, shoots were at first cultured on liquid MS medium supplemented with higher concentration of IBA (1–3 mg/l) and after 10–15 days transferred to solidified MS basal medium. However, the results of the one- and two-step methods vary from genotype to genotype in apple cultivars and should be standardized for each genotype.
11. CTAB is dissolved in the buffer at 65 °C.
12. TE buffer (1X): Consists of 1 mL 1 M Tris-HCl (pH 8), 0.2 mL 0.5 M EDTA, and 98.5 mL distilled water for a final volume of 100 mL.
13. HSP gene-specific primers were supplied in lyophilized powder. The working concentration is diluted to 10 pmol per μL using distilled water. Primers are prepared by dissolving in 0.3X TE buffer under sterile conditions. Dilute the stock solution ten times before use for polymerase chain reaction.
14. To prepare a 10 X TBE buffer, dissolve 108 g of Tris and 55 g of boric acid in 800 mL of distilled water with stirring until fully dissolved. Add 40 mL of 0.5 M Na_2EDTA, adjust the pH to 8, and bring the final volume up to 1 L with distilled water. Prepare a 1 X working solution by diluting the stock solution tenfold.
15. Response of different apple genotypes to establish in the tissue culture conditions may vary from genotype to genotype. Therefore, plant growth regulator concentration and combination needed to be standardized for the different apple genotypes. For the micropropagation of apple shoots, larger container must be used so that plantlets must have enough space to multiply and large size leaves are available for genetic transformation experiments. One can use glass jars or flasks of 150–250 mL containing 50–100 mL multiplication medium.
16. Actively grown culture must be used for preparation of the inoculum, and therefore the culture must be maintained by streaking bacteria onto fresh selective plates containing solidified LBA medium every 4 weeks. The selection agents must be

decided based on the resistance of the bacteria and selection marker present in the developed construct.

17. Optimum infection time varies from genotype to genotype and hence needs to be standardized for different apple genotypes.
18. Care must be taken while setting up large number of events at one go, so that leaf explant did not be damaged or dried out during excision and small section preparation of the leaves. The leaves must be immediately transferred to the liquid co-cultivation media after excision from the shoots and can only be taken out during the time of infection with *Agrobacterium* suspension.
19. The frequency of the transformation increases with the increase in co-cultivation time. However, overgrowth of the *A. tumefaciens* occurs in apple genotypes if we co-cultivation time is more than 2 days and later on leads to loss of the tissue as overgrowth is difficult to control even after multiple washes with the antibiotics like cefotaxime. However, the co-cultivation time may vary from genotype and needs to be standardized.
20. An A260/A280 ratio between 1.8 and 2.0 indicates high purity of nucleic acid.

Acknowledgments

We acknowledge the Director of Research, Dr. Y.S. Parmar University of Horticulture and Forestry, for essential infrastructure and research facilities. We extend our gratitude to the Director of ICAR-Central Potato Research Institute, Shimla, for additional research support. We are grateful to Dr. MV Rajam, Professor of Genetics, University of Delhi South Campus, for his valuable feedback and guidance.

References

1. Sharma JN, Sharma A, Sharma P (2004) Out-break of marssonina blotch in warmer climates causing premature leaf fall problem of apple and its management. Acta Hortic 662: 405–409. https://doi.org/10.17660/ActaHortic.2004.662.61
2. Park M, Sagong D, Kweon H, Do Y (2013) Influence of defoliation by marssonina blotch on vegetative growth and fruit quality in 'Fuji'/M.9 apple tree. Korean J Hortic Sci Technol 31:523–530
3. Sharma IM, Bhardwaj SS (2003) Efficacy and economics of different fungicide spray schedules in controlling premature defoliation disease in apple. Plant Dis Res 18:21–24
4. Tetorya M, Rajam MV (2017) RNA silencing of *PEX6* gene causes decrease in pigmentation, sporulation and pathogenicity of *Fusarium oxysporum*. Plant Pathol:67–75. https://doi.org/10.1111/ppa.12712
5. Machado AK, Brown NA, Urban M et al (2018) RNAi as an emerging approach to control Fusarium head blight disease and mycotoxin contamination in cereals. Pest Manag Sci 74:790–799. https://doi.org/10.1002/ps.4748

6. Murashige T, Skoog F (1962) A revised medium for rapid growth and bio assays with tobacco tissue cultures. Physiol Plant 15:473
7. Kumar A, Verma S, Modgil M (2021) Optimization of in vitro adventitious shoot regeneration from leaf explants of apple cultivar 'Red Chief'. Res J Biotechnol 16:180
8. Viršcek-Marn M, Javornik B, Štampar F, Bohanec B (1996) Assessment of genetic variation among regenerants from in vitro apple leaves using molecular markers. In: Eucarpia symposium on fruit breeding and genetics, vol 484, pp 299–304
9. Modgil M, Pathani P, Chauhan A (2016) Evaluation of genetic diversity and relatedness among apple cultivars using RAPD and SSR markers. Agric Res J 53:19. https://doi.org/10.5958/2395-146X.2016.00003.X
10. Modgil M, Sharma DR, Bhardwaj SV (1999) Micropropagtion of apple cv. Tydeman's Early Worcester Sci Hortic 81:179–188. https://doi.org/10.1016/S0304-4238(98)00259-3
11. Chauhan A, Modgil M (2016) Phytotoxic effects of different antibiotics on leaf explants and their potential as *Agrobacterium* counter selection agents during gene transfer in apple. Indian J Hortic 73:287. https://doi.org/10.5958/0974-0112.2016.00062.1
12. Verma S, Kumar A, Modgil M (2023) Impact of cefotaxime and kanamycin on in vitro regeneration via *Agrobacterium* mediated transformation in apple cv. Red Chief Plant Physiol Rep 28:34–42

Chapter 8

Agrobacterium-Mediated Genetic Transformation in Wheat

Satish Kumar, A. Sheena Sabatina, and Jasdeep Chatrath Padaria

Abstract

Wheat (*Triticum aestivum* L.) is an important staple food crop that significantly impacts global food security. In terms of the need for an efficient genetic transformation protocol in major cereal crops, wheat comes immediately after rice to enhance their productivity level under varied global climatic conditions. Despite the availability of various *Agrobacterium tumefaciens*-based genetic transformation techniques for wheat crops, it remains the most challenging process and efficient transformation protocol continues to be a major hurdle in the development of genetically modified crops. Here in this study, we are describing the novel modified *Agrobacterium tumefaciens*-based *in planta* genetic transformation method which targets the 2-day-old apical meristematic germinated tissue as an explant for introgression of transgene. Germinated plants infected with *Agrobacterium* are screened for the presence of transgenes using PCR with gene-specific primers at the T_0 stage. The T_1 ds obtained from PCR-positive plants at T_0 generation are subjected to hygromycin B treatment. Surviving plants are then screened for transgene presence through PCR analysis, dot blot, and Southern hybridization. The expression of the transgene in the transgenic lines is assessed using qRT-PCR. The modified *Agrobacterium*-based in planta genetic transformation method for wheat is efficient, time-saving, and less labor-intensive. By introducing this method of plant transformation, the significant challenging obstacles in the genetic modification of wheat through genome editing or transgenic approaches to wheat are successfully addressed.

Key words *Agrobacterium tumefaciens*, In planta, Introgression, Hygromycin B screening, Gene expression analysis

1 Introduction

The soil phytopathogen *Agrobacterium tumefaciens* is the natural system of genetic transformation. It can infect plant-wounded regions and deliver its T-DNA region into the host plants causing crown gall disease. During the last two decades, the *Agrobacterium tumefaciens*-based infection concept has proved its role in molecular techniques for generating transgenic plants in different crops. Over the years, researchers are extensively used the *A. tumefaciens*-based transformation system to create genetically engineered plants [1–6]. The comprehensive knowledge regarding a large number of bacterial and plant genes identified which were involved in

Nitish Kumar (ed.), *Transgenic Plants: Methods and Protocols*, Methods in Molecular Biology, vol. 2973,
https://doi.org/10.1007/978-1-0716-4803-2_8,

tumorigenesis the workflow and the interactions with host plant systems led to *Agrobacterium-mediated* tissue culture (callus)-based transformation to become the most popular technique for plant transformation [4, 7, 8]. However, researchers have developed an alternative transformation method called in planta, to avoid the problems associated with callus induction and most importantly the regeneration steps which are crucial after transformation. *Agrobacterium* based in planta transformation method was first reported for the *Arabidopsis thaliana* (dicot) model plant system, which involved dipping the floral buds into the *Agrobacterium* suspension or spraying the culture over the floral buds that creates the contact between *Agrobacterium* culture and plant germ cells directly. This simple and cost-effective method is also known as floral-dip plant transformation method [9, 10]. This method also works well in different crops such as radish [11], tomato [12], rice [13, 14], *Medicago truncatula* [15], maize [16], and wheat [17]. In planta transformation, using shoot apical meristem as a suitable target has been effectively carried out for cotton [18] and soybean [19]. The shoot tip-in planta transformation involves piercing the meristematic tissue of the shoot tip with *Agrobacterium* culture, followed by shoot induction and fertile transgenic development. Along the same lines, the particle bombardment approach has been standardized in rice and wheat [20, 21].

Wheat (*Triticum aestivum* L.) is indeed a major annual cereal crop, belonging to the Poaceae family. It is most extensively cultivated throughout the world due to its significant contribution to the human daily diet. Its adaptability to varied climatic conditions and higher yield make it a most demanding staple food crop for a large portion of the world's population. It accounts for approximately 25% protein and 20% calorific value of the world population's daily intake [22]. A biotechnological intervention in wheat improvement programs has been slow in comparison to other cereals. The hexaploid wheat has a very complex, AABBDD genome of ~17 GB which is not that much easy to transform even with the availability of various *Agrobacterium*-based transformation methods as compared to other simple, smaller, and less complex genome crops, like barley [23]. The wheat herbicide-resistant transgenic plant was developed through a microprojectile bombardment technique by generating embryogenic callus [24, 25]. The *Agrobacterium*-based biolistic gun method remained the method of choice for researchers for a longer time [23]. Previous studies and literature related to *Agrobacterium*-based wheat transformation revealed their low transformation efficiency of around 5%, but it continued to be consistent for a considerable time. To avoid this inconsistent wheat transformation efficiency, an efficient in planta *Agrobacterium*-based wheat transformation method was reported [26]. During recent times, in planta *Agrobacterium*-based methodology is becoming more acceptable [1, 27].

2 Materials

Molecular-grade chemicals are used for the sterilization and gel electrophoresis. Double-distilled autoclaved water is used to prepare all solutions related to the sterilization (sodium hypochlorite and ethanol), antibiotics (cefotaxime, acetosyringone, and hygromycin B), and media (MS broth, Luria agar, and Luria broth).

2.1 Chemicals and Experimental Materials

1. Apical meristems of germinated ds of Indian bread wheat variety (*see* **Note 1**).
2. Luria agar (*see* **Note 2**).
3. Luria broth (*see* **Note 3**).
4. Murashige and Skoog media (*see* **Note 4**).
5. Tris-acetate-EDTA (TAE) buffer: Components are 50× stock solution, mix 242 g of Tris base, 57.1 mL of glacial acetic acid, and 100 mL of 0.5 M EDTA using a magnetic stirrer, adjusting the pH of all components to 8.0. The 1× working solution is prepared fresh by adding 10 mL of the 50× TAE stock solution to 490 mL of distilled water (*see* **Note 5**).
6. Sodium hypochlorite (NaOCl): The 2% working concentration of NaOCl is prepared by adding 1 mL of sodium hypochlorite into 50 mL of distilled water (*see* **Note 6**).
7. Ethanol (EtOH): The 70% working concentration of ethanol is prepared by adding 15 mL of distilled water in 35 mL of 100% ethanol (*see* **Note 7**).
8. CTAB buffer components: 2% cetyltrimethylammonium bromide; 0.1 M Tris-HCl; 1.5 M sodium chloride; 0.5 M EDTA (*see* **Note 8**); 1% polyvinylpyrrolidone (*see* **Note 9**).
9. RNase A solution: A 10 mg/mL working solution is prepared using DNase-free water.
10. Chloroform isoamyl alcohol: The 24:1 working concentration is prepared by mixing 2 mL of isoamyl alcohol into the 48 mL of chloroform (v/v) (*see* **Note 10**).
11. Isopropyl alcohol (100%) (*see* **Note 11**).
12. Antibiotics: Kanamycin (50 mg/L); rifampicin (50 mg/L) (*see* **Note 12**).
13. Acetosyringone (100 μM).
14. Cefotaxime (250 mg/L).
15. Hygromycin B (50 mg/L).

All antibiotics are dissolved in double-distilled autoclaved water, except for rifampicin, which is dissolved in 100% molecular-grade methanol.

2.2 Equipment

1. Water bath.
2. Incubator shaker.
3. BOD incubator.
4. Growth chamber or culture room.
5. Laminar air flow chamber.
6. Thermal cycler (PCR).

2.3 Additional Materials

1. Autoclaved germination sheets (*see* **Note 13**).
2. Sterile soil rite.
3. Sterile syringe with needle (*see* **Note 14**).
4. Petri plates (*see* **Note 15**).
5. Nursery portrays (*see* **Note 16**).
6. 4-inch and 6-inch plastic pots (*see* **Note 17**).

2.4 Agrobacterium Strain

1. *EHA105* (*see* **Note 18**).

2.5 Plant Expression Vector

1. The modified *pCAMBIA 1300* vector series is used (*see* **Note 19**).

3 Methods

3.1 Plant Material Preparation

3.1.1 Seed Selection

Fresh, clean, and disease-free seeds are selected by pouring them into a tray, and then the selected seeds are placed in water to check whether they float or sink. The seeds that sink are further processed for sterilization.

3.1.2 Seed Sterilization

Carry out all the sterilization steps in a laminar hood unless otherwise specified (*see* **Note 20**).

1. Wash the healthy wheat ds in distilled water and then rinse two to three times with distilled water to further keep in a water bath for 1 h at 45 °C.
2. Subsequently sterilize the pre-treated ds with 4% (NaOCl) sodium hypochlorite and 70% ethanol treatment for 5 min to remove the presence of fungal and any other contamination.
3. Furthermore, wash the ds thoroughly with double-distilled autoclave water four to five times to remove the remnants of chemicals (NaOCl and EtOH) completely.
4. Store the ds at 4 °C.

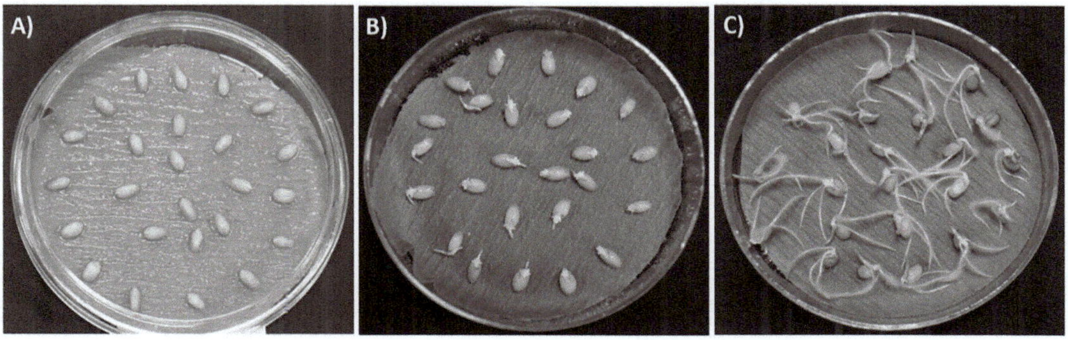

Fig. 1 Wheat ds sterilization and germination. **a** surface sterilized ds on a sterile germination sheet. **b** ds after 48 h of vernalization incubation. **c** Two-day-old germinated dlings

3.1.3 Seed Germination

1. Place the autoclave round shape cut (Petri dishes size) germination sheets in the sterile petri dishes (Fig. 1a).
2. Sprinkle the double-distilled autoclave water on the germination sheets to get wet.
3. Place the surface sterilized wheat seeds over the pre-wet germination sheets in a creeze-down position, and incubate at 4 °C for 2 days for vernalization.
4. The seeds will recover their germination viability (Fig. 1b). After vernalization, transfer the Petri dishes to the BOD incubator having 23 ± 2 °C temperature, 16 h light/8 h dark conditions, and then 2-day-old seedlings are ready for genetic transformation (Fig. 1c) (*see* **Note 21**).

3.2 Agrobacterium Culture Inoculation

1. Streak the *Agrobacterium* EHA105 strain mobilized construct on kanamycin and rifampicin antibiotics supplemented with Luria agar medium Petri dishes to proceed with only a single colony.
2. Pick the single colony and inoculate in 5 mL of Luria broth supplemented with kanamycin (50 mg L − 1) and rifampicin (25 mg L − 1) antibiotics (*see* **Note 12**).
3. Grow the secondary culture before infecting the explants, by adding 50 μL of primary culture into the 50 mL fresh Luria broth and incubate at 28 °C to reach their O.D. approximately 0.3–0.5 at 600 absorbance.
4. Incubate the culture on an incubator shaker having 200 rpm overnight (*see* **Note 22**).

3.3 Plant Transformation

1. Grow the culture till it reaches its O.D. approximately 0.3–0.5, pellet down at 3700 rpm for 15 min, and resuspend in 50 mL of half-strength broth MS medium (Murashige and Skoog 1962) (*see* **Note 23**).

2. Add the 100 μM of acetosyringone antibiotic to the resuspended culture, and mix properly (*see* **Note 24**).
3. The 2-day wheat dlings are used for genetic transformation.
4. Prick the germinated apical meristems at three sites to a depth of approximately 1 mm with a sterile needle and inject the resuspended culture into the pricked regions through the 1 mL sterile syringe (Fig. 2a).
5. Place the injected ds over the sterile soil rite (autoclaved) in the Petri dishes, and also cover with a thin layer of soil rite (Fig. 2b) (*see* **Note 25**).
6. Keep the Petri dishes under dark conditions for 2–3 days at a temperature of 23 ± 2 °C (Fig. 2c).
7. Carefully remove the excess soil rite without damaging the roots or the germinated apical meristematic region (Fig. 2d).
8. After 2 days of co-cultivation, the germinated ds are taken out from the plates and dipped into the cefotaxime solution (250 mg/L) in the conical flasks.

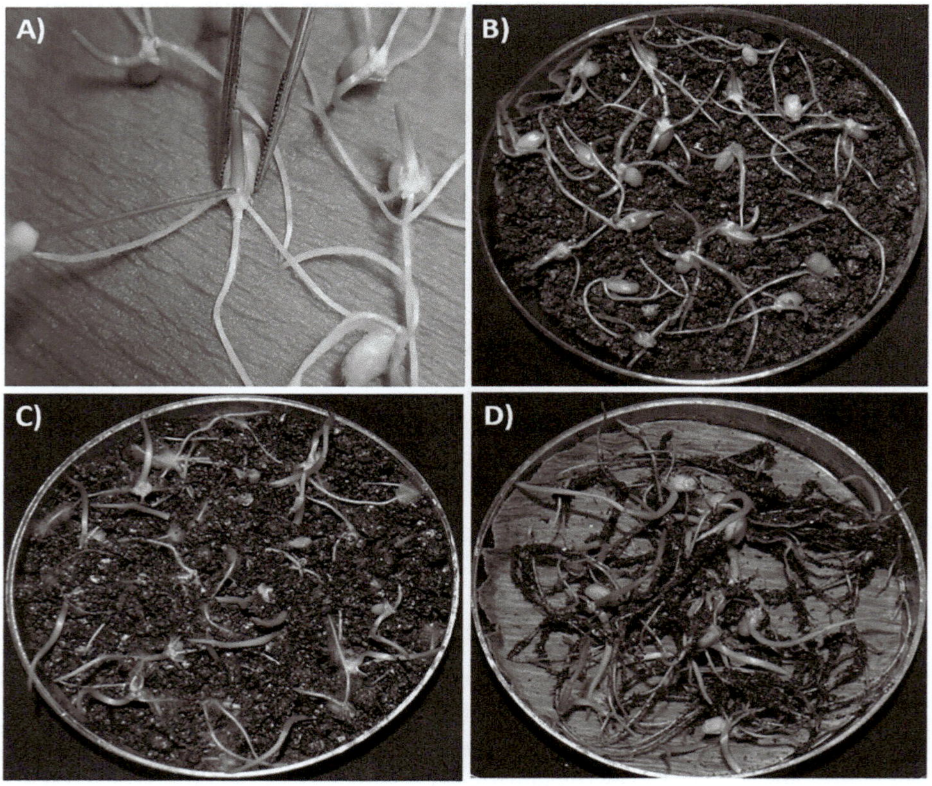

Fig. 2 *Agrobacterium tumefaciens* based *in planta* transformation steps. **a** pricking at the germinated embryonic region through a sterile needle filled with *Agrobacterium* culture. **b** Infected seedlings placed over the autoclaved soil rite mixture in the petri dish for co-cultivation. **c** Germinated dlings after 2 days of co-cultivation. **d** removed the excess amount of soil rite

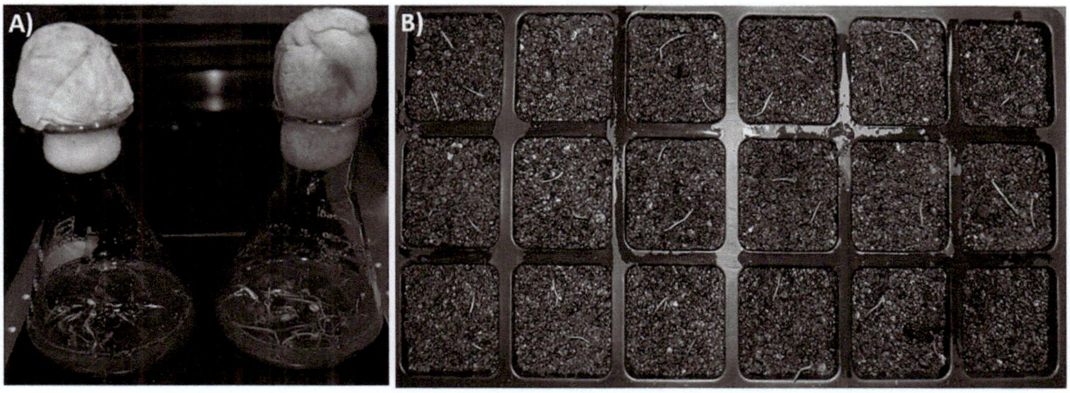

Fig. 3 Co-cultivated seedlings treatment with antibiotic and transferred to protrays. **a** Washing of germinated dlings after the co-cultivation step with acetosyringone antibiotic. **b** Seedlings transferred to the portrays filled with sterile soil rite mixture

9. Keep the conical flasks having infected dlings on an incubator shaker for 1 h having 100 rpm at 28 °C, to eliminate the overgrowth of the *Agrobacterium* (Fig. 3a).
10. Repeat the washing step with cefotaxime two to three times within 1 h, and then transfer the infected dlings to protrays having sterile soil rite (Fig. 3b) (*see* **Note 26**).
11. Maintain the trays at 23 ± 2 °C temperature in the culture room having auto set 16 h light and 8 h dark photoperiod.
12. The germinated plants in protrays were transferred to 6-inch pots and maintained in net house conditions.

3.4 Screening at T_0

1. Collect the fresh leaf samples from all generated putative wheat plants, and extract the genomic DNA using the CTAB method (*see* **Note 27**).
2. The gel electrophoresis is carried out to check the quality and quantity of extracted gDNA.
3. The gene-specific and *hpt*II primers are designed using Pearl primer software.
4. Perform the PCR to check the presence of the gene of interest in all putative plants.
5. Furthermore, the T_1 seedlings of all independent events are harvested separately and stored at 4 °C for screening at T_1 generation (*see* **Note 28**).

3.5 Hygromycin B Screening at T_1

1. Take the pre-vernalized fresh T_1 seedlings for hygromycin B screening.
2. Take the 20 ds of each independent event, wash thoroughly two to three times with tap water, and soak in distilled water.
3. Keep the ds in the water bath for 1 h at 45 °C, and then transfer to sterile 250 mL conical flasks.

4. Pour approximately 100 mL of distilled water into the flasks, supplement with hygromycin B antibiotic (50 mg/L), and keep in an incubator shaker for 1 h at 23 ± 2 °C condition.
5. After the hygromycin B treatment, transfer the seedlings to the protrays filled with sterile soil rite, and maintain the protrays in a culture room facility for 15 days.
6. Record the data of survived and dead seedlings for each independent event, transfer the survived plants further to the 6-inch pots, and maintain them in the net house facility (*see* **Note 29**).

3.6 Screening for Gene Presence in T_1 Plants

3.6.1 Genomic DNA Extraction and PCR Analysis

1. Collect the fresh leaf samples from all T_1 putative transgenic lines, and extract the genomic DNA using the CTAB method manually (*see* **Note 27**).
2. After checking the quality and quantity through electrophoresis and NanoDrop spectrophotometer, the PCR is carried out using gene-specific and *hpt*II primers.
3. The PCR-positive plants are processed for further molecular analysis.

3.6.2 Dot Blot and Southern Hybridization Blot Analysis

1. PCR-positive plants are processed to carry out dot blot hybridization.
2. Approximately 200 ng of digested genomic DNA is utilized following by protocol reported in earlier studies (Vishwakarma et al. 2018) (*see* **Note 30**).
3. Furthermore, to check the copy number of genes that carry the generated transgenic lines, Southern hybridization blot analysis is performed subsequently for all T_1 transgenic lines.
4. The PCR DIG Probe Synthesis Kit, DIG Wash and Block Buffer Set, DIG Easy Hyb, and DIG Luminescent Detection Kit (Sigma Aldrich, USA) are used to perform the whole experiment by following the blotting protocol reported by Sambrook and Russell (2001) (*see* **Note 31**).

3.6.3 Gene Expression Analysis

1. The gene expression of wheat transgenic lines is carried out using qRT-PCR.
2. The wheat *TaActin* internal control is taken to normalize the whole reaction. All the reactions are performed in triplicates.
3. Out of all transgenic lines, the line which has the lowest Ct value is set as one to calculate the fold change expression of other transgenic lines, respectively.
4. The fold change of all the transgenic lines is different because of the positional effect of the gene of interest (*see* **Note 32**).

4 Notes

1. Don't excise the germinated apical meristematic embryonic regions from the seedlings.
2. Mix the LA (Luria agar) quantity as per the required volume separately. Avoid the mixing of LA in one container and then aliquot in required volume flasks.
3. Add Luria broth (LB) powder in a glass beaker as per the total required volume. Mix the LB completely by placing it on a hot plate, and then aliquot according to the required volume.
4. Prepared 1 L MS media, mix all the components only in 800 mL of double-distilled water, set its pH between 5.8 and 6.0, and then make the volume to 1000 mL.
5. Prepare all the components separately, set their pH to 8.0, and then use as required. Add glacial acetic acid in an open area as it releases fumes.
6. Shake the sodium hypochlorite solution before preparing the working solution as it is suspended in the bottom mostly.
7. Use only 70% ethanol during washing the genomic DNA, and keep it at 4 °C before using.
8. Prepare all the components separately in BOD bottles, and then use them as per required. Set the pH of Tris- HCl and EDTA at 8.0. Do not shake the Cetyl-trimethylammonium bromide solution during the mixing. Place the Cetyl-trimethylammonium bromide solution bottle on a hot plate to mix completely.
9. Do not autoclave the polyvinylpyrrolidone. Add after autoclaving all the components.
10. Mix well both the components before use.
11. Use pre-chilled isopropyl alcohol as it increases the precipitation rate.
12. First dissolve the rifampicin powder in 2 mL 100% methanol, and then make up the volume using double-distilled autoclaved water.
13. Use sterile forceps to handle germination sheets.
14. Use a fresh syringe for each *Agrobacterium* culture, and do not exceed the depth of pricking during the infection.
15. Use large-size Petri dishes of 25 mm × 25 mm.
16. Keep the square piece of blotting sheet at the bottom of each column in protrays.
17. Use plastic pots for maintaining the plants in net house conditions because they were easy to transfer in case of any experiment plan.

18. The genotype background of the *EHA105* strain is C58 Ti pEHA105 (pTiBo542DT-DNA) succinamopine, and prepare rifampicin selection LBA plates to screen colonies.
19. The *pCAMBIA1300* vector is modified by cloning gene *GFP* in between *Bam*HI and *Sac*I restriction sites, and the available *35S* promoter is replaced by inserting ***ubiquitin*** promoter in between *Bam*HI and *Spe*I restriction sites. Amplify the GFP gene from the *pCAMBIA-1302-GFP* vector using gene-specific primers (Tarafdar et al. 2019, Kumar et al. 2024).
20. During the sterilization steps, the 50 mL tube-containing seedlings are continuously shaken for 5 min for sodium hypochlorite and 70% ethanol treatment.
21. Cover the Petri dishes with aluminum foil if possible, during the vernalization step for 2 days. Do not add too much water to the Petri dishes containing sterilized seedlings.
22. Before inoculating the secondary culture, the volume of the primary culture should be checked to obtain an accurate optical density (O.D.) at 600 nm absorbance. Do not use the secondary culture if its O.D. crosses the value of 1.0. Avoid the dilution of secondary culture which will be used for infection.
23. Do not exceed the rpm of the centrifuge for pelleting the *Agrobacterium* cells, and use Oakridge tubes for handling *Agrobacterium* culture during the centrifugation step.
24. After adding the acetosyringone into the resuspended *Agrobacterium* cells in half-strength MS broth medium, incubate it at least for 30 min at a 28 °C incubator shaker having 200 rpm.
25. Do not damage the germinated embryonic region apical meristems during pricking and covering with soil rite in protrays.
26. Remove the excess amount of soil rite from the roots of wheat seedlings before washing them with cefotaxime solution and avoid damaging the seedlings.
27. Dip the collected fresh leaf samples into liquid nitrogen, and then store them in a −80 °C deep freezer. Use 100 mg of leaf samples for DNA extraction. Mix all the components for 100 mL, by the following: 8.18 g of NaCl, 2 g of CTAB, 4 mL of 0.5 M EDTA, and 10 mL of 1 M Tris, and then autoclave the samples.
28. The gene-specific and *hpt*II primers are designed using Pearl primer software. The PCR-positive plants were kept separately in the net house (avoid mixing or crossing with normal wheat plants during the anthesis and harvesting period.
29. In this step, wash the seedlings in vivo conditions, and soak them in water for 30 min before incubation in a water bath. The hygromycin B concentration of 50 mg/L is decided after

performing the dosage effect of hygromycin B on normal wild-type wheat varieties of the same cultivar in which we have generated the transgenic lines. Do not let the seedlings in protrays for a longer time and avoid exceeding 15 days period.

30. Dot blot analysis was carried out to narrow down the total number of PCR-positive plants because it is not possible to do Southern hybridization blot analysis for a higher number of plant samples.
31. Wear gloves while handling the buffers, solutions, and blots. Avoid the air bubbles in the agarose gel which might be entrapped during the separation of fragments. Do not add too much buffer in the tray (fill half of the total volume of the tray) during the transfer of bands from agarose gel to nitrocellulose membrane. We used square-shaped cutting of blotting sheets in place of Whatman filter paper as it showed good results. Handle the nitrocellulose membrane from the corners (one fixed place), and avoid touching in between the blot as it showed the hand prints on the X-ray film.
32. The fold change in all transgenic lines is calculated using the 2^ (−ΔΔCt) method described by Livak and Schmittgen (2001).

Acknowledgments

Authors are thankful to Director, NIPB, for providing facilities to carry out the work mentioned here. Dr. G.P. Singh, Director, ICAR-NBPGR, and Dr. P.K. Singh, Principal Scientist, Division of Genetics, ICAR-IARI, are duly thanked for providing the material to carry out transformation studies.

References

1. Kumar S, Vishwakarma H, Ghosh G, Singh J, Padaria JC (2024) *In planta* transformation in wheat: an improved protocol to develop wheat transformants. Mol Biol Rep 51(1):407
2. Christie PJ (1997) *Agrobacterium tumefaciens* T-complex transport apparatus: a paradigm for a new family of multifunctional transporters in eubacteria. J Bacteriol 179:3085–3094
3. Riva D, Cabrera JG, Padron RV, Pardo AC (1998) *Agrobacterium tumefaciens*: a natural tool for plant transformation. Electron J Biotechnol 1:1–16
4. Gelvin SB (2000) *Agrobacterium* and plant genes involved in T-DNA transfer and integration. Annu Rev Plant Physiol Plant Mol Biol 51:223–256
5. Tzfira T, Citovsky V (2000) From host recognition to T-DNA integration: the function of bacterial and plant genes in the *Agrobacterium*-plant cell interaction. Mol Plant Pathol 1:201–212
6. Zupan J, Muth TR, Draper O, Zambryski P (2000) The transfer of DNA from *Agrobacterium tumefaciens* into plants: a feast of fundamental insights. Plant J 23:11–28
7. Tzfira T, Rhee Y, Chen MH, Kunik T, Citovsky V (2000) Nucleic acid transport in plant-microbe interactions: the molecules that walk through the walls. Ann Rev Microbiol 54:187–219
8. Tzfira T, Citovsky V (2002) Partners-in-infection: host proteins involved in the transformation of plant cells by *Agrobacterium*. Trends Cell Biol 12:121–128

9. Clough SJ, Bent AF (1998) Floral dip: a simplified method for *Agrobacterium-mediated* transformation of *Arabidopsis thaliana*. Plant J 16:735–743
10. Bechtold N, Jaudeau B, Jolivet S, Maba B, Venzon D, Voisin R, Pelletier G (2000) The maternal chromosome set is the target of the T-DNA in the *in planta* transformation of *Arabidopsis thaliana*. Genetics 155:1875–1887
11. Curtis IS, Nam HG (2001) Transgenic radish (*Raphanus sativus* L. *longipinnatus Bailey*) by floral-dip method: plant development and surfactant are important in optimizing transformation efficiency. Transgenic Res 10:363–371
12. Yasmeen A, Mirza B, Inayatullah S, Safdar N, Jamil M, Ali S, Choudary MF (2009) *In Planta* transformation of tomato. Plant Mol Biol Report 27:20–28
13. Rod-in W, Sujipuli K, Ratanasut K (2014) The floral-dip method for rice (*Oryza sativa*) transformation. J Agric Technol 10:467–474
14. Ratanasut K, Rod-In W, Sujipuli K (2017) In planta Agrobacterium-Mediated transformation of rice. Rice Sci 24:181–186
15. Trieu AT, Burleigh SH, Kardailsky IV, Maldonado-Mendoza IE, Versaw WK, Blaylock LA, Shin H, Chiou TJ, Katagi H, Dewbre GR et al (2000) Transformation of *Medicago truncatula* via infiltration of dlings or flowering plants with *Agrobacterium*. Plant J 22:531–541
16. Mu G et al (2009) Genetic transformation of maize female inflorescence following floral dip method mediated by *Agrobacterium*. Biotechnology 11:178–183
17. Zale JM, Agarwal S, Loar S, Steber CM (2009) Evidence for stable transformation of wheat by floral dip in *Agrobacterium tumefaciens*. Plant Cell Rep 28:903–913
18. McCabe DE, Martinell BJ (1993) Transformation of elite cotton cultivars via particle bombardment of meristems. Nat Biotechnol 11: 596–598
19. McCabe DE, Swain WF, Martinell BJ, Christou P (1988) Stable transformation of soybean (*Glycine Max*) by particle acceleration. Nat Biotechnol 6:923–926
20. Supartana P, Shimizu T, Shioiri H, Nogawa M, Nozue M, Kojima M (2005) Development of simple and efficient *in planta* transformation method for rice (*Oryza sativa* L.) using *Agrobacterium tumefaciens*. J Biosci Bioeng 100: 391–397
21. Supartana P, Shimizu T, Nogawa M, Shioiri H, Nakajima T, Haramoto N, Nozue M, Kojima M (2006) Development of simple and efficient *in planta* transformation method for wheat (*Triticum aestivum* L.) using *Agrobacterium tumefaciens*. J Biosci Bioeng 102:162–170
22. Shewry PR (2009) Wheat. J Exp Bot 60(6): 1537–1553
23. Harwood WA (2012) Advances and remaining challenges in the transformation of barley and wheat. J Exp Bot 63(5):1791–1798
24. Vasil V, Castillo AM, Fromm ME, Vasil IK (1992) Herbicide resistant fertile transgenic wheat plants obtained by microprojectile bombardment of regenerable embryogenic callus. Nat Biotechnol 10:667–674
25. Cheng M, Fry JE, Pang S, Zhou H, Hironaka CM, Duncan DR, Conner TW, Wan Y (1997) Genetic transformation of wheat mediated by *Agrobacterium tumefaciens*. Plant Physiol 115(3):971–980
26. Risacher T, Craze M, Bowden S, Paul W, Barsby T (2009) Highly efficient *Agrobacterium-mediated* transformation of wheat via *in planta* inoculation. Methods Mol Biol 478: 115–124
27. Tarafdar A, Vishwakarma H, Gothandapani S, Bhati M, Biswas K, Prakash A, Chaturvedi U, Solanke AU, Padaria JC (2019) A quick, easy and cost-effective *in planta* method to develop direct transformants in wheat. 3 Biotech 9(5): 180

Chapter 9

Development of Hairpin RNAi Tobacco Plants for Management of Insects and Pests

Janani Sree Sivakumar, Selvi Chellamuthu, Vasanth Krishnan, and Dhandapani Gurusamy

Abstract

Sustainable agriculture is severely affected by insect as well as pest attack. The manipulation of numerous genes to control pests by RNAi involves post-transcriptional gene silencing that is sequence-specific. This chapter is the first to describe the use of hairpin RNAi in tobacco plants as a novel approach to target insect pests and the potential impact this method could have on future pest management. Here we describe the generation of hpRNA constructs in tobacco for controlling insect pests. First, it describes how to make and test hpRNA constructs aimed at essential pest genes such as those required for development or survival. In this part of our study, the chapter on RNAi design explains in depth the procedures to create these RNAi vectors using cloning strategies in vector structure. It also deals with the use *of Agrobacterium*-mediated transformation for both tobacco plants and selection of transgenic plants. We look at approaches for measuring the impact of RNAi in these plants, such as gene expression analysis and evaluation of host resistance in laboratory or field conditions. This chapter is a comprehensive guide by giving an in-depth view on the steps from construct designing to in field applications; thus it will serve as reference for all future workers of RNAi technology looking towards effective and eco-friendly ways of pest management in agriculture.

Key words RNAi, Insect and pest resistance, hpRNA, Tobacco transformation

1 Introduction

Gene silencing is enhanced by means of a highly promising strategy, RNA interference (RNAi) that targets mRNA degradation in late 1990s. The RNAi is achieved by an enzyme dicer, enables to cleave the double-stranded RNA (dsRNA) into small interfering RNAs (siRNAs). Followed by the incorporation of "siRNAs with in the RNA-induced silencing complex (RISC)" results in invading the complex that degrades mRNA [1, 2]. RISC plays a significant role for regulating the gene, viral defense in addition to silencing the transposon cascade [3].

Nitish Kumar (ed.), *Transgenic Plants: Methods and Protocols*, Methods in Molecular Biology, vol. 2973,
https://doi.org/10.1007/978-1-0716-4803-2_9,

"Hairpin RNA (hpRNA) constructs" are employed as a most significant tool for gene silencing mediated by RNAi. Unlike single-strand siRNAs, hairpin RNAs are transcribed off dsRNA constructs that fold into a stem-loop structure. This stem region of the hairpin RNA has complementary sequences, and so when it is processed by Dicer, mature siRNAs are produced. The advantage of this approach is that it creates many siRNA molecules from one hairpin construct and therefore increases the silencing effect [4].

One of the most popular model systems for plant genetic engineering has been tobacco (*Nicotiana tabacum*) because of its high transformation efficiency, and it is easy to grow [5, 6].

Research on RNAi using tobacco plants can lead to the understanding of gene function and possibly the creation of new pest control methods. Tobacco has large leaves and grows vigorously so it should be a good subject for this kind of work, both molecular and agronomic. It has recently received wide recognition that RNAi can be used as a tool in pest management because it offers a very targeted and eco-friendly means of control. RNAi allows for the control of agricultural pests through the turning off of specific genes that are necessary for pest survival or development, thus avoiding the broad-spectrum effects of the traditional chemical pesticides used [1].

2 Materials

Total RNA, cDNA (*see* **Note 9**), vector, restriction enzymes and DNA ligase, kits for extraction of plasmid/gels, PCR kit, gel electrophoresis system, plasmid DNA, *E. coli* and *Agrobacterium*-competent cells, ice box with ice, liquid nitrogen, centrifuge, centrifuge tubes, sterile whatman 1 and 3 filter paper, micro tips, pipettes, conical flasks, incubator, LB and YEP medium, laminar flow chamber, sterile glass spreader, rifampicin kanamycin and cefotaxime antibiotics, MS medium (Table 1), tobacco plants, cork borer,

Table 1
Media essential for tobacco transformation

Media	Description
YEP agar	1.6 g Agar in 100 mL
MSO	MS medium without hormones and antibiotics
MSAS	MS medium with 0.8% Agar + Antibiotics
MSAP	MS medium with 0.8% Agar + Phytohormones (*see* **Note 10**)
MSAPS	MS medium with 0.8% Agar + Phytohormones (*see* **Note 10**) + Selection antibiotics
Antibiotics	100 mg/L Kanamycin + 500 mg/L Cefotaxime
Phytohormones	2 mg/L BAP + 0.5 mg/L NAA

forceps, centrifuge, Petri plates, double-distilled water (DDW), autoclave, phytohormones, forceps, scalpel holder, scalpel blade, 70% ethanol.

3 Methods

3.1 Selection of Target Region and Development of Target hpRNA Construct

3.1.1 Selection of Target Region

1. The target gene sequence was obtained from the National Center for Biotechnology Information (NCBI) via nucleotide search.
2. Acquire sequence length that corresponds to the target mRNA sequence.
3. Blast the sequence to identify the region.
4. Utilize the selected fragment for designing the construct.

3.1.2 Development of Target hpRNA Construct (See Note 1)

1. Refer Chapter 14 for total RNA isolation (*see* **Notes 6**, **11**, and **12**).
2. From the total RNA, synthesize cDNA (refer Chapter 14) (*see* **Notes 6**, **7**, **8**, **11**, **12**, and **13**).
3. The "Snapdragon—dsRNA design" Link: https://www.flyrnai.org/snapdragon is a web-based tool for designing the double-stranded RNA (dsRNA) for RNA interference. Paste the sequence and adjust the size based on the requirement to obtain a good region for RNAi.
4. The sense, antisense, and intron regions should be PCR-amplified and separately cloned in to pGEM T-Easy vector (*see* **Notes 1** and **2**).
5. After being separately cloned into the pGEM T-Easy vector, the amplified targeted sense, antisense, and then the intron fragment of the "potato granule-bound starch synthase" (GBSS) gene were verified by restriction digestion examination and sequencing (*see* **Note 2**).
6. The fragments underwent additional subcloning and integration within a pUC19 vector. Prior to creating the pUC19-target gene, the sense fragment should be removed using restriction from the pGEM T-Easy vector and cloned onto the pUC19 vector at the appropriate region.
7. After that, *Pst*I-*Bam*HI integrated the GBSS intron into pUC19-target to create target-GBSS-Int; that restriction analysis should be confirmed.
8. To generate pUC19-target-GBSS-Int-target, the antisense segment has subsequently been subcloned onto pUC19-target-GBSS Int at the *Bam*HI-*Eco*RI site.
9. After the hpRNA construct has been subcloned into the pBI121 binary vector under the transcriptional control of the

CaMV 35S promoter along with NOS terminator, restriction digestion and sequencing examination have been performed to ensure the existence of target sense, antisense, and GBSS intron in the correct orientation and sequence.

10. This resulted in the binary vector cassette pBI121::target hpRNA shown in Fig. 1. Then, using the freeze-thaw

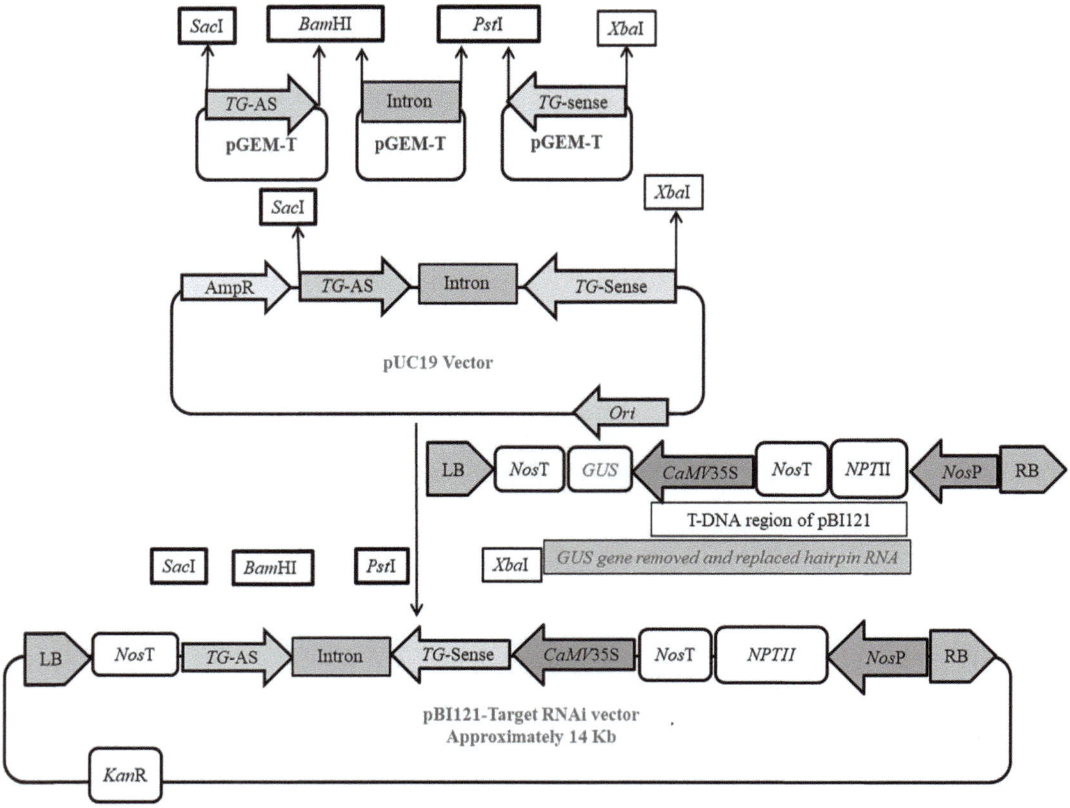

Fig. 1 Construction of pBI121-Target RNAi vector. Specific lengths of (400–700 bp) sense and antisense fragment have PCR amplified using cDNA of respected target species with sense and antisense specific primers contains specific restriction sites for cloning to the vector. Meanwhile approximately 100–150 bp fragmets of suitable intron (PDK, GPSS and etc) have individually PCR amplified and cloned into pGEM T-Easy vector (by employing TA cloning) and confirmed by restriction digestion analysis (restriction sites present in the primer) and sequencing. These fragments were further subcloned and assembled into the pUC19 vector where the target gene antisense fragment were first taken out from the pGEM T-Easy vector with HindIII - SalI restriction and cloned onto the pUC19 vector at the corresponding site to create pUC19- TG-AS. Subsequently, the intron was incorporated into "pUC19::TG-AS" by PstI-BamHI to get "TG-AS-Int" and confirmed by restriction analysis. Next, the sense fragment was subcloned into TG-AS-Int at the BamHI-SacI site to develop the "pUC19::TG-AS-Intron-TG-S". Restriction digestion and sequencing analysis have to be done to confirm the presence of all the fragments. Further, "pUC19::TG-AS-Intron-TG-S" have been subcloned into pBI121 binary vector (at Xba1- Sac1 restriction sites) under the transcriptional control of CaMV 35S promoter and NOS terminator to get the binary vector cassette pBI121::TGhpRNA. The binary vector then mobilized into *A. tumefaciens* strain EHA105 by freeze-thaw method [7]. The colony PCR confirmed *Agrobacterium* strain harboring pBI121::TGhpRNA have then used for the tobacco transformation [8]

procedure, the binary vector has been mobilized into *Agrobacterium tumefaciens* strain, EHA105.

11. The *A. tumefaciens* colony containing pBI121 binary vector has been confirmed by PCR, perform the tobacco transformation.

3.2 Preparation of Agrobacterium tumefaciens-Competent Cells and Mobilization of Binary Vector into Agrobacterium-Competent Cells by Freeze and Thaw Method (See Note 3)

1. *Agrobacterium* strain (EHA105/LBA4404) has streak from the -80 °C deep freezer glycerol stock on a YEP agar plate containing rifampicin (10 mg/L) and incubates at 28 for 2–3 days (*see* **Note 14**).
2. A loop of colony should be picked from the above plate and inoculated into 5 mL liquid YEP medium (1% yeast extract, 1% peptone, 0.5% NaCl, (pH 7.1–7.2); for solid medium 1.5% agar was added) containing required antibiotic and incubated at 28 °C with a vigorous shaking at 200 rpm for overnight.
3. A 2 mL of the culture has been taken and added into 50 mL of YEP medium containing required antibiotic (in a 500 mL conical flask) and incubated at 28 °C with a vigorous shaking at 250 rpm for 5 h (to OD600 = 0.5–1.0).
4. Keep the culture keeps ice for chilling for 30 min, and pellet by centrifugation at 10,000 rpm for 5 min at 4 °C.
5. Discard the supernatant, and resuspend the pellet with 30 mL of 20 mM ice-cold $CaCl_2$ solution (*see* **Note 4**).
6. Repeat the steps for two more times, resuspend the pellet with 1 mL of 20 mM ice-cold $CaCl_2$ (*see* **Note 4**) and 750 μl of 100% glycerol, and keep on 4 °C refrigerator with ice for overnight.
7. The next day, the competent cells have dispense into sterilized, chilled 1.5 mL Eppendorf tubes (100 μl each) (*see* **Note 4**), and keep the tubes at -80 °C deep freezer after quick freeze in liquid nitrogen.

3.3 Mobilization of Binary Vector into Agrobacterium-Competent Cells [9]

1. Take the *Agrobacterium*-competent cells out from -80 °C deep freezer, and thaw on ice (*see* **Note 5**).
2. Add 2.0 to 5.0 μl of plasmid DNA (approximately 500–1000 ng) to the thawed competent cells, mix it gently, and then freeze the cells immediately in liquid nitrogen for 2 min followed by thawing at 37 °C water bath for 2 min.
3. 1.0 mL of liquid YEP medium without antibiotics should be added to the cells.
4. Then incubate the cells in the orbital shaker at 28 °C for 4 h with 200 rpm.
5. Centrifuge the cells that were harvested at 28 °C for 30 s, and discard the supernatant.

6. Resuspend the pellets in 100 μl of YEP medium (without antibiotics), and spread the cells on YEP agar plates containing appropriate antibiotics (10 mg/L rifampicin, 50 mg/L kanamycin) using sterile glass spreader.
7. Incubate the plates at 28 °C for 2–3 days for colonies to appear.

3.4 Tobacco (Nicotiana tabacum Var. Petit Havana) Transformation

Day 1

1. Streak the *A. tumefaciens* stain harboring the binary plasmid in YEP agar medium with 10 mg/L rifampicin +50 mg/L kanamycin

DAY 2

2. Preparation of all necessary media and other things for infection and co-cultivation.

DAY 3

3. Cut healthy young leaves of axenic tobacco plants in the transfer laminar hood. Place the leaves in a Petri dish with sterile H_2O (*see* **Note 7**).
4. Place two to three Whatman 1 circles (sterile) in a sterile Petri disc. Wet them with sterile H_20 and place the leaves over the filter paper.
5. Cut out discs (about 7.5 mm in diameter) is using a pre-sterilized cork borer, and gently push the leaf discs from inside the cork borer using a glass rod into the Petri disc containing sterile water.
6. Repeat this procedure until sufficient amount of leaf discs are cut.
7. Using a spatula, transfer the discs from water to a Petri dish with a piece of sterile Whatman 1 paper.
8. Blot the discs dry and then place them inverted (lower epidermis facing upward) on MSP (shooting) medium.
9. Incubate for 2 days in the tissue culture room (*pre-incubation*).
10. Culture the *A. tumefaciens* stain EHA105 harboring the binary plasmid, pBI121, in 50 mL of YEM medium with selective antibiotics (10 mg/L rifampicin +50 mg/L kanamycin) at 28 °C with 220 rpm shaking for 16–20 hrs at by inoculating a loop of colony with a sterile loop.

DAY 4

11. In the evening (after 16–20 hrs), *Agrobacterium* culture (approximately 0.8–1.0 OD at 600 nm) is transferred to sterile 50 mL tube and centrifuged at room temperature for 10 min at 5000 rpm.

12. Discard the supernatant completely, and resuspend the pellet in 10 mL of MS liquid medium.
13. *Infection*: The preincubated leaf disc has been taken from the plate and mixed with *Agrobacterium* culture which contains in the 250 mL conical flask (*see* **Note 15**).
14. Gently mix the leaf discs with the bacterial suspension by gentle swirling for 10–20 min (*see* **Note 8**).
15. Blot dry the infected leaf discs with sterile tissue paper contains in the Petri dish (*see* **Note 8**).
16. *Co-cultivation*: Place the infected leaf discs upside down in fresh MSPA plates for 2 days in a culture room at dark place (*see* **Notes** 8 and 15).

DAY 6

17. *Washing of co-cultivated leaf discs*: After 2 days, co-cultivated leaf discs have been washed with 200 mg/L cefotaxime solution in DDW, followed by blot dry the leaf discs with sterile Whatman 1 paper contains in the Petri dish (*see* **Note 9**).
18. *Selection*: Transfer the blot tried leaf discs onto a MSP medium contain 100 mg/L kanamycin in which only plant cells transformed with *neomycin phosphotransferase* (*NPT*II) gene would grow. This medium also contains 500 mg/L cefotaxime, which would kill *Agrobacterium* which grows outside the cell (*see* **Note 9**).

Subculture

19. Once in 2 weeks, subculture the regenerated leaf MSPAS medium (*see* **Note 10**).
20. Proliferation of callus induction observed in the edge of the leaf discs under selection medium.
21. Three to four weeks after infection, shoots begin to develop.
22. At this stage segments with shoots are cut out and transferred to rooting medium (MSA) with kanamycin and cefotaxime.
23. After roots initiate from the shoots and elongated, the plants are further subcultured into 250 mL conical flask containing the selection media. Once the plants are matured with broadened leaves and thick shoots, they will be subcultured for multiple clones.
24. Finally, one batch of plants have acclimatized in the transgenic greenhouse for seed collection, while the remaining clones are still maintained under tissue culture conditions for further insect bioassays.

4 Notes

1. Researchers can use the commercially available pHANNIBAL and pKANNIBAL vectors (https://www.novoprolabs.com/) to construct the hairpin RNAi vector. These vectors contain an "intron," and it is simple to clone the sense and antisense fragments between it with the appropriate restriction sites.
2. The quality of the plasmids and the yield of the gel products are enhanced by using commercially available kits for plasmid isolation and gel purification, such as QIAGEN (https://www.qiagen.com/) and others.
3. The process of competent cells preparation should be in clean and contaminated-free zone; use laminar air flow chamber while opening/closing the tubes.
4. Centrifuge tubes, 1.5 mL Eppendorf tubes for storage, and $Cacl_2$ solution should be properly autoclaved and prechilled prior to use.
5. Avoid excessive thawing of competent cells.
6. All the materials used for isolation RNA should be treated with DEPC water and autoclaved.
7. The leaf disc explants used for the transformation should be taken and prepared for in vitro culture tobacco plants.
8. Avoid excessive wounding during co-cultivation because it may cause mechanical injury to the explant.
9. Prolonged existence of *Agrobacterium* within the host cells results in mortality of the transformed crop.
10. Addition of cytokinin improves the regeneration potential in several crops [10].
11. Wearing gloves and mask is mandatory to get rid from denaturation of RNA.
12. To obtain good quality RNA, avoid extend duration of vortex as it may mix the genomic DNA [11].
13. The dilution of cDNA and template concentration should be based on our experiment need.
14. Utilizing auxotrophic *Agrobacterial* strains such as LBA4404/EHA105 elevates the efficiency of plant transformation [12].
15. No need acetosyringone (AS) and any other specific medium for the infection and co-cultivation.

References

1. Baum JA, Bogaert T, Clinton W, Heck GR, Feldmann P, Zeng Q (2007) Control of coleopteran insect pests through RNA interference. Nat Biotechnol 25(11):1322–1326
2. Bernstein E, Caudy AA, Hammond SM, Hannon GJ (2001) Role for a bidentate ribonuclease in the initiation step of RNA interference. Nature 409(6818):363–366
3. Cullen BR (2004) RNA interference: Insights from viral infections. Nat Rev Microbiol 2(2): 1–7
4. Elbashir SM, Lendeckel W, Tuschl T (2001) RNA interference is mediated by 21- and 22-nucleotide RNAs. Genes Dev 15(2): 188–200
5. Fire A, Xu S, Montgomery MK, Kostas SA, Driver SE, Mello CC (1998) Potent and specific genetic interference by double-stranded RNA in *Caenorhabditis elegans*. Nature 391(6669):806–811
6. Horsch RB, Fry JE, Hoffmann NL, Rogers SG, Fraley RT (1985) A simple and general method for transferring genes into plants. Science 227(4691):1229–1231
7. Hofgen R, Willmitzer L (1988) Storage of competent cells for *Agrobacterium* transformation. Nucl Acids Res 16:9877
8. Jaiswal S, Saini RP, Tilgam J, Gurusamy D, Sreevathsa R, Pattanayak D (2022) Host-delivered RNAi mediated silencing of HaAce1, encoding the major isoform of acetylcholinesterase, imparts resistance against the polyphagous insect-pest, *Helicoverpa armigera*. J Plant Biochem Biotechnol:1–12
9. Gurusamy D (2014) Transcriptome analysis of cotton (*Gossypium hirsutum* L.) during boll development under drought condition and functional characterization of drought tolerance genes. PhD thesis. https://hdl.handle.net/10603/215568
10. Sivakumar JS, Gurusamy D, Subramaniam S, Krishnan V (2024) Enlightening the biotechnological approaches of legumes: present and past scenario to mitigate the biotic and abiotic stress. Plant Biotechnol Rep 18(5):605–623
11. Farrell RE Jr (2009) RNA Methodologies: laboratory guide for isolation and characterization. Aca Press
12. Aliu E, Ji Q, Wlazlo A, Grosic S, Azanu MK, Wang K, Lee K (2024) Enhancing *Agrobacterium*-mediated plant transformation efficiency through improved ternary vector systems and auxotrophic strains. Front Plant Sci 15: 1429353

Chapter 10

Characterization of RNAi Tobacco Plants and Detection of siRNA

Janani Sree Sivakumar, Selvi Chellamuthu, Vasanth Krishnan, and Dhandapani Gurusamy

Abstract

In this chapter we focus for characterizing the RNA interference (RNAi) tobacco plants and detecting existence of the small interfering RNAs (siRNAs). RNAi, the most powerful gene silencing approach that engages in introducing double-stranded RNA (dsRNA) homologous to the target gene. The popular model plant system, tobacco, is employed for studying RNAi because of easy transformation as well as fast growth. This chapter emphasizes the steps that include total RNA extraction, cDNA synthesis, qRT-PCR amplification, as well as northern blotting. To segregate the DNA contamination from isolated total RNA extracted using TRIzol method, *DNase* treatment was carried out. PCR amplification was performed to check for the target gene and housekeeping gene. Northern blot experiment was done to examine the existence of siRNAs. The experiment worked in that RNAi-mediated gene silencing occurred in the tobacco plants, as proven by lower expression of the target gene and detection of siRNAs. This research offers a great protocol for the analysis of RNAi tobacco plants and the detection of siRNAs and can be utilized in many different research fields, such as plant biotechnology and functional genomics.

Key words Total RNA, Northern blot, Double-stranded RNA (dsRNA), Small interfering RNA (siRNA), α-32^{P}, Digoxigenin (DIG) labeling, Autoradiography

1 Introduction

RNA interference (RNAi) is an evolutionarily phenomenon below which one can find certain sequence-specific gene silencing due to the degradation of respective mRNA molecules, exposed to dominance by tiny RNA species, discovered initially in the late 1990s [1]. RNAi itself owns its attributes or features as initiated by double-stranded RNA (dsRNA), which is then further processed into small interfering RNAs (siRNAs) by the enzyme Dicer, generally composed of 21–24 nucleotides in size [2]. The siRNA finally gets delivered into RNA-induced silencing complex, which gets directed toward mRNA targets, resulting in the cleavage of corresponding mRNA and its subsequent destruction [3]. The

Nitish Kumar (ed.), *Transgenic Plants: Methods and Protocols*, Methods in Molecular Biology, vol. 2973,
https://doi.org/10.1007/978-1-0716-4803-2_10,

length of siRNA should range between 20 and 25 nucleotides in the initial step of RNAi. Derived from longer dsRNA precursors, siRNA duplexes are processed by Dicer into siRNA duplexes. The duplexes are unwound to produce single-stranded siRNAs [4]. RNA-induced silencing complex (RISC) recognizes mRNA targets and destroys it along with the guidance of siRNA to downstream events leading to gene expression silencing. In addition, this mechanism is critical for gene regulation as well as defense against viral infection and transposons [5].

The tobacco (*Nicotiana tabacum*) plant is becoming gradually more established as a model system for RNAi in plant systems because of its higher efficiency of transformation and relative ease in cultivation. Because of several benefits, the use of RNAi in plants has mainly focused on tobacco. An introduction of RNAi constructs may be far easier because these are effectively transformed into tobacco using *Agrobacterium*-mediated methods [6]. There is this higher transgene expression that gives an amazing model to study the potential impact of RNAi on a gene in other hosts [7]. Knowing that with their high transgene expression, tobacco constitutes an interesting model to study RNAi effects on a gene in other hosts, it becomes blessed with such holders that should otherwise be under close watch. Such is the case for molecular assays and phenotypic analysis [8]. This chapter focuses on characterization of RNAi tobacco plants and detection of siRNA (*see* **Note 1**).

Refer Chapters 14 and 16 for the Total RNA Isolation and Analysis.

2 Materials

Roche DIG Luminescent Detection Kit (Sigma-Aldrich), Amersham Megaprime DNA labeling system (GE Healthcare UK Ltd.), Ambion ULTRAhyb-oligo Hybridization Buffer (Thermo Fisher Scientific), TBE Electrophoresis Buffer (Table 1).

Prepare using RNase-free water and store up to 6 months at room temperature [9].

Table 1
TBE electrophoresis buffer (10×)

Reagent	Quantity (for 1 L)
Tris base (1 M)	121.1 g
Boric acid (1 M)	61.8 g
EDTA (disodium salt) (0.02 M)	7.4 g

2.1 Gel Preparation Buffer (100 mL)

Take 70 mL of RNase-free water in beaker and add 48 g of urea, and then add 15 mL of 40% acrylamide solution, followed by adding 10 mL of 10× TBE electrophoresis buffer, and make up the final volume to 100 mL [10] (*see* **Note 12**).

2.2 1-Ethyl-3-(3-Dimethylaminopropyl) Carbodiimide EDC Cross-Linking Solution

1. EDC (0.16 M) prepared using 1-methylimidazole (0.13 M) at pH 8.
2. Add 245 mL of 12.5 M 1-methylimidazole to 9 mL DEPC-treated water. Then, adjust the pH to 8.0 with 1 M HCl.
3. It should be prepared 1–2 h prior to use. Keep it in room temperature.
4. Immediately prior to use, add 753 mg EDC, and make up the final volume to 24 mL with DEPC water.
5. This provides the working solution of about 0.16 M EDC in the 0.13 M 1-methylimidazole at pH 8.

3 Methods

3.1 Non-radioactivity Method [11] (see Note 12)

1. 50–90 μg of total RNA sample should be resolved using gel preparation buffer (denaturing urea PAGE in 0.5 × TBE buffer) (*see* **Notes 2**, **3**, **4**, **5**, **6**, and **7**).
2. Perform the electro blotting on a nylon membrane which is positively charged. Perform at 4 °C for 4 h (*see* **Notes 13** and **14**).
3. Then, the blot should be treated using 1-ethyl-3-(3-dimethylaminopropyl) carbodiimide (EDC) [11, 12].
4. The target fragment DNA should be amplified in PCR using specific primers to utilize it as probe followed by DIG labeling.
5. Probe preparation, pre-hybridization, hybridization, and band detection were performed by Roche DIG Luminescent Detection Kit as per the manufacturer protocol (*see* **Notes 8**, **9**, **10**, **15**, and **16**).
6. Hybridization should be performed using (Ambion ULTRAhyb-oligo hybridization buffer) (*see* **Note 10**).
7. Then, 0.3 nM of DIG end-labeled probe should be washed at 37 °C for 16 h in the "low stringency conditions" (2× SSC + 0.1% SDS) at 28–32 °C.
8. The blot has been developing by exposing the membrane to the X-ray film in intensifying cassette under the dark conditions (*see* **Notes 17**, **18**, **19**, and **20**).
9. The cassette should be kept for 2 h and expose it to the X-ray film to visualize the bands after developing and fixing (*see* **Notes 17**, **18**, **19**, and **20**).

3.2 Radioactivity Method [13] (see Note 12)

1. Extract the total RNA and the small RNAs must fractionated using the RNA clean up protocol of the Midi RNeasy Kit (Qiagen GmBH, Hilden, Germany) (*see* **Notes 2**, **3**, **4**, and **6**).
2. Small RNA fraction (20 μg) should be loaded onto a 15% polyacrylamide [w/v] gel with 8 M urea and was electrophoresed at 300 V (*see* **Notes 5**, **6**, and 7).
3. Small RNAs must blotted onto the positively charged nylon membrane using the Trans blot-SD semidry transfer apparatus (Bio-Rad, Hercules, USA) in 1 × TBE at 7 V for 45 min (*see* **Notes 13** and **14**).
4. The membrane is cross-linked twice in a "UV cross linker" (Hoefer Scientific Instruments, San Francisco, USA).
5. Approximately 300–500 bp fragment of the gene of interest (used as probe) should be labeled using (α-32^{P}) dCTP using the Amersham Megaprime DNA labeling system (GE Healthcare UK Ltd., Little Chalfont, UK) (*see* **Notes 8**, **9**, and **10**).
6. The hybridization should be performed at 37 °C for 16 h (*see* **Notes 15** and **16**).
7. In post-hybridization, washes should be done for four times (duration of each wash is 20 min) with 2 × SSC/0.2% SDS at 50 °C (*see* **Note 16**).
8. The washed membrane is fixed in the intensifying cassette and exposed X-ray film under the dark conditions.
9. The cassette should be keep for 2 h to overnight in the -80^{0} deep freezer and develop the X-ray film to visualize the bands (*see* **Notes 17**, **18**, **19**, and **20**).
10. The autoradiography shows 21 bp of siRNA band (Fig. 1) (*see* **Note 11**).

4 Notes

1. The quality of total RNA extraction, siRNA probe design, sample preparation, gel electrophoresis conditions, normalization techniques, and possible off-target effects are important factors to consider when detecting high-quality siRNA in RNAi plants.
2. Make sure that siRNA is efficiently produced and delivered to the target tissue.
3. Optimize the RNA extraction procedure, including tissue selection and homogenization conditions, to prevent low RNA yield. Additionally, do an RNA integrity number (RIN) analysis on the extracted RNA to look for possible RNA degradation.

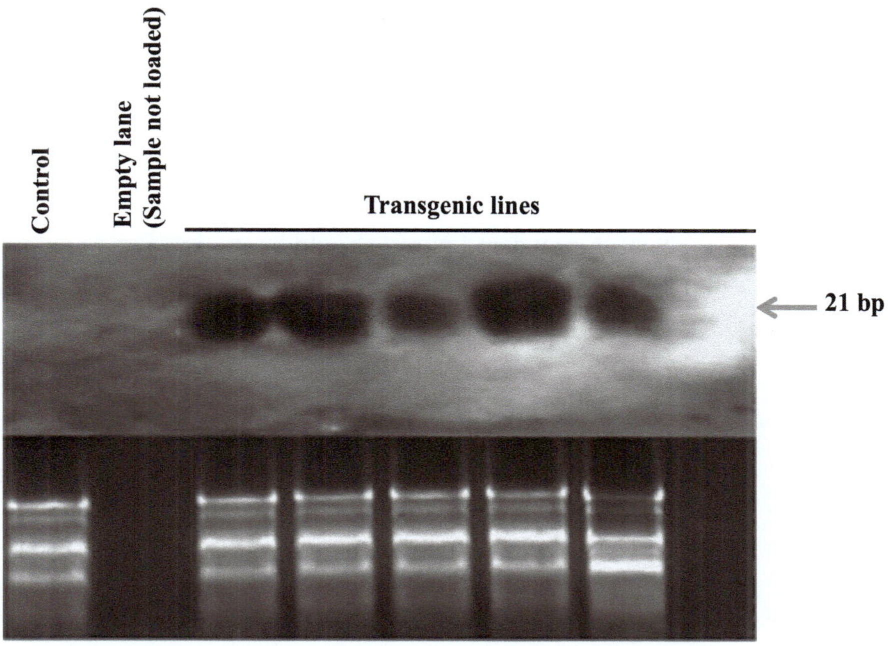

Fig. 1 Northern blot analysis of RNAi tobacco plants and detection of siRNA production. **a** Autoradiography shows detection of 21 bp siRNA production in hairpin RNAi delivered tobacco plants, whereas no siRNA detection observed in the control plant. **b** Image shows ethidium bromide-stained Northern gel shows distinct RNA band [11]

4. Make sure that all genomic DNA is removed using *DNase* treatment, and if RNA contamination does occur, use the right RNA separation kits developed to reduce contamination from other molecules.
5. Optimize the gel concentration (usually 15–20% denaturing urea-PAGE) to efficiently separate small RNA fragments and increase the amount of RNA loaded onto the gel to improve detection sensitivity.
6. To prevent the damage of RNA, cut tips are used while pipetting the aqueous phase [14].
7. Double-check running conditions (voltage and running time) to ensure proper siRNA migration.
8. Design probes that are highly specific to the target siRNA sequence, avoiding possible off-target binding sites. To increase the probe's specificity, employ sequence analysis tools to look for any secondary structures that might be present.
9. Make sure the probe length is suitable for the size of the siRNA being detected, which is usually between 21 and 23 nucleotides.

10. Use a more sensitive detection technique, such as improved chemiluminescence or radioactive labeling, and optimize probe binding to the target siRNA by adjusting hybridization settings to prevent low signal intensity.
11. By evaluating the expression of related genes, do additional investigations to rule out possible off-target silencing.
12. Non-powdered gloves should be used to avoid contamination.
13. The membrane used for the blotting should be always be slightly larger compared to gel size.
14. To prevent contamination, avoid handling the membrane directly even if the hands are gloved, instead use forceps (blunt end).
15. Formation of air bubbles and heat may be due to unsaturated filters. Make sure the filters are fully saturated using buffer.
16. To avoid short-circuit the apparatus should be wrapped with plastic material [15].
17. The X-ray film box should "never" be opened in light. It contains X-ray sheets with inter-leaves which together are enclosed in a black polythene cover.
18. The box should be opened only in dark and after using the required number of X-ray films, it "has to be" replaced back into the box "in dark."
19. The opening of the cover has to be folded properly and inserted back into the box in such a way that it faces the bottom of the box. The lid is put back and fastened with rubber bands to avoid the lid from releasing, accidently.
20. The X-ray film should be handled on only one corner and never touched in any other portion.

References

1. Fire A, Xu S, Montgomery MK, Kostas SA, Driver SE, Mello CC (1998) Potent and specific genetic interference by double-stranded RNA in *C. elegans*. Nature 391(6669): 806–811
2. Bernstein E, Caudy AA, Hammond SM, Sharp PA (2001) Role for a bidentate ribonuclease in the initiation step of RNA interference. Nature 409(6818):363–366
3. Hammond SM, Bernstein E, Beach D, Hannon GJ (2001) An RNA-directed nuclease mediates post-transcriptional gene silencing in *Drosophila* cells. Nature 404(6775):293–296
4. Elbashir SM, Lendeckel W, Tuschl T (2001) RNA interference is mediated by 21- and 22-nucleotide RNAs. Genes Dev 15(2): 188–200
5. Cullen BR (2004) RNA interference: antiviral defense and regulation of gene expression. Mol Cell 16(1):15–19
6. Horsch RB, Fry JE, Hoffmann NL, Eichholtz D, Rogers SG, Fraley RT (1985) A method for detecting expression of genes transferred into plant cells using a binary vector system. Science 227(4691):1229
7. Sambrook J, Russell DW (2001) Molecular cloning: a laboratory manual. Cold Spr Harbor Lab Press
8. Graham TL, Miller JC, Smith RH (2003) Tobacco: a versatile system for the production

of recombinant proteins. In: Methods in molecular biology, vol 207. Humana Press, pp 265–278
9. Rio DC (2014) Northern blots for small RNAs and microRNAs. Cold Spr Harbor Protoc 2014(7):080838
10. Rio DC, Ares M, Hannon GJ, Nilsen TW (2010) Polyacrylamide gel electrophoresis of RNA. Cold Spr Harbor Protoc 2010(6):5444
11. Jaiswal S, Raman KV, Saini RP, Tilgam J, Saakre M, Dhandapani G, Pattanayak D (2024) Host-delivered RNAi-mediated seilencing of HaAce1, encoding the major isoform of acetylcholinesterase, imparts resistance against the polyphagous insect-pest, *Helicoverpa armigera*. J Plant Bio Biotech:1–12
12. Pall GS, Hamilton AJ (2008) Improved northern blot method for enhanced detection of small RNA. Nat Protoc 3(6):1077–1084
13. Shanmugapriya G, Das SS, Veluthambi K (2015) Transgenic tobacco plants expressing siRNA targeted against the Mungbean yellow mosaic virus transcriptional activator protein gene efficiently block the viral DNA accumulation. Virusdisease 26:55–61
14. Sahu SK, Thangaraj M, Kathiresan K (2012) DNA extraction protocol for plants with high levels of secondary metabolites and polysaccharides without using liquid nitrogen and phenol. ISRN Mol Biol 2012(1):205049
15. Brown T (2001) Analysis of DNA sequences by blotting and hybridization. Curr Protoc Mol Bio

Chapter 11

Development of Transplastomic Tobacco Plants

Kanakachari Mogilicherla, Janani Sree Sivakumar, Selvi Chellamuthu, Vasanth Krishnan, and Dhandapani Gurusamy

Abstract

Chloroplasts, originating from ancestral cyanobacteria, are essential plastids involved in photosynthesis and other key metabolic processes, making them vital components of plant cells. Chloroplast transformation is a preferred alternative to nuclear transformation in plant genetic engineering due to its numerous advantages and positive outcomes. This chapter explores the available various chloroplast transformation vectors, cloning the gene of interest (GOI) in the chloroplast transformation vector with step-by-step protocol, and preparation of the vector for chloroplast transformation of tobacco using leaf disc explants through a biolistic method for efficient gene delivery. Additionally, it highlights how crucial some factors are for the effective integration and expression of the chloroplast genome by DNA-coated particle bombardment, including microcarrier type, DNA concentration, helium accelerating pressure, leaf layout and firing distance, and duration of pretreatment for explants. Further explored the generation of homoplasmic plants through re-regeneration with antibiotic selection confirmed correct cassette insertion and homoplasmy using PCR and Southern blot analyses and verified maternal inheritance of chloroplast-transferred genes by examining green and albino seedlings from reciprocal crosses between control and transplastomic tobacco plants.

Key words *Nicotiana tabacum*, Chloroplasts vector, Microcarrier, Biolistic transformation, Spectinomycin resistance, Transplastomic plants

1 Introduction

Plastids are organelles found in plant cells, algae, and some parasites, believed to have originated from cyanobacteria through an ancient symbiotic relationship with eukaryotic cells [1, 2]. Chloroplasts, originating from proplastids, are essential for photosynthesis and also contribute to amino acid production, nitrogen metabolism, secondary metabolite synthesis, and fatty acid production, with genomes that have reduced to 100–200 kb from their ancestral sizes [3, 4]. Chloroplast transformation, developed 20 years ago as an alternative to nuclear transformation, is a highly effective method for expressing transgenes and preventing their spread due

Nitish Kumar (ed.), *Transgenic Plants: Methods and Protocols*, Methods in Molecular Biology, vol. 2973,
https://doi.org/10.1007/978-1-0716-4803-2_11,

to the maternal inheritance of chloroplast genomes in most angiosperms, making it ideal for generating genetically modified plants [5, 6]. Chloroplast transformation offers higher protein yields, with studies showing up to 46% foreign protein expression, and produces proteins free from viral or human pathogens, making it a promising method for protein production research [5, 6].

The fundamental components of the double-stranded DNA that make up chloroplast genome are usually 120–220 kb and are organized in both linear molecules as well as multimeric rings. In order to enable double recombination, "homologous flanking sequences" are incorporated into the design of chloroplast transformation vectors on both sides of the transgene cassette. Targeting sequences are typically 1 kb in size and usually homologous with the selected target region; they possess no other unique characteristics [7].

Integration of the transforming DNA into the plastid genome occurs through homologous recombination, followed by selective enrichment on a medium containing spectinomycin or streptomycin, which inhibits cell growth and regeneration. Transplastomic clones, initially heteroplastomic due to a mix of wild-type and transformed cells, are subjected to repeated shoot regeneration until homoplasmy is confirmed through PCR, DNA gel blot analysis, and stable transmission in seed progeny [8]. Plastid transformation, first performed in *Nicotiana tabacum* cv. Petit Havana due to its short life cycle and fast regeneration allows for stable homo-transplastomic plant generation in 4–6 months. Different methods are used to manipulate the plastid genome, including insertion of marker genes and linked genes of interest, marker gene excision after transformation, replacement of wild-type genes with mutant forms, gene knockouts using selectable markers, and co-transformation using two plasmids to recover a non-selectable gene [9]. This chapter describes the chloroplast transformation vector, the construction of the gene of interest (GOI) into the vector, the protocol for plastid transformation in *Nicotiana tabacum* using spectinomycin resistance for selection, and incorporating advancements in plastid transformation [10–12].

2 Materials

Chloroplast expression vector (*see* **Note 1**), gene of interest, restriction enzymes, DNA ligase, competent cells, antibiotics, LB medium, micropipettes, sterile microtips, centrifuge tubes, water bath, 37 °C shaker and incubator, Petri plates, laminar airflow chamber, inoculation/spreading loops, competent cells, ligated mixture, Eppendorf tubes (*see* **Note 2**), ice box, preparation LB medium (Table 1), gold or tungsten particles, biolistic gene gun,

Table 1
Composition of LB medium

LB medium	Quantities (for 1 L)
Tryptone	10.0 g
NaCl	10.0 g
Yeast extract	5.0 g

vacuum pump, Whatman No. 4 filter paper, 70% ethanol, vortexer, tobacco leaf disc, MS medium, BAP, NAA, streptomycin, DNA construct (recombinant plasmid), macro-carrier, rupture disc, calcium chloride, isopropyl alcohol, glycerol, and Helium gas.

Add all the components given in the tables and dissolve using double-distilled water. Adjust the pH to 7.0. Finally make up the volume to 1litre and autoclave it.

3 Methods

3.1 Chloroplast Transformation Vector and Cloning the Gene of Interest (GOI) [15]

3.1.1 Designing Chloroplast Vector (Fig. 1) and Genes for Chloroplast Transformation Vector (Table 2)

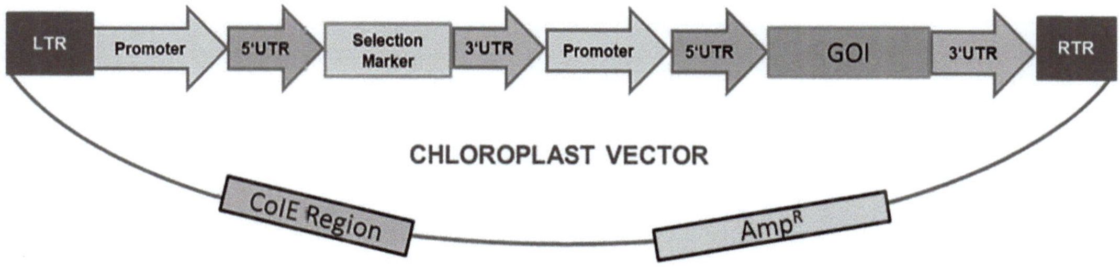

Fig. 1 Design and structure of a common chloroplast transformation vector. Linear map of chloroplast transformation vector contains chloroplast specific promoter, 5' UTR, selectable marker gene, 3' UTR, gene of interest (GOI) [7]

Table 2
List of genes for chloroplast transformation vector [7]

Right flanking sequence	trnA, accD, trnL, psbJ, trnV, 3'rps7/12, trnG, trnR, trnS
Left flanking sequence	trnL, rbcL, rpI32, petA, 3'rps7/12, rrn16, trnV, trnfM, trnN, ycf3
Promoter	Prm, PpsbA, PrbcL
5′ regulatory elements (RE)	ggagg, *psbA*, rbcL, atpB, Cry2a, T7gene 10
Selectable marker (SM)	*aad*A, *npt*II, *aph*A6, badh
3'UTR	*rps*16, *rbc*L, *psb*A, *pet*D
Gene of interest (GOI)	Based on the requirement

3.1.2 Restriction Digestion and Ligation

1. The gene of interest (GOI) should be selected and primers to be designed with suitable restriction sites followed by PCR amplification and gel purification of the gene product.
2. The chloroplast transformation vector also streak in the LB agar medium and inoculate in the LB broth with suitable antibiotic and pure plasmid should be isolated.
3. Both vector (cloning vector digested at the specific *Pvu*II site within the region which has been retained in all higher plants studied using a "clone chloroplast transformation cassette") and gene of insert should be restricted with suitable specific restriction sites (37 °C water bath for 4 h to overnight), and purify the restricted products.
4. Ligation should be performed with 3:1 ratio of insert and vector with T4 DNA ligase at 4 °C for overnight or 22 °C for 4 h.
5. To avoid the self-ligation of vector, alkaline phosphatase should be used to treat the vector dephosphorylate for 30–60 min at 37 °C.
6. The ligated mixture should be transformed to the *Escherichia coli*, DH5 alpha competent cells by heat-shock method and plate to the LB agar medium with vector specific selective antibiotic and incubate for 12–16 h at 37 °C to appear the colonies.
7. The colony PCR have to performed to confirm the positive clones and isolate pure plasmid DNA by using Qiagen Plasmid Extraction Midi Kit.
8. The pure plasmid DNA is delivered to the tobacco leaf disc by gene gun process method (*see* **Note 15**).

3.2 Preparation of Competent Cells (DH5 Alpha)

3.2.1 Preparation of 0.1 M Calcium Chloride Solutions (for 100 mL)

1. Weigh 11 g of $CaCl_2$ and dissolve it using 90 mL of double-distilled water.
2. Then make up the final volume to 100 mL and then autoclave it (*see* **Note 3**).

3.2.2 Preparation of DH5 Alpha Competent Cells

1. Inoculate an LB DH5 alpha culture with cells (directly from the frozen stock without thawing), and grow overnight at 37 °C.
2. Add 5 mL of this overnight culture to 100 mL of LB medium in a 500 mL flask.
3. Grow cells to an OD between 0.4 and 0.6. Do not exceed 0.6.
4. Chill culture on ice for 10 min.
5. Transfer culture to a sterile 50 mL centrifuge tubes (*see* **Note 3**), and pellet the cells at 4000 g for 5 min at 4 °C. Decant off the LB medium.
6. Gently resuspend the pellet in 50 mL of ice-cold 100 mM $CaCl_2$ solutions, and incubate the cell suspension on ice for 15–30 min (*see* **Note 2**).
7. Pellet the cells at 4000 g for 5 min at 4 °C. Decant off the $CaCl_2$ solution.
8. Repeat the **steps 6** and 7 one more time.
9. Gently resuspend the cells with 1.2 mL of 100 mM $CaCl_2$ solution (*see* **Note 2**).
10. Incubate the cell suspension on ice for 5 min. Then add 0.8 mL of 90% glycerol and gently mix it and keep it in ice. The ice box should be kept overnight at 4 °C refrigerator.
11. The next day, resuspend the suspension into sterile microcentrifuge tubes (100 μL aliquots is generally appropriate) (*see* **Notes 2** and **3**), and flash-freeze in liquid nitrogen and store at −80 °C (*see* **Note 4**).

3.3 Transformation of Ligate Product to the DH5 Alpha Competent Cells by Heat-Shock Methods [13]

1. Thaw the competent cells using ice (*see* **Note 4**).
2. Add 1–5 μL of ligate mixture (*see* **Note 3**), and incubate in ice for about 30 min.
3. Heat-shock the sample by placing it in the 42 °C water bath for 30 s.
4. Keep cells in ice for 2 min.
5. Then, add 1 mL pre-warmed LB media and incubate at 37 °C of 200 rpm for 1 h.
6. Make a spread plate and then incubate at 37 °C for 16 h.

3.4 Particle Bombardment (see Note 15)

3.4.1 Gold Particle Preparation (see Note 11)

1. Weigh 30 mg of 0.6 μm gold microcarriers in a 1.5 mL Eppendorf tube, and add 1 mL of ice-cold 70% ethanol.
2. Vortex for 5 min, and then let particles settle for 15 min at room temperature.
3. Spin the tubes at 3000 rpm for 1 min to sediment the gold, remove ethanol, and wash twice with 1 mL ice-cold sterile distilled water, repeating the spinning step each time.
4. Spin at tubes at 5000 rpm for 15 s, remove water, add 500 μL of 50% glycerol (60 mg/mL of gold will be present), and vortex for 1 min.
5. Store at room temperature for up to 2 weeks.

3.4.2 Plasmid DNA Coating

1. Vortex stored gold particles for 5 min before use (*see* **Note 11**).
2. Shake 50 μL aliquots of gold in Eppendorf tubes, and add 5 μL of plasmid DNA (1 μg/μL) (*see* **Note 8**), 50 μL of 2.5 M $CaCl_2$, and 20 μL of 0.1 M spermidine (*see* **Note 9**) in sequence.
3. Vortex the tubes for 5 min, spin at 3000 rpm for 1 min to sediment, remove supernatant, add 140 μL of 70% ethanol, and lightly tap or pipet to dissolve the pellet.
4. Spin at 3000 rpm again, remove supernatant, add 140 μL of ice-cold 100% ethanol, and tap or pipet to dissolve.
5. Spin at 5000 rpm for 15 s, resuspend the pellet in 50 μL of 100% ethanol, and shake tubes while waiting for bombardment.
6. Replace ethanol if sitting for a long period.

3.4.3 Plasmid DNA Delivery by the Biolistic Process

1. Place sterile Whatman No. 4 filter papers on 20 mL solid MS medium in a Petri plate (100 × 15 mm size).
2. Position sterile leaves abaxial side up (flat upper surface is in contact with the medium) on the filter papers (*see* **Notes 5** and **13**).
3. They can be stored like this for up to a day.

3.4.4 Biolistic Gun Setup (see Note 14)

1. In a sterile laminar flow hood, sterilize the chamber, rupture-disc cap, launch assembly, etc., with 70% ethanol.
2. Sterilize macro-carriers, macro-carrier holders, and stopping screens with 100% ethanol, and sterilize rupture discs (1100 psi) in isopropanol.
3. Set up the helium tank to a pressure of 1300 psi, ensuring its 200–300 psi above the rupture disc rating (*see* **Note 12**).
4. Adjust the vacuum pump and gene gun settings with a vacuum rate of 7 and a venting rate of 2.
5. Coat the macro-carriers with 10 μL of DNA-coated gold, and allow them to dry for 5 min.

3.4.5 Bombardment

1. Secure the ruptured disc tightly into the retaining cap (*see* **Note 6**).
2. Set up the stopping screen and flying disc in the microcarrier launch assembly.
3. Place the assembly in a chamber below the rupture disc.
4. Put a leaf (*see* **Note 5**) on a thin MS medium plate into the chamber (*see* **Note 13**), positioning it 9 cm below the microcarrier launch assembly (*see* **Note 10**).
5. Close the chamber door, and activate the vacuum by pressing the vacuum button until it reaches 28 in Hg.
6. And then hold down the fire button until you hear the pop of the ruptured disc breaking.

3.4.6 Post-bombardment

1. Turn off equipment and release pressure.
2. Incubate bombarded leaves in clear plastic sleeves in a culture room.
3. After 2 days, cut leaves into small pieces (1 cm^2).
4. Place them on a selective MS medium with spectinomycin (500 mg/L spectinomycin).
5. Incubate at 26 °C with 16 h of light.

3.5 Transformation of Tobacco Leaf Disc (Nicotiana tabacum Var. Petit Havana) by Biolistic Method

1. Select appropriate leaf disc for transformation, and place them in a Petri dish for bombardment (*see* **Note 7**).
2. Clean gold or tungsten metal particles by washing them with isopropanol to remove impurities.
3. Wash the particles again using glycerol to ensure they are free from contaminants.
4. Mix the purified metal particles with the DNA construct (recombinant plasmid) (*see* **Note 8**). Overall workflow is enclosed in Fig. 2.
5. Use the precipitation principle where the negatively charged DNA coats the positively charged metal particles.
6. Add calcium chloride to facilitate binding between the DNA construct and the particles.
7. Use isopropyl alcohol to complete the formation of the DNA-metal complex, ensuring an even coating on the particles.
8. Transfer the DNA-coated metal particles onto a macro-carrier (plastic bullet).
9. Load the macro-carrier into the gene gun's chamber (ensuring it is positioned correctly around 7 inches from the rupture disc).

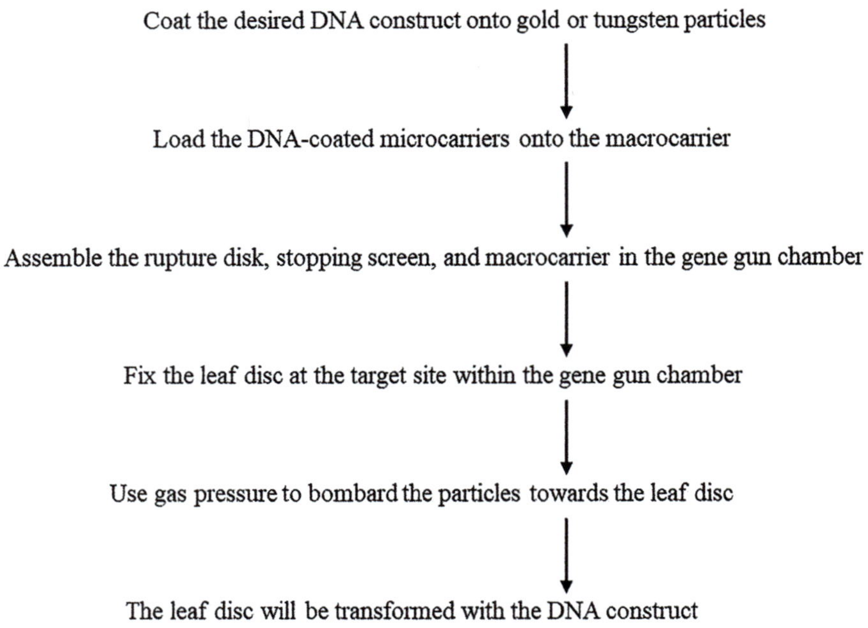

Fig. 2 Overview of the biolistic gene delivery process

10. Place the Petri dish containing the plant leaf explants or target cells at the bottom of the biolistic chamber.
11. Adjust the pressure of the chamber based on the cell type and transformation distance.
12. Ensure the optimal distance between the rupture disc and the Petri dish for precise delivery of the particles (*see* **Note 1**).
13. Activate the system, applying high pressure to the gas chamber.
14. The pressurized helium will propel the DNA-coated metal particles into the target cells at high velocity, penetrating the cell wall and delivering the DNA to the chloroplast (*see* **Note 12**).
15. After bombardment, incubate the transformed plant leaf disc on MS medium supplemented with 2 mg/L BAP and 0.5 mg/L NAA for 2 days before selection process.
16. *Selection*: Transfer the leaf discs onto a MS medium [14] supplemented with 2 mg/L BAP and 0.5 mg/L NAA + 100 mg/L streptomycin in which only plant cells transformed with *aad*A gene (*see* **Note 14**).
17. Observe for callus formation in Fig. 3, indicating that the cells are healing and integrating the introduced DNA.
18. *Subculture*: Once in 2 weeks, subculture the regenerated leaf disc to the fresh MS medium supplemented with 2 mg/L BAP and 0.5 mg/L NAA + 100 mg/L streptomycin.

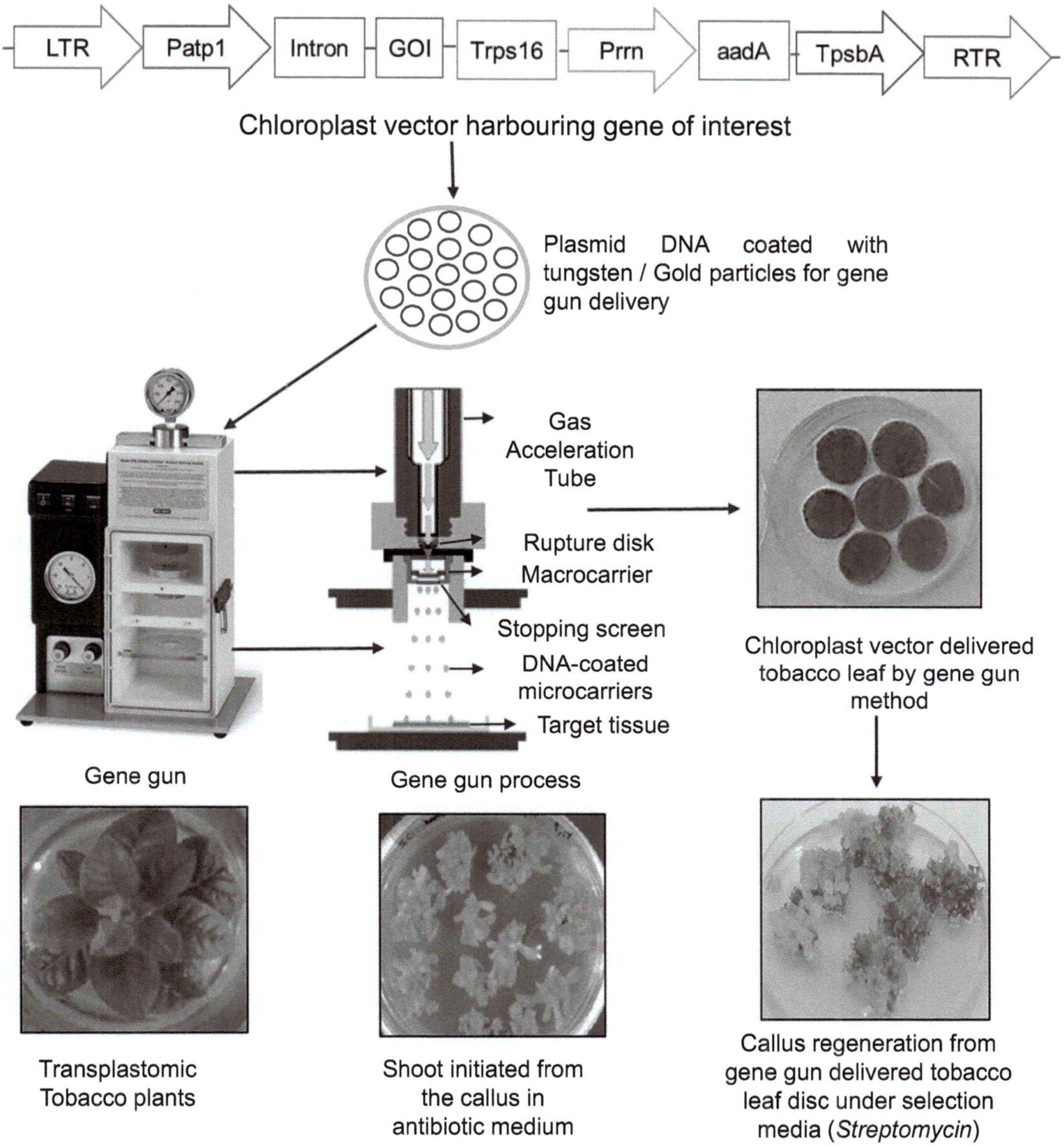

Fig. 3 Development of transplastomic tobacco plants. Schematic representation of the delivery of chloroplast vector DNA harboring the gene of insert (GOI) to the tobacco leaf disc by gene gun method and tissue culture process to develop the complete plant lets of transplastomic tobacco plants

19. Proliferation of callus induction observed in the edge of the leaf discs under selection medium.
20. Three to four weeks after infection, shoots begin to develop.
21. At this stage segments with shoots are cut out and transferred to rooting with MS medium (without phytohormones) + 100 mg/L streptomycin.
22. After roots initiate from the shoots and elongated, the plants are further subcultured into 250 mL conical flask containing

the selection media. Once the plants are matured with broadened leaves and thick shoots, they will be sub-cultured for multiple clones.

23. Finally, one batch of plants acclimatized to transgenic greenhouse to collect the seeds from remaining clones still maintained in the tissue culture condition for further various insect bioassays.
24. Later, perform the molecular analysis such as PCR, Southern, Northern, and Western blotting to confirm the presence of the transgene, integration and copy number detection, expression of the transgene, and detection of the protein, respectively.

4 Notes

1. A few chloroplast transformation vectors are available in the Addgene (https://www.addgene.org/).
2. The centrifuge tubes, $CaCl_2$ solution, etc. should be properly autoclaved and prechilled prior to use.
3. Culture aliquots should be performed in the laminar air flow chamber.
4. To check the viability of cells, transform at least 1–10 ng of the target plasmid.
5. Ensure employing healthy tissue to prevent from post-bombardment damage of tissues.
6. Employ rupture leaf discs for tobacco explants because rupture discs enhance optimal transformation in several plant species.
7. Avoid over-matured leaves for transformation. Use four to five leaf stages of young plant for good results.
8. To avoid low transformation efficiency, use purified plasmid DNA at a concentration of 1–2 μg per shot.
9. Refreezing followed by thawing of spermidine results in lack of gene integration.
10. Inadequate delivery of DNA may also be a reason for the lack of gene integration, so optimize the delivery of DNA precipitation parameters including the distance of the target and pressure in the rupture discs.
11. Use the right size of gold or tungsten particles for the experiment to address the microparticle problems, and optimize the DNA precipitation parameters, such as spermidine concentration, ethanol volume, etc., to minimize poor coating efficiency and guarantee thorough mixing of DNA and microparticles to prevent uneven coating.

12. Optimize the helium pressure; if it is too high, that may cause excessive tissue damage and reduce pressure gradually. If it is too low, DNA delivery may not be successful.
13. One of the reasons for poor regeneration is that the bombarded surface of explant has not contacted the medium.
14. Regularly clean and sterilize the gene gun components, properly check the stopping screen for potential blockages, and optimize the culture conditions after bombardment (light, humidity, and other parameters in the culture room).
15. To determine the ideal parameters for your particular tobacco variety and transformation technique, optimize your experiment by varying its parameters. If this is your first time, make sure the gene gun is working correctly by including positive control samples. You should also do numerous duplicates to guarantee accurate results.

References

1. Kato Y, Sakamoto W (2010) New insights into the types and function of proteases in plastids. Int Rev Cell Mol Biol 280:185–218
2. Inaba T, Yazu F, Ito-Inaba Y, Kakizaki T, Nakayama K (2011) Retrograde signaling pathway from plastid to nucleus. Int Rev Cell Mol Biol 290:167–204
3. Jensen PE, Leister D (2014) Chloroplast evolution, structure and functions. F1000 Prime Rep 6:40
4. Sabater B (2018) Evolution and function of the chloroplast. Current investigations and perspectives. Int J Mol Sci 19(10):3095
5. Wang HH, Yin WB, Hu ZM (2009) Advances in chloroplast engineering. J Genet Genomics 36(7):387–398
6. Kumar AU, Ling APK (2021) Gene introduction approaches in chloroplast transformation and its applications. J Genet Eng Biotechnol 19:1–10
7. Verma D, Daniell H (2007) Chloroplast vector systems for biotechnology applications. Plant Physiol 145(4):1129–1143
8. Lutz KA, Maliga P (2008) Plastid genomes in a regenerating tobacco shoot derive from a small number of copies selected through a stochastic process. Plant J 56:975–983
9. Maliga P (2012) Plastid transformation in flowering plants. In: Bock R, Knoop V (eds) Genomics of chloroplasts and mitochondria, Advances in photosynthesis and photorespiration, vol 35. Springer, Dordrecht, pp 393–414
10. Bock R (2015) Engineering plastid genomes: methods, tools, and applications in basic research and biotechnology. Annu Rev Plant Biol 66:211–241
11. Fuentes P, Armarego-Marriott T, Bock R (2018) Plastid transformation and its application in metabolic engineering. Curr Opin Biotechnol 49:10–15
12. Martin-Avila E, Lim YL, Birch R, Dirk LMA, Buck S, Rhodes T, Sharwood RE, Kapralov MV, Whitney SM (2020) Modifying plant photosynthesis and growth via simultaneous chloroplast transformation of Rubisco large and small subunits. Plant Cell 32(9):2898–2916
13. Chang AY, Chau V, Landas JA, Pang Y (2017) Preparation of calcium competent *Escherichia coli* and heat-shock transformation. JEMI Methods 1:22–25
14. Murashige T, Skoog F (1962) A revised medium for the growth and bioassay with tobacco tissue culture. Physiol Plant 15:473–497
15. Verma D, Samson NP, Koya V, Daniell H (2008) A protocol for expression of foreign genes in chloroplasts. Nat Protoc 3(4): 739–758

Chapter 12

Development of Transgenic Cotton Plants for Insect Resistance

Janani Sree Sivakumar, Dhandapani Gurusamy, Selvi Chellamuthu, Sakthi Ambothi Rathnasamy, and Vasanth Krishnan

Abstract

Genetically engineered plants are emerging as an increasingly popular tactics for combating insect pests, offering the possibility of increased crop yields with reduced reliance on toxic pesticides. A notable instance is the emergence of Bt-transgenic cotton that incorporates a gene obtained from the bacterium *Bacillus thuringiensis* which releases a toxin lethal for certain insect pests especially bollworms. This strategy has become extensively utilized because Bt-cotton has enabled for significant rises in cotton production. Concerns are being expressed regarding the possibility for insect resistance to Bt toxins along with the chance of adverse impacts to organisms that are not targets. This work seeks to provide an optimized manual approach for producing Bt-transgenic cotton cultivars that can successfully manage insect pests while reducing potential negative consequences. The main procedures for introducing genes, transforming plants, and choosing transgenic lines exhibiting desired characteristics will all be covered in the protocol. The resultant transgenic cotton will undergo extensive investigation and assessment in order to determine their overall agronomic performance, potential effects on nontarget organisms, and effectiveness against target pests. This research aims to actively contribute toward the ethical creation and implementation of insect-resistant genetically modified cotton by offering a strong and well-characterized manual process, balancing the need for improved agricultural output and preserving the environment. This manual provides a step by step protocol for successful transformation of transgenic cotton.

Key words Transgenic cotton, Insecticide, Insect resistance, *Agrobacterium*, Textile industry

1 Introduction

A vital worldwide commodity crop, cotton (*Gossypium* spp.) provides oil and fiber to the food and textile industries [1]. Because of its economic importance, it is referred as "white gold" [2]. Moreover half of the garments in the world is made primarily of cotton fiber. In addition, it is the second-most important oilseed crop in the world [3]. Lack of water limits the growth and advancement of the crop, which drastically lowers the amount of cotton produced and the profit generated from it [4]. However, cotton production

Nitish Kumar (ed.), *Transgenic Plants: Methods and Protocols*, Methods in Molecular Biology, vol. 2973, https://doi.org/10.1007/978-1-0716-4803-2_12,

faces significant challenges from insect pests, which can cause substantial yield losses and quality degradation [5]. It has been documented that 162 different insect species are pests of cotton. Fifteen species stand out among them because of their serious occurrence. The pest complex that affects cotton consists of fragile and soft stem feeders, sap eaters, and an array of lepidopteron insects whose bodies directly feed on cotton as well as spread a variety of diseases. Farmers primarily make use of pesticide sprays for managing insect pests, while frequent application of chemical sprays generates a risk to both targeted in addition to nontargeted wildlife. Although nontargeted insects, notably the predatory organisms, fall extinct as a result of the deadly effects of insecticides, target insect pests show tolerance [6]. Insect populations are becoming more resistant to traditional pest management measures, particularly chemical pesticides; the effectiveness of these approaches has been declining over time due to mounting environmental challenges. Most artificial pesticides have more detrimental effects on the natural environment [4]. Due to increasing health as well as environmental issues, traditional pest management methods particularly chemical pesticides are losing some of their effectiveness as insect populations become more resilient to them [7]. This has led to researchers examining and utilizing genetic engineering techniques in order to develop insect resistant cotton cultivars. Later twentieth century brought a new approach to creating pest-resistant crops with the advent of genetic engineering, and transgenic cotton was one of the initial examples [8]. The Bollgard cotton carrying the Bt toxin molecules provides insect resistance, a manufactured variation of one half dozen or so toxic proteins made by the bacterium *Bacillus thuringiensis* hence Bt that attack lepidopteran insects. The most successful strategy has been the deployment of gene constructs from *Bacillus thuringiensis* (*Bt*). In addition, Bt biosynthesizes insecticidal proteins which are harmful to certain groups of insects in the larval stage and have been shown not affect humans or other nontarget organisms [9]. These *Bt* genes, when integrated into the cotton genome, let the plants to manufacture these proteins; offering a very good protection against pests. The *Bt* genes, when introduced into the cotton genome, enable the plants to produce these proteins, providing an effective defense mechanism against pests. The Bt proteins function by binding to receptors in the gut cells of susceptible insects, causing cell lysis and death. This mode of action is highly specific to certain insect orders, making Bt cotton an environmentally friendly alternative to broad-spectrum chemical pesticides [10]. The introduction and commercial release of Bt cotton in the United States in 1996 heralded a new era for pest management [11]. Bt cotton development and commercialization requires identification and isolation of insecticidal genes genetically engineering the cotton genome followed by evaluation of transgenic plants for resistance to insects as well as their agronomic performance. The discovery of advanced molecular tools like *Agrobacterium*-mediated transformation and gene-editing has paved the way for developing transgenic cotton varieties, which are resistant to a wide range of insect pests [12].

The worldwide commercial release of Bt cotton provides a case in point. Bt cotton has benefited in major cultivation countries including India, China, and the United States with declined uses of insecticides which resulted to lowering cost production as well increase yieldings [13]. Bt cotton reduced pesticide applications by 30–50% with beneficial effects on the environment and economy [14]. Even with the success of Bt cotton, there are still some challenges associated with resistance management first is the risk that insect populations could evolve resistance. As with anything alive, insects can evolve resistance to control measures over time. To prevent this from happening, growers should implement an integrated pest management (IPM) plan. These include preventing planting, where non-Bt cotton is grown alongside Bt-cotton to keep a population of susceptible insects alive, and utilizing two or more genes derived from different sources into the background [15]. Additionally, ongoing research aims to develop transgenic cotton with multiple modes of action and to explore new genes and technologies for improved insect resistance. Reduced pesticide use translates to lower chemical runoff into water bodies, decreased soil contamination, and reduced harm to nontarget organisms, including beneficial insects like pollinators and natural predators of pests [16]. Economically, Bt cotton has contributed to increased farmer incomes due to higher yields and lower input costs [17]. In India, for example, the adoption of Bt cotton has been associated with a substantial increase in cotton production and a corresponding rise in farmer incomes. However, the benefits are not uniform across all regions and farming systems, highlighting the need for context-specific approaches to transgenic crop adoption [18].

Different methodologies have been employed to hinder or prevent the rise of insect populations that have developed resistance to Bt, like stacking two or more *Bt* genes. To decline the development of insects that are resistant to Bt, it is crucial to use various *Bt* genes that work in different modes. *Bt* genes that are chimeric often effective for controlling insect toxicity and even resistance. Transgenic cotton provides resistance to lepidopteron insects by using the unique *cry* gene [19]. This chapter provides a clear, step-by-step procedure to establish transgenic cotton crop in embryogenic calli by employing *Agrobacterium*-mediated transformation harboring a unique *Bt cry* gene.

2 Materials

1. *Explant:* Cotyledons and hypocotyls (*see* **Note 8**).
2. *Surface sterilization:* 70% ethanol, 0.1% mercuric chloride and sterile distilled water.
3. *MS basal medium:* [20].
4. *Preparation of stock solutions:* Prepare the following stock solutions, and store in a refrigerator for further use (media preparation) (Tables 1, 2, 3, 4, 5, and 6).

   Stock MS-I: Weigh the specified quantities of salts in Table 1, and dissolve one after the other in about 300 mL of

Table 1
Stock MS-I (*see* Note 3)

Salt	g/500 mL
NH_4NO_3	33.0 g
KNO_3	38.0 g
KH_2PO_4	3.4 g
$MgSO_4.7H_2O$	7.4 g

Table 2
Stock MS-II (*see* Note 3)

Salt	g/250 mL
$CaCl_2.2H_2O$	11.0 g

Table 3
Stock MS-III (*see* Note 3)

Salt	g/250 mL
H_3BO_3	155.00 mg
$MnSO_4.4H_2O$	422.50 mg
$ZnSO_4.7H_2O$	215.00 mg
KI	20.00 mg
$Na_2MoO_4.2H_2O$	6.25 mg
$CuSO_4.5H_2O$	0.63 mg
$CoCl_2.6H_2O$	0.63 mg

Table 4
Stock MS-IV (*see* Note 3)

Salt	g/500 mL
Na_2.EDTA	1.835 g
$FeSO_4.7H_2O$ (*see* **Note 8**)	1.390 g

Table 5
Stock MS-V (*see* Note 3)

Vitamins	gm/100 mL
Nicotinic acid	10 mg
Pyridoxine HCl	10 mg
Thiamine HCl	2.0 mg
Glycine	40 mg

Table 6
MS media preparation

Stock	Quantity/liter
MS-I	25 mL
MS-II	10 mL
MS-III	10 mL
MS-IV	10 mL
MS-V	5.0 mL
Sucrose	30 gm
Myoinositol (*see* **Note 11**)	100 mg

double-distilled water (DDW), and then make the final volume up to 500 mL. Label the reagent bottle as MS-I with date of preparation and store at 4–8 °C.

Stock MS-II: Dissolve 11 g of $CaCl_2.2H_2O$ (Table 2) in about 100 mL of DDW, and then make the final volume to 250 mL. Label the reagent bottle as MS-II with date of preparation and store at 4–8 °C.

Stock MS-III: Weigh the specified quantities of salts in Table 3, dissolve one after other in about 150 mL of distilled, and then make the final volume up to 250 mL with DDW. Label the reagent bottle as MS-III with date of preparation and store at 4–8 °C.

Stock MS-IV: Boil around 300 mL of DDW in a glass beaker wrapped with brown paper on magnetic stirrer cum heater. Add 1.835 g of Na_2.EDTA and 1.390 g of FeSO4.7H_2O (Table 4) to the boiling water with constant stirring. Make up the final volume to 500 mL with DDW, and store in amber (brown color) reagent bottle. Label the reagent bottle as MS-IV with date of preparation and store at 4–8 °C.

Stock MS-V: Weigh the specified quantities of these vitamins (Table 5), dissolve one after other in about 50 mL of distilled, and then make up the final volume to 100 mL with DDW. Label the reagent bottle as MS-V with date of preparation and store at 4–8 ° C.

5. *MS Media preparation:* Medium should be prepared in a beaker which is double the volume of required medium. For the preparation of 1000 mL medium, use 2000 mL capacity beaker. Take 500 mL of DDW in a 2 L capacity beaker, and place it on a magnetic stirrer. Put a magnetic bead inside the beaker. While stirring add stock solutions of MS-I (25 mL), MS-II (10 mL), MS-III (10 mL), MS-IV (10 mL), and MS-V (5 mL) to the beaker (Table 6). Weigh 30 g of sucrose and 100 mg of Myoinositol and add them in to the beaker. After complete dissolution of the stocks and sucrose, make up the final volume to 1000 mL with DDW. Add required quantities of hormones to the medium (*see* **Note 2**). Set the medium pH to 5.8 with 0.1 N NaOH or 0.1 N HCl. Add agar (0.8%) to the required culture vessel (test tube, conical flask, culture bottle), and pour the medium in to the culture vessel, and plug with nonabsorbent cotton. Sterilize the medium by autoclaving at 121 °C, 15 PSI for 15 min. Allow the medium to cool to RT and pour into sterile Petri plates before it solidifies.
6. *(50 mg/mL) Kanamycin stock preparation:* Dissolve 5.0 grams of kanamycin sulfate in 100 mL distilled water, and vortex to make sure completely soluble. Then, filter through a 0.2 μm filter, aliquot in the Eppendorf tubes, and store at −20 °C for up to 6 months (*see* **Notes 1**, **2**, and **3**).
7. *(10 mg/mL) Rifampicin stock preparation:* Weigh 10 mg of rifampicin and dissolve with 10 mL of 100% methanol. Store the stock at −20 °C deep freezer for 1 year. Rifampicin may precipitate in the freezer. Dissolve again before use (*see* **Notes 1**, **2**, and **3**).
8. *(100 mM) Acetosyringone stock preparation:* Weigh 0.3924 g of acetosyringone (*see* **Note 7**), and dissolve with 12 mL 95% ethanol, once completely dissolved; adjust the final volume to 20 mL with Milli-Q water to equal 20 mL. Filter-sterilize and store the stock solution at −20 °C (*see* **Notes 1**, **2**, and **3**).

9. *(100 mg/mL) Cefotaxime stock preparation:* Weigh 1.0 g of cefotaxime, dissolve with Milli-Q water, and make the final volume to 10 mL. Filter-sterilize and store the stock solution at −20 °C (*see* **Notes 1**, **2**, and **3**).
10. *Hormone preparation.*

    (1 mg/mL) GA3 stock preparation: To prepare a 1 mg/mL stock solution, 100 mg of the GA3 powder was dissolved in 2–5 mL of ethanol, once completely dissolved; adjust the final volume to 100 mL with Milli-Q water. Store the stock solution at 2–8 °C (*see* **Notes 1**, **2**, and **3**).

    (1 mg/mL) Indole-3-acetic acid stock preparation: To prepare a 1 mg/mL stock solution, dissolve 100 mg of the IAA powder in 2–5 mL of ethanol/1 N NaOH. Once completely dissolved, adjust the final volume to 100 mL with milli-Q water, and store the stock solution at 2–8 °C (*see* **Notes 1**, **2**, and **3**).

    (1 mg/mL) Kinetin stock preparation: To prepare a 1 mg/mL stock solution, 100 mg of kinetin powder was dissolved in 2–5 mL of 1 N NaOH. Once completely dissolved, adjust the final volume to 100 mL with Milli-Q water, and store the stock solution at 2–8 °C (*see* **Notes 1**, **2**, and **3**).

    (1 mg/mL) 2,4-Dichlorophenoxyacetic acid (2,4-D) stock preparation: To prepare a 1 mg/mL stock solution, dissolve 100 mg of 2,4-D in minimum volume of Milli-Q water, and once it dissolves make up the final volume to 100 mL, and store the stock solution at 2–8 °C (*see* **Notes 1**, **2**, and **3**).

3 Methods

3.1 Preculture

1. Immerse cotton seeds and wash for a minute using 70% ethanol. It is followed by sterilizing the seeds using 0.1% mercuric chloride of about 10 min. Finally wash three times in autoclaved distilled water.
2. Dry the surface-sterilized seeds on a sterile tissue paper, and plate on half-strength MS medium containing sucrose (15 g/L) along with a solidifying agent agar (8 g/L) (*see* **Note 3**).
3. Excise the cotyledons as well as hypocotyls explants from the in vitro seedlings.
4. Inoculated the explants in MS medium containing kinetin (0.5 mg/L) and 2,4-D (0.1 mg/L).
5. Utilize calli that were 6 months for preculture.

6. Transfer the callus into MS media containing potassium nitrate (1.9 g/L), maltose (30 g/L) (*see* **Notes 1** and **5**), and Phytagel (4 g/L) prior a week to co-cultivation.
7. Incubate the cultures at 26 °C with a photoperiod (16/8 h) under white fluorescent lamp, and light intensity is 100 μmol/m^2 s inside the growth chamber.

3.2 Co-cultivation (see Note 9)

1. *Agrobacterium* strains (*see* **Note 12**) such as LBA4404/EHA105 harboring the pCAMBIA1300-*cry* gene cassettes (Figs. 1 and 2) were streaked in a Petri plates containing YEP medium provided with kanamycin (100 mg/L) and rifampicin (20 mg/L) (*see* **Note 4**).
2. Incubate the cultures overnight under 28 °C.
3. Take loop full of bacteria from the streak plate, and inoculate in liquid medium containing YEP supplemented with kanamycin (100 mg/L) and rifampicin (20 mg/L) (*see* **Note 4**).
4. Then keep it overnight in a rotary shaker at 28 °C with 220 rpm.
5. Check the optical densities for the cultures at 600 nm (*see* **Notes 14** and **15**).
6. Take the cell cultures with (0.5–0.6) cell density (*see* **Notes 14** and **15**), and centrifuge for 10 min at 3220 g. Then decant the supernatant, and dissolve the pellets in AAM (media containing glutamine, aspartic acid, arginine, and glycine) glycine (mg/L), sucrose (20 g/L), 2,4-D (1 mg/L), kinetin (0.2 mg/L), and GA3 (0.1 mg/L) (*see* **Notes 2** and **6**) broth [21] (*see* **Note 3**).
7. Swirl the embryogenic calli (*see* **Notes 9** and **16**) inside a sterile glassware in *Agrobacterium* suspension of about 30 mL.
8. Later, blot-dry the callus for 20 min using sterilized filter paper.
9. Subsequent to blot dry, inoculate the callus on co-culture medium containing MS salts supplemented with KNO3 (1.9 g/L), maltose (30 g/L) (*see* **Notes 1** and **5**), and

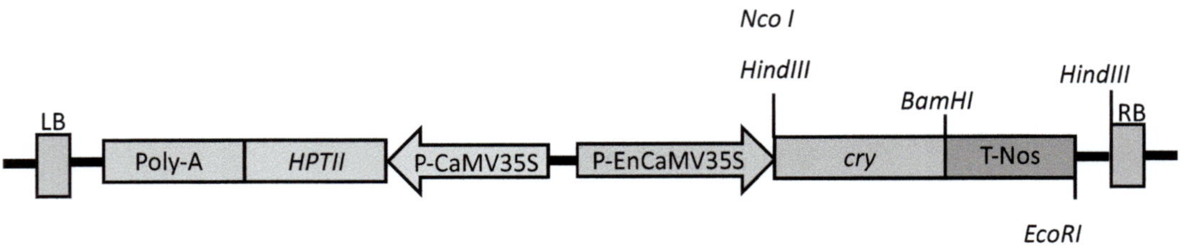

Fig. 1 Linear map of binary vector pCAMBIA1300 carrying the *cry* gene(s) driven by the enhanced CaMV35S promoter and *HPT*II gene as plant selectable marker, LB and RB: left and right borders, respectively. (Adapted from Ref. [19])

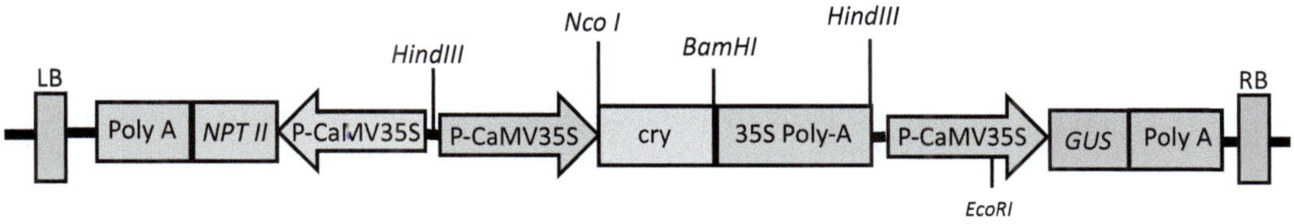

Fig. 2 Linear map of binary vector pCAMBIA2301 carrying the *cry2AX1* driven by the CaMV35S promoter, *NPT*II as a plant selectable marker gene, and *GUS* as a reporter gene. LB, left border; RB, right border; *NPT*II, neomycin phosphotransferase; GUS-β, glucuronidase. (Adapted from Ref. [19])

Fig. 3 *Agrobacterium*-mediated genetic transformation of cotton. **a** Leaf explants, **b** callus initiation from inoculated explant, **c** matured callus used for co-cultivation somatic embryo, **d** co-cultivated cotton calli on selection medium after 6 weeks showing a large number of globular embryo clusters, **e** enlarged view of one of the globular embryo clusters from **d** and **f** regeneration of plantlets, **g** rooted plantlets, **h** acclimatize the putative transgenic (kanamycin resistance) cotton plants established in the transgenic greenhouse. (Adapted from Refs. [19, 22])

acetosyringone (100 μM) (*see* **Note 7**) along with Phytagel (100 μM).

10. Maintain at 26 °C in dark for 48 h.

3.3 Selection (see Notes 10 and 13)

1. Clean the explants with sterile double-distilled water, followed by washing it using cefotaxime solution (500 mg/L).
2. Then wash it using sterilized distilled water for two times. Then blot dry the callus, and inoculate on selection medium MS salts, KNO_3 (1.9 g/L), maltose (30 g/L) (*see* **Note 5**), kanamycin (25 mg/L) (*see* **Note 13**), and Phytagel (4 g/L) (*see* **Note 3**).
3. Transfer the matured somatic embryos (*see* **Note 16**) to medium provided with MS salts, GA3 (0.1 mg/L) (*see* **Note**

6), IAA (1.0 mg/L), sucrose (30 g/L), and Phytagel (4 g/L) enclosed in Fig. 3.

4. Keep the transgenic plantlets containing four to six leaves along with adequate root into pots with sterilized soil. Then cover the pots by polythene bags.
5. Keep it for 15 days inside the culture room. Then well-grown plantlets must be transferred in the greenhouse.
6. Analyze the putative transgenic line by PCR, Southern, Northern, and Western blots, followed by insect bioassay.

4 Notes

1. All hormone and antibiotic stocks should be prepared in a laminar air flow chamber using newly autoclaved, sterile DDH_20, or Milli-Q H_20. The stocks should only be opened and used in this chamber.
2. Prepare the bare minimum of hormones and antibiotics; do not prepare more, and keep them in the refrigerator for an extended period of time.
3. Use newly purchased culture media, hormones, and antibiotics; do not use ones that have expired.
4. The *Agrobacterium* strain harboring the binary vector using for the genetic transformation should be freshly streaked/ cultured, and don't use one that has been stored in the refrigerator for too long.
5. Maltose enables to avoid browning the cells. Moreover, it leads to during the development stage that enhances the formation of embryonic calluses and the germination of somatic embryos [23].
6. GA3 triggers the development of somatic embryos [24].
7. If the exhibiting wound responsible molecules is merger quantity for activating virulence genes, pre-treatment of *Agrobacterium* using synthetic Acetosyringone results in the increasing transformation rate [25].
8. Iron in addition to copper chelators inhibits specific enzymes, which leads to increased T-DNA transformation. The rise in transformation emerges through thiol-inhibiting peroxidase together with polyphenol oxidase within the explant.
9. Avoid excessive wounding during co-cultivation because it may cause mechanical injury to the explant and results in declined vigorous GUS stains.
10. Prolonged existence of *Agrobacterium* within the host cells results in mortality of the transformed crop.

11. Myoinositol, an antioxidant supressess the oxidative burst in defense mechanism of the crop against *Agrobacterium*-mediated genetic transformation [26].
12. An auxotrophic Agrobacterial strains enhance the efficiency of plant transformation [27].
13. Adjust the concentration of selection antibiotics to balance between eliminating non-transformed cells and preventing toxicity to transformed cells.
14. Generally, longer duration of infection and prolonged exposure to elevated cell density leads to explant necrosis.
15. Eliminating the *Agrobacterium* from the explant becomes difficult when exposed with excessive cell density [28].
16. Optimize the process of somatic embryogenesis to increase the conversion of transformed calli into plantlets.

References

1. Wendel J, Cronn R (2003) Polyploidy and the evolutionary history of cotton
2. Din SU, Azam S, Rao AQ, Shad M, Ahmed M, Gul A, Shahid AA (2021) Development of broad-spectrum and sustainable resistance in cotton against major insects through the combination of *Bt* and plant lectin genes. Plant Cell Rep 40:707–721
3. Farooq J, Farooq A, Riaz M, Shahid MR, Saeed F, Iqbal MS, Mahmood A (2014) Cotton leaf curl virus disease a principle cause of decline in cotton productivity in Pakistan (a mini review). Can J Plant Prot 2:9–16
4. Carriere Y, Dennehy TJ, Pedersen B, Haller S, Ellers-Kirk C, Antilla L, Tabashnik BE (2001) Large-scale management of insect resistance to transgenic cotton in Arizona-can transgenic insecticidal crops be sustained. J Econ Entomol 94(2):315–325
5. Dhillon MK, Sharma HC (2013) Comparative studies on the effects of Bt-transgenic and non-transgenic cotton on arthropod diversity, seed cotton yield and bollworms control. J Environ Biol 34:67–73
6. Sarwar M, Sattar M (2016) An analysis of comparative efficacies of various insecticides on the densities of important insect pests and the natural enemies of cotton, *Gossypium hirsutum* L. Pakistan J Zool 48(1)
7. Ahmad S, Raza I, Ali H, Shahzad AN, Atiq-ur-Rehman SN (2014) Response of cotton crop to exogenous application of glycinebetaine under sufficient and scarce water conditions. Braz J Bot 37(4):407–415
8. Perlak FJ, Deaton RW, Armstrong TA, Fuchs RL, Sims SR, Greenplate JT, Fischhoff DA (1990) Insect resistant cotton plants. Biotechnology 8(10):939–943
9. Aronson AI, Beckman W, Dunn P (1986) *Bacillus thuringiensis* and related insect pathogens. Microbiol Rev 50(1):1–24
10. Barton KA, Whiteley HR, Yang NS (1987) *Bacillus thuringiensis* § -endotoxin expressed in transgenic *Nicotiana tabacum* provides resistance to lepidopteran insects. Plant Phys 85(4):1103–1109
11. Galanopoulou-Sendouca S (1998) Towards a more competitive and sustainable cotton production. In: Freer trade, sustainability, and the primary production sector in the Southern EU: unraveling the evidence from Greece. Springer, Dordrecht, pp 112–127
12. Ahmad S, Cheema HMN, Khan AA, Khan RSA, Ahmad JN (2019) Resistance status of *Helicoverpa armigera* against Bt cotton in Pakistan. Trans Res 28:199–212
13. Ahmad S, Abbas Q, Abbas G, Fatima Z, Atique-ur-Rehman NS, Younis H, Khan RJ, Nasim W, Habib ur Rehman M, Ahmad A, Rasul G, Khan MA, Hasanuzzaman M (2017) Quantification of climate warming and crop management impacts on cotton phenology. Plants 6(7):1–16
14. Ali H, Afzal MN, Ahmad F, Ahmad S, Akhtar M, Atif R (2011) Effect of sowing dates, plant spacing and nitrogen application on growth and productivity on cotton crop. Int J Sci Eng Res 2(9):1–6

15. Bailey JC, Hanny BW, Meredith WR (1980) Combinations of resistant traits and insecticides: effect on cotton yield and insect populations. J Econ Entom 73(1):58–60
16. Ali AA, Watson TF (1982) Effects of *Bacillus thuringiensis* var. kurstaki on tobacco budworm (Lepidoptera: Noctuidae) adult and egg stages. J Econ Entomol 75:596–598
17. Ahmad A, Ramani R, Sharma KK, Vidyarthi AS, Ramamurthy VV (2014) Distinction of Indian commercial lac insect lines of *Kerria spp* (Homoptera: Coccoidea) based on their morphometrics. J Insect Sci 14(1):263
18. Ahmad S, Iqbal M, Muhammad T, Mehmood A, Ahmad S, Hasanuzzaman M (2018) Cotton productivity enhanced through transplanting and early sowing. Acta Sci Biol Sci 40:1–7
19. Sakthi AR, Naveenkumar A, Deepikha PS, Balakrishnan N, Kumar KK, Devi EK, Balasubramanian P (2015) Expression and inheritance of chimeric cry2AX1 gene in transgenic cotton plants generated through somatic embryogenesis. In Vitro Cell Dev Biol-Plant 51:379–389
20. Murashige T, Skoog F (1962) A revised medium for rapid growth and bio assays with tobacco tissue cultures. Physiol Plant 15(3)
21. Toriyama K, Hinata K (1985) Cell suspension and protoplast culture in rice. Plant Sci 41: 179–183
22. Leelavathi S, Sunnichan VG, Kumria R, Vijaykanth GP, Bhatnagar RK, Reddy VS (2004) A simple and rapid *Agrobacterium*-mediated transformation protocol for cotton (*Gossypium hirsutum* L.): embryogenic calli as a source to generate large numbers of transgenic plants. Plant Cel Rep 22:465–470
23. Pereira AVC, Takamori LM, Yaguinuma DH, de Oliveira AM, Ribas AF (2019) Maltose in culture media improves the *in vitro* regeneration of *Urochloa brizantha* cv. 'Marandu'plants. Biotecnol Veg 19:205–213
24. Montero-Cortes M, Saenz L, Cordova I, Quiroz A, Verdeil JL, Oropeza C (2010) GA 3 stimulates the formation and germination of somatic embryos and the expression of a knotted-like home box gene of *Cocos nucifera* (L.). Plant Cel Rep 29:1049–1059
25. Sheikholeslam SN, Week DP (1987) Acetosyringone promotes high efficiency transformation of *Arabidopsis thaliana* explants by *Agrobacterium tumefaciens* plant. Mol Biol 8: 291–298
26. Sivakumar JS, Gurusamy D, Subramaniam S, Krishnan V (2024) Enlightening the biotechnological approaches of legumes: present and past scenario to mitigate the biotic and abiotic stress. Plant Biotech Rep 18(5):605–623
27. Aliu E, Ji Q, Wlazlo A, Grosic S, Azanu MK, Wang K, Lee K (2024) Enhancing *Agrobacterium*-mediated plant transformation efficiency through improved ternary vector systems and auxotrophic strains. Front Plant Sci 15: 1429353
28. Prakash DP, Ramachandra YL, Hanur VS (2016) Effect of *Agrobacterium* infection time, co-cultivation and cell density on *in vitro* response in hypocotyl of eggplant (*Solanum melongena* L.). J Hort Sci 11(1): 21–26

Chapter 13

Characterization of Transgenic Cotton Plants by GUS Histochemical Staining and GUS Fluorometric Assay

Janani Sree Sivakumar, Dhandapani Gurusamy, Selvi Chellamuthu, Sakthi Ambothi Rathnasamy, and Vasanth Krishnan

Abstract

Transgenic cotton plants are responsible of a greater breakthrough in the field of agricultural biotechnology which focuses on improving plant characteristics such as pest resistance, drought tolerance, and fiber quality. The precise molecular characterization of these genetically transformed plants is fundamental in assessing the insertion, inheritance, and expression of transgenes. The β-glucuronidase (GUS) assay is among the conventional tools to test transgenic cotton that offer a consistent and productive process for determining gene expression. This chapter offers the exploration of the GUS assay using transgenic cotton plants, which may be useful for gene development. The enzyme that gives the blue product is β-glucuronidase, which in turn cleaves its substrate (5-bromo-4-chloro-3-indolyl β-D-glucuronide [X-Gluc]). It signals a change in color which can be seen directly as an indicator for the transgene expression. The histochemical assay significance is based on tissue fixation time, incubation periods, staining, and decolorizing of chlorophyll procedures for reproducible results of putative transgenic lines. It also covers approaches for optimizing assay conditions, such as changing the levels of substrates and incubation times to improve staining clarity and intensity. This chapter provides the step-by-step protocol for initial analysis of transgenic cotton by reporter assay by GUS histochemical staining.

Key words Transgenic cotton, *β-glucuronidase* (*GUS*) gene, GUS histochemical assay, X- Gluc, Blue color, Methanol

1 Introduction

Genetic engineering has created revolution agricultural biotechnology by the production of transgenic plants with superior characteristics [1]. The genetic manipulations used to improve stress tolerance; insect resistance and fiber quality in the economically important crop species such as cotton are common examples of these techniques [2]. Properly characterizing such transgenic plants is essential for the efficiency of genetic modifications and so that science can advance further in our understanding of plant biology. The β-glucuronidase (GUS) assay is commonly used and efficient

Nitish Kumar (ed.), *Transgenic Plants: Methods and Protocols*, Methods in Molecular Biology, vol. 2973,
https://doi.org/10.1007/978-1-0716-4803-2_13,

tool for the qualitative as well as quantitative measurement of gene expression. Modern agriculture is predicated on the use of these transgenic plants that include one or a small number of new genes which were recently introduced by genetic engineering techniques. These plants are genetically engineered to have traits that do not exist in their wild counterparts, including resistance to insects or herbicides, improved nutritional profiles, and increased tolerance of environmental stresses. In the case of transgenic cotton varieties that could provide a key to enhance yield and quality, transgenic cottons that express resistance genes to pests such as the cotton bollworm or tolerance of herbicide, namely, glyphosate, have already been clearly successful [3, 4]. Nevertheless, thorough characterization is needed for successful genetic manipulations to ensure that the genes added are expressed in a proper manner and confer the desired traits [5]. For this reason the GUS assay is very useful in this system because it provides an indication of transgene functional activity as well as presence.

The GUS (β- glucuronidase) assay detects the conversion of a synthetic substrate 5-bromo-4-chloro-3-indoly-glucuronic acid into X-Gluc by an enzyme reporter. X-Gluc is a substrate for GUS, and when acted on by GUS, it produces an insoluble, blue-colored precipitate that can be seen either directly or under the microscope [6]. This colorimetric reaction allows researchers to visually track the expression of the GUS gene, which is often used as a reporter in transgenic constructs [7].

The GUS assay also numerous characteristics that it a desirable technique like its high sensitivity, ease, and adaptability. It can be used for gene expression profiling in different plant tissues, at various developmental stages and under multiple environmental conditions [3]. Moreover, the assay is reproducible, economical, and straight-forward that allows an accessible high-throughput screening for routine usage in plant genetic researches and breeding programs. In addition, the assay is useful for detecting possible problems with transgene expression (e.g., silencing) which may be crucial in pursuing an optimal line of a transgenic event existing at gene level traits. It is currently applied not only to basic research but also to breeding and biotechnology industries as an indispensable key technology for development of genetically modified crops. However, over the past decade or so, both sensitivity and utility have been enhanced by improvements in GUS assay methodology resulting from improved substrate formulations [8, 9] as well as more accurate detection methods [10].

New advances in transgenic technology and cotton biotechnology document the ongoing importance of the GUS assay. Researchers, for instance, have used GUS assay to toward monitoring abiotic stress in cotton resistance genes leading the development of tolerant sweet varieties [11]. Furthermore, GUS has been important when combined with other molecular techniques such as quantitative PCR and high-throughput sequencing to have a broader assessment of gene expression and function in transgenic cotton.

2 Materials

1. *Plant materials:* Leaf, stem, bud, boll, and fibers of transgenic cotton.
2. *Requirements:* Mortar and pestle, centrifuge, centrifuge tubes, (substrate) 1 mM 4-methyl umbelliferyl glucuronide (MUG) (*see* **Notes 8**, **10**, and **14**).
3. *(50 mM) Sodium phosphate buffer pH 7.0:*
   A. *0.2 M stock monobasic sodium phosphate:* (Volume: 200 mL). Dissolve 6.24 g $NaH_2PO_4.2H_2O$ in 150 mL Milli-Q water, and then make up the volume to 200 mL. Autoclave the reagent autoclave and store at room temperature.
   B. *0.2 M stock dibasic sodium phosphate:* (Volume: 200 mL): Dissolve 5.678 g Na_2HPO_4 in 150 mL of Milli-Q water, and then make up the volume to 200 mL. Autoclave the reagent and store at room temperature. Take 39 mL from *stock A*, mix with 61 mL from *stock B*, and make up the volume with autoclaved Milli-Q water to the final volume of 400 mL (the pH should read 7.0).
4. *GUS histochemical staining reagents:* Sodium phosphate buffer, 5.0 M EDTA, 100 mM potassium ferricyanide, 100 mM potassium ferrocyanide (Tables 1 and 2).
5. *5.0 M EDTA: 100 mL (pH 7.0):* Dissolve 18.6 g of EDTA (M. Wt: 372.0) in 60 mL of Milli-Q water. Adjust the pH to 7.0 using 10 N NaOH. Make up the volume to 100 mL.
6. *100 mM potassium ferricyanide 100 mL:* Dissolve 3.3 g of potassium ferricyanide in 100 mL of 50 mm phosphate buffer (pH 7.0).

Table 1
GUS histochemical reagent: 50 mL. Make up the volume to 50 mL using 50 mM phosphate buffer (pH 7.0)

Name of the chemicals	Final volume	Final concentration
50 mM sodium phosphate buffer (pH 7.0)	50 mL	50 mM
0.5 M EDTA (pH 7.0)	100 μL	1 mM
100 mM potassium ferricyanide (*see* **Note 7**)	50 μL	1 mM
100 mM potassium ferrocyanide (*see* **Note 7**)	50 μL	1 mM

Table 2
X- Gluc stock (2 mM): 5 mL. Dissolve 5.2 mg X-gluc in 50 μL of DMSO and add 4.95 mL of histochemical reagent

Name of the chemicals	Final volume	Final concentration
Methanol	2.0 mL	20%
X- Gluc (2 mM stock)	5.0 mL	1 mM
Phosphate buffer (100 mM)	2.5 mL	50 mM
Milli-Q water	0.5 m	–

7. *100 mM potassium ferrocyanide 100 mL:* Dissolve 4.22 g of potassium ferrocyanide in 100 mL of 50 mM phosphate buffer (pH 7.0).
8. *50 mM sodium phosphate (pH 7.0):* Provide a stable pH environment for enzyme activity.
9. *0.1% Triton X-100:* A detergent that helps to lyse cell membranes and release the GUS enzyme.
10. *10 mM 2-mercaptoethanol:* A reducing agent that helps to protect the enzyme from oxidative damage.
11. *10 mM 1,2-diaminocyclohexane-N,N,N,N-tetra-acetic acid (DCTA):* A chelating agent that binds divalent metal ions which could otherwise interfere with enzyme activity.
12. *0.1% sodium lauryl sarcosine:* Another detergent that assists in cell lysis.
13. *GUS extraction buffer:* 50 mM phosphate buffer, pH 7.4;10 mM β mercaptoethanol, 10 mM EDTA, 0.1% sodium lauryl sarcosine, and 0.1% Triton X-100.

3 Methods (*see* Note 10)

3.1 GUS Histochemical Staining

1. Wash the explants (*see* **Note 11**) in 50 mM phosphate buffer (pH 7.0).
2. Place the explants in phosphate buffer containing 1% Triton X-100, and incubate for 1 h in a 37 °C incubator.
3. Transfer the explants to X-Gluc staining solution (1 mM), and place in a desiccator. Subject to vacuum infiltration for about 5 min (*see* **Notes 6**, **9**, and **12**).
4. Incubate the explants in a 37 °C incubator for 16–24 h (*see* **Note 12**).
5. Observe the blue spots/staining in the explants enclosed in Fig. 1, and store the samples with 50% glycerol (*see* **Notes 1**, **2**, **3**, **4**, and **5**).

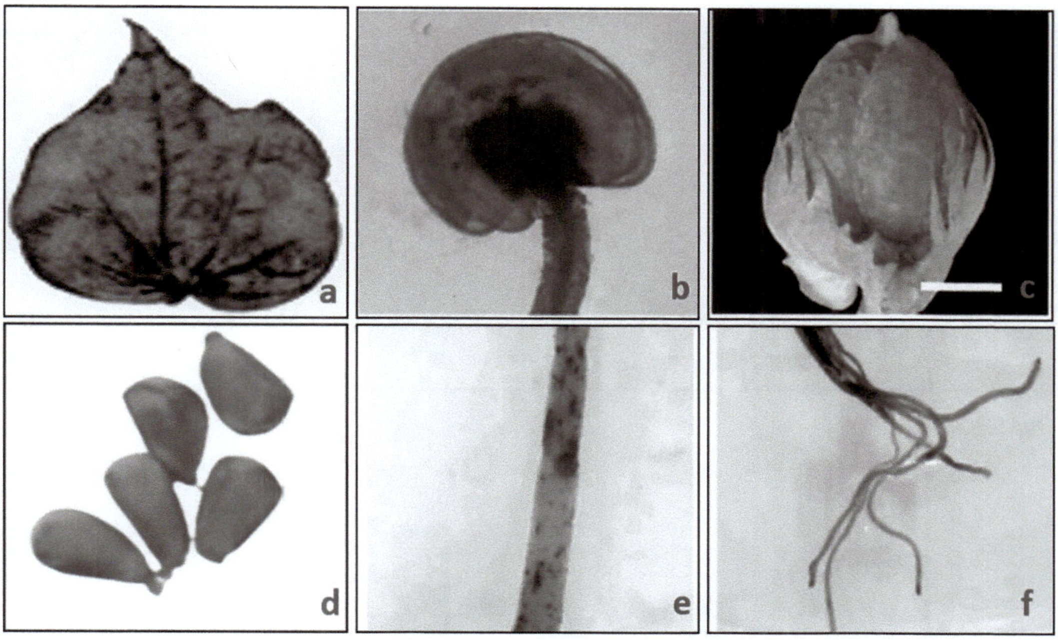

Fig. 1 Histochemical staining of GUS expression in the transgenic cotton plant. **a** Leaf samples, **b** anther, **c** matured boll, **d** ovule, **e** stem and **f** root [2, 12, 13, 14]

3.2 GUS Fluorometric Assay [15]

3.2.1 Lysis Conditions

1. 200 mg of fresh tissues (the same which were used for the histochemical staining) were ground with 500 μL of GUS extraction buffer using mortar and pestle.
2. Centrifuge the samples at 12000 × g for 15 min at 4 °C.
3. Transfer10 μL of supernatant to 500 μL of substrate (MUG) for GUS activity assay (one more 10 μL of supernatant were used to determine the protein concentration using Bradford's assay) (*see* **Note 13**).
4. After incubation at 37 °C for 2 h, the reactions were stopped by the addition of 2 mL of 0.2 M Na_2CO_3, and the fluorescence was measured with a fluorometer at 360 nm excitation and 455 nm emissions (*see* **Notes 13**, **14**, and **15**).
5. Estimate the total proteins using Bradford method, and calculate the GUS protein presence in the total protein (*see* **Note 13**).
6. The GUS activity is expressed as "pmol of 4-MU/min/μg of total protein" shown in Fig. 2 (*see* **Note 16**).

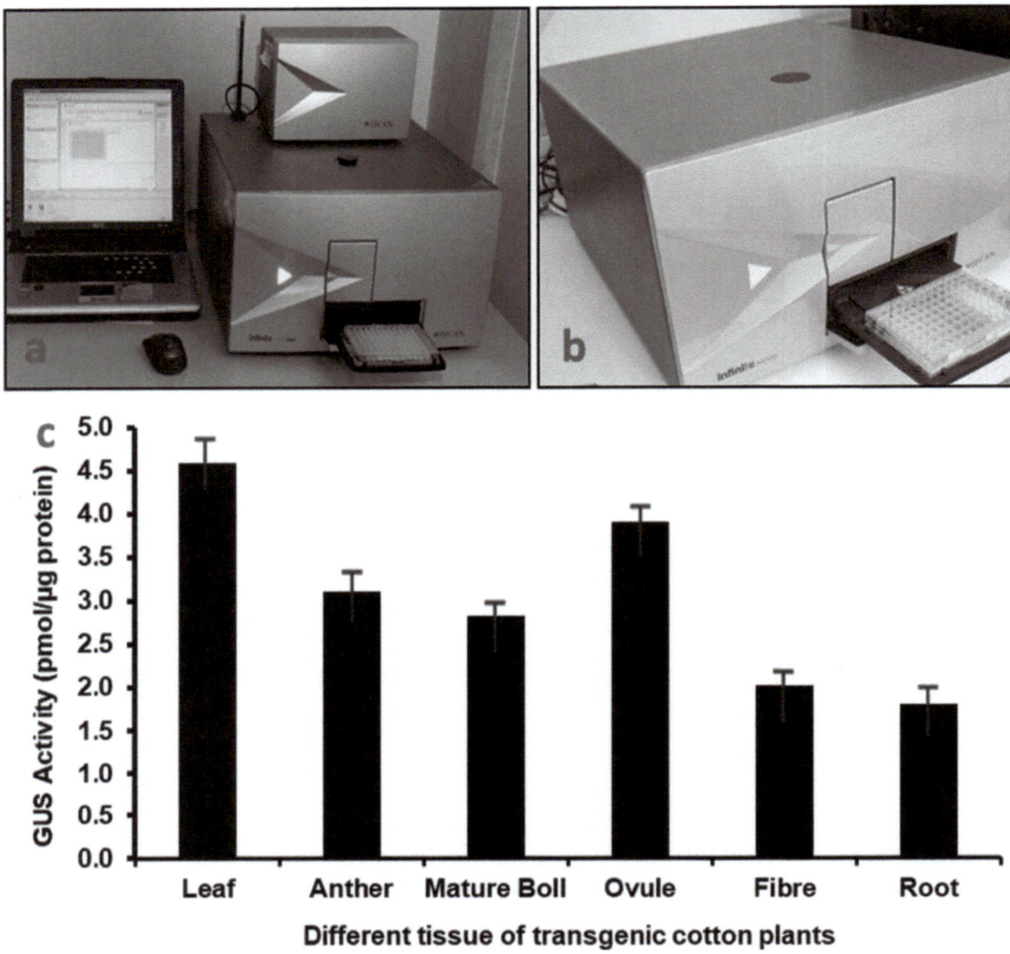

Fig. 2 Fluorometric quantification of GUS activity measured in different tissues of transgenic cotton. **a** and **b** Fluorometer instrument; **c** graphs show GUS activity of different tissues of transgenic cotton and GUS expression represents in pmol/μg protein

4 Notes

1. Chlorophyll clearing should be done if explants with chlorophyll are used for GUS staining.
2. Chlorophyll clearing of GUS-stained leaf discs: Remove the explants from X-Gluc solution and wash twice with sterile Milli-Q water (100 mL conical flask).
3. Wash once with 70% ethanol, and then transfer the explants to acetone/methanol mixture (1:3 volume) to remove chlorophyll and keep the vials at 4 °C for 1 h.
4. After that, transfer the samples in the fresh solution of acetone/methanol mixture (1:3 volume), and repeat till the chlorophyll is completely removed.

5. After the tissues become clear, wash with sterile Milli-Q water twice, and store in 50% glycerol and keep it at 4 °C.
6. It is advised to utilize incubation media along with an increased concentration of methanol (20% v/v), to raise the pH to 7.5, or to optimize the temperature to 60 °C if difficulties with intrinsic GUS expression are found.
7. It is recommended to prepare the potassium ferricyanide and ferrocyanide with an optimal concentration of 1 mM [16].
8. Employing the substrate (MUG) enables for the quantifying the activity of GUS exactly even in the intact seedlings [17].
9. Try different fixation methods such as vacuum infiltration or acetone fixation to improve penetration of the staining solution, especially for thick tissues to avoid poor tissue penetration.
10. If there is no staining at all or very little staining in the tissues, make sure your construct is expressing the GUS gene at a level that can be detected. Make sure the tissue is properly sliced and fixed, check the pH of your staining buffer, adjust the amount of X-Gluc in the staining solution, and, if your plant species has natural GUS activity, think about applying a blocking agent to prevent it.
11. Try using different fixation techniques or tissue sectioning techniques if the tissues exhibit patchy or uneven staining. If the tissues have cell wall interference, use enzymes to partially digest the cell walls to improve penetration. After staining, make sure to follow thorough washing procedures to remove excess substrate.
12. To lessen background staining, lower the amount of X-Gluc in the staining solution, and optimize the incubation period.
13. In GUS fluorometric assay: If the samples exhibit a low fluorescence signal, make sure the fluorometer's settings (such as excitation and emission wavelengths) are optimized for the selected substrate, check the purity and expiration date of your fluorogenic substrate, and optimize the protein extraction procedure to guarantee effective protein solubilization.
14. To reduce high background fluorescence, ensure proper cleaning of pipettes and tubes to avoid contamination, and test for autofluorescence from your plant tissue extract.
15. Maintain a constant temperature during the assay, and calibrate the fluorometer frequently using a standard curve to prevent inconsistent readings.
16. Perform the experiment with proper controls such as both negative controls (untransformed tissue) and positive controls (known GUS expressing tissue) to validate the results.

References

1. Babinard J (2001) A short history of agricultural biotechnology. In: Genetically modified organisms in agriculture. Aca Press, pp 271–274
2. Amarasinghe BHRR, Faivre-Nitschke E, Wu Y, Udall JA, Dennis ES, Constable G, Llewellyn DJ (2006) Genomic approaches to the discovery of promoters for sustained expression in cotton (*Gossypium hirsutum* L.) under field conditions: expression analysis in transgenic cotton and Arabidopsis of a Rubisco small subunit promoter identified using EST sequence analysis and cDNA microarrays. Plant Biotech 23(5):437–450
3. Gong H, Zhang H, Chen W (2013) Transgenic cotton: current status and future prospects. Plant Biotech J 11(6):715–732
4. Zhang Z, Li M, Liu B (2019) Biotechnology in cotton: a review. J Biotechnol 299:58–66
5. Ovchinnikov DA (2020) Genetically-modified cell lines: categorisation and considerations for characterisation. Stem Cell Res 49:102103
6. Jefferson RA, Burgess SM, Hirsh D (1987) β-glucuronidase from *Escherichia coli* as a gene-fusion marker. Proc Natl Acad Sci 84(22):7413–7417
7. De Ruijter NCA, Verhees J, Van Leeuwen W, Van Der Krol AR (2003) Evaluation and comparison of the GUS, LUC and GFP reporter system for gene expression studies in plants. Plant Bio 5(02):103–115
8. Hegde SS, Belyavskaya NA, Muthukrishnan S (2000) Improved β-glucuronidase assays for plant genetic transformation. Plant Mol Bio Rep 18:379–387
9. Wangou S, Sticklen MB (1993) Enhanced β-glucuronidase assay for plant research. Plant Cell Rep 12(1):26–29
10. Meyer P, Gatz C, Miesfeld R (2018) Advanced β-glucuronidase assays for plant research. Methods Mol Bio 1706:41–50
11. Liu X, Zhang W, Wang L (2020) Monitoring abiotic stress response in cotton using β-glucuronidase (GUS) assay. Front Plant Sci 11:1100
12. Yadav VK, Yadav VK, Pant P, Singh SP, Maurya R, Sable A, Sawant SV (2017) GhMYB1 regulates SCW stage-specific expression of the GhGDSL promoter in the fibres of *Gossypium hirsutum* L. J Plant Biotech 15(9): 1163–1174
13. Shangguan XX, Xu B, Yu ZX, Wang LJ, Chen XY (2008) Promoter of a cotton fibre MYB gene functional in trichomes of *Arabidopsis a*nd glandular trichomes of tobacco. J Exp Bot 59(13):3533–3542
14. Li H, Li K, Guo Y, Guo J, Miao K, Botella JR, Miao Y (2018) A transient transformation system for gene characterization in upland cotton *(Gossypium hirsutum)*. Plant Met 14(1):50
15. Song P, Heinen JL, BurnsT H, Allen RD (2000) Expression of two tissue-specific promoters in transgenic cotton plants. J Cot Sci 4(4):217–223
16. Vitha S (2007) Histochemical localization of β-glucuronidase (GUS) reporter activity in plant tissues. Microscopy and Imaging Center
17. Dedow LK, Oren E, Braybrook SA (2022) Fake news blues: a GUS staining protocol to reduce false-negative data. Plant Dir 6(2):367

Chapter 14

Characterization of Transgenic Cotton Plants by PCR

Janani Sree Sivakumar, Dhandapani Gurusamy, Selvi Chellamuthu, Sakthi Ambothi Rathnasamy, and Vasanth Krishnan

Abstract

The most significant traits in agricultural biotechnology are those resulting from expression of transgenes, introduced to plant species such as cotton which has incorporated genes encoding resistance against biotic and abiotic stress. This chapter emphasizes on characterization of transgenic cotton plants in terms of polymerase chain reaction (PCR) techniques. This chapter describes in detail the protocol for primer design, genomic DNA and total RNA extraction, cDNA synthesis from RNA, master mix preparation, thermo cycler procedure, and finally analysis of PCR amplified product by agarose gel electrophoresis. This chapter describes not only the basic foundations of PCR, for instance, primer design, thermal cycling, and DNA amplification. It also covers the aspect of using reverse transcription PCR (RT-PCR) which is a type to measure gene expression amounts by transcribing RNA as complementary DNA (cDNA) before amplification. One needs to use RT-PCR for a more functional readout, especially if trying to measure the change caused by the transgene insertions.

Key words Putative transgenic cotton, *NPT*II, Primers, Gene expression

1 Introduction

Transgenic crops undoubtedly form the basis of modern agriculture, affording traits like pest resistance, drought tolerance, and fiber quality improvement. The cotton is one of the major crops across the globe and has made remarkable achievements with genetic engineering. The cotton bollworm resistance and herbicide tolerance genes in the list increased yields while reducing chemical use [1]. This application of genetic modifications tackles economic as well as environmental problems associated with cotton [2]. Genetic analysis of transgenic plants is important to secure and phenotypically characterize the presence of appropriate modifications. This approach comprises confirming the existence, integration, and expression of introduced genes. Polymerase chain reaction (PCR) is a widely used molecular tool in this respect to its high sensitivity and specificity [3]. PCR also enables researchers

Nitish Kumar (ed.), *Transgenic Plants: Methods and Protocols*, Methods in Molecular Biology, vol. 2973,
https://doi.org/10.1007/978-1-0716-4803-2_14,

to amplify and detect the presence of a specific transgene-based DNA construct in the transgenic cotton [4]. There are several advantages in the PCR-based characterization for transgenic cotton plants. The plant's genome has been correctly inserted, so it does not disturb any essential genes [5]. Additionally, these PCR-based methods can be used to directly measure the expression level of transgenes a critical element for evaluating whether genetic modifications are functioning as expected [6]. This thorough characterization is also essential for regulatory compliance as a prerequisite evidence to claims of genetic stability and safety in the context of an environmental release or commercialization.

PCR tools provide different techniques to serve these needs, such as qualitative and quantitative methods. While genomic PCR can detect the presence, transgene and qPCR will allow the quantification of gene expression providing a more holistic view of their implications [7, 8]. The methods are relevant in both research and practical contexts, including breeding programs and regulatory assessments. They are important tools for both research and practical purposes such as breeding programs, regulatory assessments, etc. PCR techniques help for initial characterization of putative transgenic cotton plants as regards whether transgenes are inserted or not. This chapter will describe the methodologies and protocols used to define via PCR-based technology and guidelines for researchers and practitioners in this field.

2 Materials

1. *Requirements:* Nucleotides sequence, software, Diethyl Pyrocarbonate (DEPC) water (*see* **Note 5**), 70% ethanol, gloves, mask, mortar and pestle, liquid nitrogen, spatula, microfuge tubes and micropipettes, microtips, water bath, ice packet with ice, micro spin, thermo cycler, genomic DNA, TAE buffer (Table 1), total RNA, 70% ethanol, gloves, centrifuge, PCR buffer, dNTPs, Taq DNA polymerase, primers, template DNA (*see* **Note 8**), nuclease-free water, PCR reaction mix (Table 3), 0.2 mL PCR tubes, lysis buffer, (2- mercaptoethanol), binding solution, wash solutions I and II, elution buffer, MOPS, DEPC

Table 1
TAE buffer preparation for 1000 ml (pH 7.6 to 7.8)

Chemicals required	Volume
Tris Base (2 M)	242.2 g
Glacial acetic acid (1 M)	57.1 mL
EDTA (0.5 M)	100 mL

water/nuclease-free water, sodium acetate, disodium EDTA, dNTPs, random primers, oligo primers, dNTP, reaction mix, DTT, M-MLV RT enzyme. For 50X TAE preparation: Weigh 242.2 g of Tris-base and dissolve it in 700 mL of Milli-Q water. Once completely dissolved, add 57.1 mL of 100% glacial acetic acid, followed by 100 mL of 0.5 M EDTA should be added. Make up the solution to a final volume of 1000 mL with Milli-Q water and ensure the pH is between 7.6 to 7.8. Autoclave the solution and store the stock at room temperature.

2. *(10 mg/mL) Ethidium bromide preparation:* Weigh 10 mg of ethidium bromide (Etbr) and dissolve it in 1.0 mL of DDW/MQW.
3. *10× MOPS Buffer (1 L):* Weigh 41.86 g of 3-(N-morpholino) propanesulfonic acid (MOPS) free acid, then 4.1 g of sodium acetate, and 3.72 g of disodium EDTA; dissolve it in 800 mL of DEPC water. Make up the final volume to 1000 mL after the chemicals have been completely dissolved, and autoclave. The bottles containing MOPS buffer should be covered properly with aluminum foil to avoid exposure to light.

3 Methods

3.1 Primer Designing

3.1.1 Sequence

Obtain desired sequence for gene of interest from the National Center for Biotechnology Information (NCBI) for designing the primers.

3.1.2 Designing Primer (see Notes 1, 2, 3, 4, 5, 6, and 7)

1. A wide range of tools along with software are accessible to design PCR primers.
2. The majority of them are Internet-based freeware programs.
3. Some of the software is as follows (Table 2).
4. Download the software, register with email ID, and perform the program as per the guidelines given for particular software.

Table 2
List of software's used for primer designing [9] (*see* Notes 1, 2, 3, 4, 5, 6, and 7)

S. no.	Name of the software	Link
1.	IDT PrimerQuest Tool	(https://sg.idtdna.com/PrimerQuest/Home/Index)
2.	Primer3	(http://primer3.wi.mit.edu/)
3.	Net Primer	(http://www.premierbiosoft.com/netprimer/index.html)
4.	Primer BLAST	(http://www.ncbi.nlm.nih.gov/tools/primer-blast/)
5.	PCR Designer	(http://primer1.soton.ac.uk/primer.html)
6.	Eprimer3	(http://emboss.bioinformatics.nl/cgi-bin/emboss/eprimer3)

3.2 DNA Extraction by CTAB Method [10, 11] (see Note 8)

1. Weigh 1 gram of putative transgenic cotton leaves, and freeze it in liquid nitrogen (*see* **Note 9**).
2. Grind the frozen tissue into fine powder with the aid of a chilled mortar and pestle in the presence of liquid nitrogen.
3. Approximately 350 g of the powder should be distributed in each of three micro fuse tubes (micro fuse tubes must be chilled in liquid nitrogen before and after weighing. Spatula should be pre-chilled and used to transfer the ground powder). Equal volume (w/v) of pre-hot (65 °C) 2× CTAB buffer should be added (*see* **Note 9**).
4. Mix well by inversion in vigorously and equal volume of (700 μL v/v of ice-cold chloroform: Isoamyl alcohol (24:1) should be added.
5. Mix thoroughly by inversion to form an emulsion, and centrifuge for 5 min at 14,000 rpm.
6. Then, using the cut tips, upper aqueous content should be transferred to new microfuge tubes.
7. Measure the volume, 1/5 volume of 5% CTAB solution should be added, and mix well by gentle inversion.
8. Equal volume of chloroform: Isoamyl alcohol (24:1) should be added. Then mix thoroughly to form emulsion and centrifuge for 5 min at 14,000 rpm.
9. Using cut tips, upper aqueous content should be transferred to new centrifuge tubes, then add equal volume of CTAB Precipitation buffer, and mix gently by inversion.
10. Centrifuge for 1 min at 14,000 rpm, discard the supernatant, and keep the tube by inverting in tissue paper and dry it for 30 min.
11. Dissolve the pellet in 200 μL of high-salt TE buffer and gently tap the tubes. Keep it overnight at 4 °C fridge.
12. Next day, tap the tube gently and make sure the pellet dissolved completely, and then 2.5 volumes of ice-cold 95% ethanol should be added, and mix gently by inversion (*see* **Note 14**).
13. Centrifuge for 10 min at 14,000 rpm at 4 °C and discard the supernatant.
14. Add 500 μL of 70% ethanol to the pellet, and centrifuge for 5 min at 14,000 rpm at 4 °C (*see* **Note 14**).
15. Air-dry or vacuum dry the pellet and dissolve in 50 μL of 0.1×TE (*see* **Note 13**).
16. Leave it for some time at room temperature or 4 °C then analyze the isolated genomic DNA by 0.8% agarose gel electrophoresis (Fig. 1).

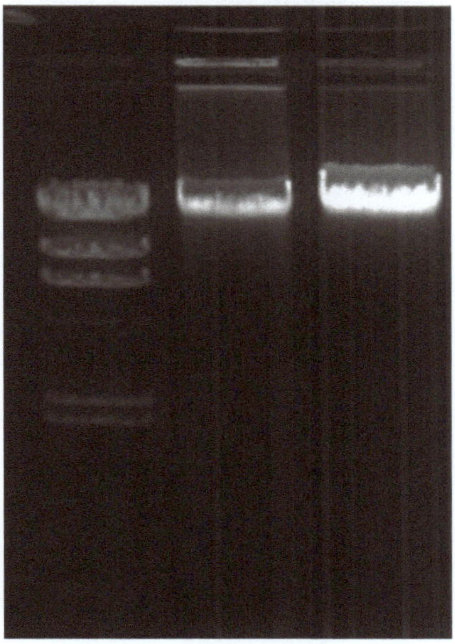

Fig. 1 Agarose (0.8%) gel electrophoresis analysis of genomic DNA isolated from the leaves of putative transgenic cotton plants

17. In order to eliminate protein contamination and produce more pure DNA, the extracted genomic DNA have further digest using proteinase K and purified using a phenol/chloroform mixture (*see* **Notes 9**, **10**, **11**, **12**, and **13**).

3.3 PCR Analysis

1. Add all PCR components from Table 3 in the following order as nuclease-free water, 10× buffer, dNTP mix forward and reverse primers, Taq DNA polymerase, templates (genomic DNA/plasmid for positive control/water (for negative control) in a aseptic condition.
2. Set the thermocycler as mentioned in Table 4.
3. The PCR-amplified sample will be analyzed in agarose gel electrophoresis and visualize in the gel documentation system (Fig. 2).

3.4 Agarose Gel Electrophoresis

1. Prepare enough quantity of the 1× TAE buffer (running buffer), and fill it in electrophoresis tank.
2. Weigh 0.5 g of agarose, and add it into 50 mL of TAE buffer (1×) in 150 mL conical flask.
3. Heat the mixture in microwave oven until the agarose completely dissolves.

Table 3
Prepare the PCR reaction mix (50 μL reaction)

Requirements	Volume
10× PCR buffer (with $MgCl_2$)	5.0 μL (1×)
dNTP mix (10 mM)	1.0 μL
Forward primer (10 μM)	1.0 μL
Reverse primer (10 μM)	1.0 μL
Taq DNA polymerase	1.0 μL (1 Unit)
Template DNA (*see* **Notes 7**, **20**, and **25**)	1.0 μL (50–100 ng)
Nuclease-free water	40.0 μL

Table 4
PCR reaction conditions [12]

Steps	Condition
Initial denaturation	94 °C for 5 min
Denaturation	94 °C for 1 min
Annealing (*see* **Notes 20** and **25**)	50–60 °C for 1 min
Extension	72 °C for 1 min (1 min/kb)
Final extension	72 °C for 7–10 min
No. of. cycles	**30–35** (*see* **Note 26**)

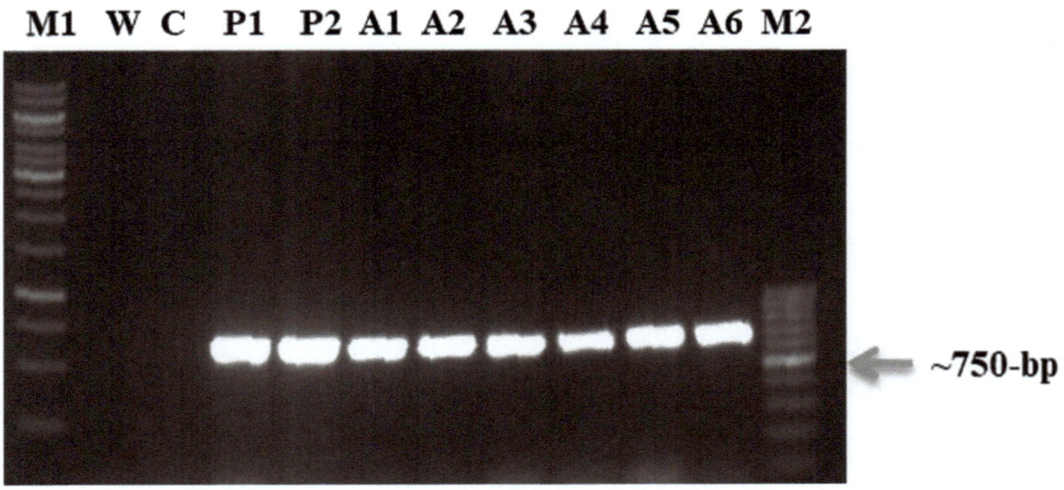

Fig. 2 PCR analysis of putative transgenic cotton plants by using *NPT* II primer. *Lanes M1 and M2*, 1 Kb plus DNA ladder and 100 bp ladder, respectively; *Lane W*, negative control (water used as a template); *Lane C*, negative control (non-transgenic cotton DNA used as a template); *Lanes P1 and P2*, positive control (plasmid used as a template); *Lanes A1–A6*, difference transgenic cotton lines

4. Allow to cool the mixture. Prior solidifying, add 5 μL of Ethidium bromide to the mixture, and mix well.
5. Place the comb in the casting tray and pour the mixture in it.
6. After solidification (approximately 30–45 min) carefully remove the comb without disturbing the gel.
7. Gels in the casting tray move to electrophoresis tank which contains 1× TAE buffer.
8. Take 25 μL of PCR product from the PCR reaction tubes, add 5 μL of 6× loading dye, and mix it by pipetting the substance.
9. Load the sample in the respective wells.
10. Run the gel in 1× TAE buffer for 30–60 min in 80 volts until the dye migrated to a three-fourth distance of the gel.
11. To visualize the DNA bands, keep it in gel documentation system and capture the image [13].

3.5 Total RNA Isolation (see Notes 8, 15, and 16)

1. Weigh 1 g of genomic PCR-positive transgenic cotton leaves, and freeze with liquid nitrogen and ground well using mortar and pestle.
2. Transfer the frozen ground powder to the 1.5 centrifuge tubes (DEPC-treated/RNase-free).
3. Pipette 1000 μL of the lysis solution with 2-mercaptoethanol (add 10 μL of 2-ME into 1000 μL of the lysis solution) for 500 mg of tissue powder, and vortex immediately and vigorously for at least 30 s. Incubate the sample at 56 °C for 5 min, and then bring to the room temperature (*see* **Note 17**).
4. Centrifuge the sample at 10,000–12,000 rpm for 5 min to pellet cellular debris.
5. Pipette the lysate supernatant into a filtration column (blue retainer ring) seated in a 2.0 mL collection tubes by using cut tips.
6. Close the cap and centrifuge at 10,000–12,000 rpm for 1 min to remove residual debris. Save the clarified flow-through lysate.
7. Pipette 500 μL of binding solution into the clarified lysate and mix immediately and vortex gently for 1 min.
8. Pipette 700 μL of the mixture into a binding column (red retainer ring) seated in a 2.0 mL collection tube. Close the cap and centrifuge at 12,000 rpm for 1 min to bind RNA.
9. Decant the flow-through liquid, and tap the collection tube (upside down) briefly on a clean absorbent paper to drain the residual liquid. Return the column to the collection tube, pipette the remaining mixture to the column, and repeat the centrifugation and decanting steps. Repeat **Step 8**.

10. Pipette 500 μL of wash solution I into the column. Close the cap and centrifuge at 12,000 rpm for 1 min. Decant the flow-through liquid, and tap the collection tube (upside down) briefly on a sterile filter paper to drain the residual liquid. Return the column to the collection tube.
11. Pipette 500 μL of the diluted wash solution II into the column. Close the cap and centrifuge at maximum speed for 30 s. Decant the flow-through liquid, and tap the Collection Tube (upside down) briefly on a sterile filter paper to drain the residual liquid. Return the column to the collection tube (*see* **Note 18**).
12. Again pipette 500 μL of the diluted wash solution II into the column, close the cap, and centrifuge at maximum speed for 30 s. Decant the flow-through liquid, and tap the collection tube (upside down) briefly on a sterile filter paper to drain the residual liquid. Return the column to the collection tube.
13. Centrifuge the column at 12,000 rpm for 1 min to dry. Remove the column from the tube.
14. Transfer the column to a new, clean 2.0 mL collection tube. Pipette 10 μL of elution solution directly onto the center of the binding matrix inside the column. Close the cap and let the tube sit for 1 min. Centrifuge at 12,000 rpm for 1 min to elute (elution 1) (*see* **Note 19**).
15. Pipette 10 μL of elution solution directly onto the center of the binding matrix inside the column. Close the cap and let the tube sit for 1 min. Centrifuge at 12,000 rpm for 1 min to elute in separate tubes (elution 2) (*see* **Note 19**).
16. Analyze the total RNA samples by 1× MOPS gel (formaldehyde denaturing) (*see* **Note 27**).

3.6 Denaturing Agarose Gel for RNA Analysis

1. Weigh 1.2 g of agarose, and add to the 250 mL conical flask that contains 97 mL of 1× MOPS buffer.
2. Heat the mixture in the microwave oven until the agarose completely dissolves.
3. Add 3 mL of 37% formaldehyde.
4. Mix well and pour into the gel casting tray inserted with the comb.
5. After solidification (approximately 30–45 min), carefully remove the comb without disturbing the gel.
6. Casting tray with gel move to electrophoresis tank which contains 1× MOPS buffer.
7. Mix all the components from the Table 5 in 0.5 mL RNase-free Eppendorf tube.
8. Incubate in water bath/heating block at 70 °C for 5 min, and keep the tubes in ice (*see* **Note 21**).

Table 5
Sample preparation for gel electrophoresis analysis

Sample	Volume (10 μL)
RNA (1.0 μg)	2.0 μL
2× RNA gel loading dye (Thermo Fisher Scientific)	5.0 μL
Nuclease-free water	3.0 μL

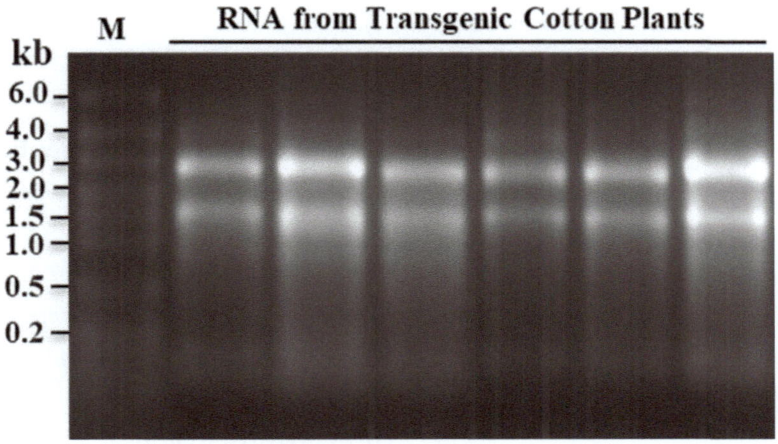

Fig. 3 Analysis of total RNA isolated from transgenic cotton plants using a denaturing formaldehyde agarose gel (1.2%) with 1× MOPS buffer. Lane M: RNA ladder

9. Briefly spin the samples and load it in the respective wells of the gel.
10. Run the gel with 1× MOPS buffer for 40–60 min in 60 volts until the dye migrated to a three-fourth distance of the gel.
11. Visualize the separated RNA under UV transilluminator of gel documentation system and capture the image (Fig. 3).

3.7 First-Strand cDNA Synthesis (for Troubleshoot, Follow the Manufacturer's Instruction) (see Notes 23 and 29)

Take the following components out from −20 °C deep freezer, and thaw in the ice to set up the cDNA synthesis (Tables 6 and 7):

1. Mix and briefly centrifuge each component before use. Preheat the thermal cycler to 65 °C (*see* **Note 27**).
2. Prepare the component given in Table 6 to 0.2 mL PCR tubes on ice.
3. The PCR tubes contain reaction mixture incubate in a thermal cycler at 65 °C for 5 min, and then immediately place on ice for at least 1 min. Collect the contents of the tube by a short spin.
4. Next, add the following to the tube.
5. Mix the sample and collect by a brief spin. Incubate in thermocycler at 25 °C for 10 min followed by 50 °C for 50 min (*see* **Note 27**).

Table 6
Components for first-strand cDNA synthesis (*see* Note 23)

Components	Volume (13.0 μL)
Total RNA (2.0 μg) (*see* **Note 28**)	2.0 μL
10 mM dNTP mix (*see* **Note 27**)	1.0 μL
Random Primer (10.0 μM) (*see* **Note 24**)	0.5 μL
Oligo (dT) Primer (10.0 μM) (*see* **Note 24**)	0.5 μL
Nuclease-free water	11.0 μL

Table 7
cDNA preparation component mixture (*see* Note 23)

Components	Volume (7.0 μL)
5× first-strand reaction mix *(Thermo Fisher Scientific)*	4.0 μL
Dithiothreitol (DTT) *(Thermo Fisher Scientific)*	2.0 μL
M-MLV Reverse Transcriptase (Thermo Fisher cientific)	1.0 μL

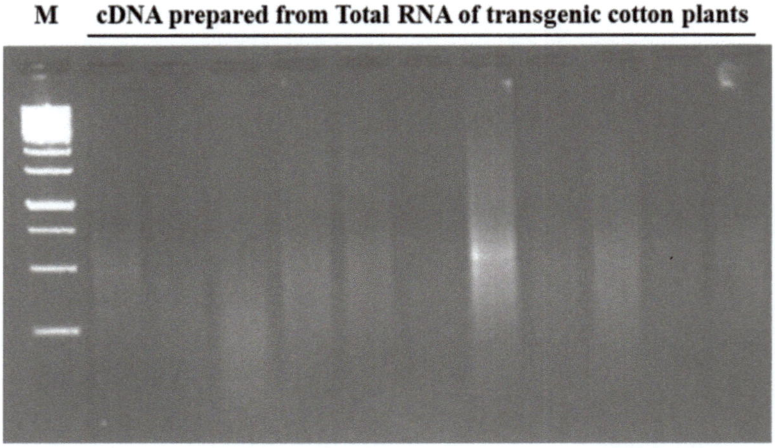

Fig. 4 Agarose (1%) gel electrophoresis analysis of cDNA prepared from total RNA extracted from the transgenic cotton. Lane M: 1 kb plus DNA ladder

6. Terminate the reactions at 70 °C for 15 min followed by hold at 4 °C.
7. Dilute the synthesized cDNA with one volume of nuclease-free water (Fig. 4).
8. Then 1.0 μL from diluted cDNA will be used for the RT-PCR reaction.

3.8 Reverse Transcriptase Polymerase Chain Reaction (RT-PCR)

Perform the reverse transcriptase PCR like genomic PCR; instead of genomic DNA, cDNA (*see* **Note 22**) will be used as a template for PCR reaction (Fig. 5). The reaction conditions are given as follows (Table 8).

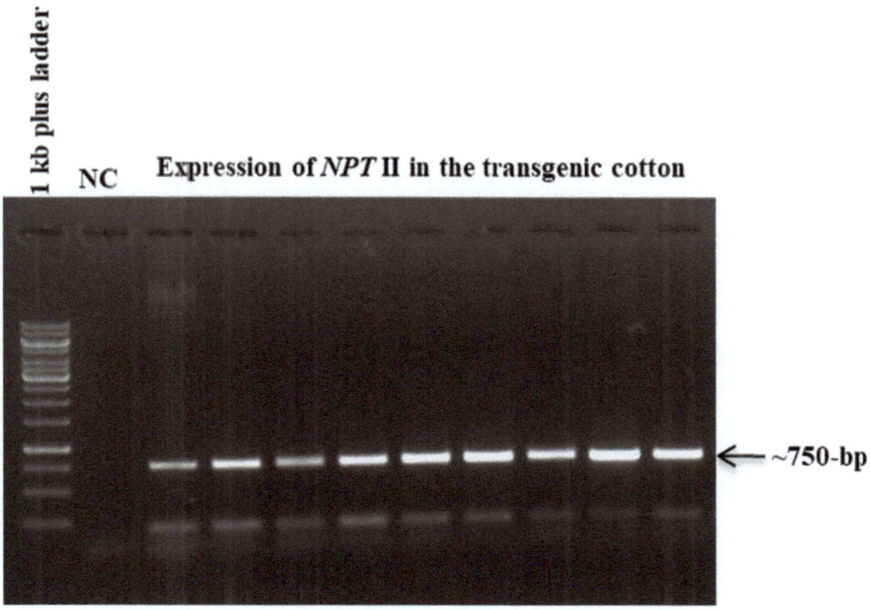

Fig. 5 Agarose gel electrophoresis analysis of RT-PCR product of *NPT* II gene expression in the transgenic cotton

Table 8
RT-PCR reaction conditions

Steps	Condition
Initial denaturation	93 °C for 3 min
Denaturation	93 °C for 30 s
Annealing	60 °C for 40 s
Extension	72 °C for 7 min (1 min per kb)
Final extension	72 °C for 7 min
Pause	4 °C
No. of cycles	**30**

4 Notes

1. To steer clear of problems such as Tm mismatch, self-complementarity, and secondary structures, use primer design tools.
2. Design primers with an ideal length (usually 18–25 nucleotides) since too short primers may have poor specificity and too lengthy primers may be prone to secondary structures.
3. For obtaining best results, primers with annealing temperature ranges from 60 to 65 °C should be considered.
4. Too high or too low GC content can affect primer binding stability, so aim for a balanced GC content (around 40–50%) depending on the target sequence [14].
5. Carefully choosing primer sites to avoid troublesome regions—think about utilizing touchdown PCR—is necessary to prevent nonspecific amplification caused by repeating sequences or high GC content inside the target region.
6. The amplification effectiveness is decreased if the primer produces dimer structures or hairpin loops; hence caution should be taken when designing the primer to prevent primer self-complementarity at the 3′ end.
7. Determine the ideal annealing temperature for the primer set using a gradient PCR. Then, test various primer concentrations to find the most effective amplification. Finally, make sure the DNA template is high-quality and concentrated. If problems continue, design new primer sets that target different regions of the target sequence.
8. The working place should be thoroughly cleaned with 75% ethanol/RNaseZap solution and always free with RNase contamination.
9. Increase the weight of the tissue sample, make sure the plant tissue is thoroughly ground with a mortar and pestle or a tissue homogenizer, modify the volume of CTAB buffer to match the amount of tissue, extend the incubation period with proteinase K, or use a higher concentration to prevent incomplete protein digestion if the DNA yield is lower.
10. Add an additional phenol-chloroform extraction step to remove protein contamination completely.
11. Increase the centrifugation speed and duration during the phenol-chloroform extraction steps, add more polyvinylpyrrolidone (PVP) to the CTAB buffer to bind to the polysaccharides present in the samples, and make sure that the aqueous and organic phases are properly mixed during the phenol-chloroform extraction to prevent poor phase separation in the event that DNA is contaminated with polysaccharides.

12. Make sure to always adhere to the correct laboratory safety protocols when handling chemicals such as phenol and chloroform.
13. To prevent PCR inhibition, perform a thorough wash step with ethanol after DNA precipitation to avoid residual plant metabolites, dialyze the DNA or adjust the salt concentration in the CTAB buffer, and carefully remove any phenol traces during the phenol-chloroform extraction.
14. Use high-quality ethanol for effective DNA precipitation and perform multiple washes with 70% ethanol to remove residual contaminants.
15. All the materials used for isolation RNA should be treated with DEPC water and autoclaved.
16. Wearing gloves and mask is mandatory to get rid from denaturation of RNA.
17. To obtain good-quality RNA, avoid extended duration of vortex as it may mix the genomic DNA [15].
18. Wash solution II should be used twice if we take more quantity of sample.
19. Elutions I and II should be collected in the individual tubes and quantified.
20. Faint/nonspecific/no amplification is rectified by optimizing the annealing temperature and template concentration.
21. Prior to load the isolated RNA for analysis, the total RNA mixed with RNA loading dye should be incubated with 70 °C for 5 min then immediately kept it in ice.
22. The dilution of cDNA and template concentration should be based on our experiment need.
23. Researchers can buy the complete cDNA synthesis kit from different manufacturers and followed their protocol as per the manufacturer instruction.
24. We can use both or anyone of the primers of random and oligo (dT) based on the requirement.
25. If any additional, nonspecific bands appear on the PCR gel, lower the concentration of the template, and adjust the temperature in the thermal cycler to match the primer's annealing temperature.
26. Reduce the number of cycles and adjust the thermal cycler's temperature to prevent too much smearing of the amplified DNA on the PCR gel.
27. Check the RNA quality, make sure there are no cDNA synthesis inhibitors, raise the cDNA synthesis temperature to prevent secondary structures on the RNA, and apply the ideal quantity of dNTP if the cDNA yield is low or nonexistent.

28. Before utilizing your sample for cDNA synthesis, make sure to treat the total RNA with *DNase* I because the extracted RNA may be contaminated with genomic DNA.

29. Follow these guidelines if you intend to repeat the experiment: use filter tips for RNA extraction, clean the workspace with RNase decontamination products beforehand, only use RNase-free water for extraction, begin cDNA synthesis right away after RNA extraction, and assess the quality of extracted RNA before beginning cDNA synthesis.

References

1. James C (2023) Global Status of Commercialized Biotech/GM Crops ISAAA Brief Inter Serv for the Acqu of Agri-biotech Appl 56
2. Ellstrand NC (2019) The evolution of plant genetic modification: a historical perspective. Camb Univ Press
3. Mullis K, Faloona F (1987) Specific synthesis of DNA in vitro via a polymerase-catalyzed chain reaction. Methods Enzymol 155:335–350
4. Holland PM. Detection of specific DNA sequences using the polymerase chain reaction. Nucl Acid Res. 19(15):4193–4197
5. Kumar S (2018) Molecular techniques for characterization of transgenic plants. Plant Biotech Rep 12:489–501
6. Meyer P et al (2004) PCR analysis of genetically modified crops. J Agri Food Chem 52(3): 782–788
7. Robertson LJ, Clark CG, Debenham JJ, Dubey JP, Kvac M, Li J, Tsaousis AD (2019) Are molecular tools clarifying or confusing our understanding of the public health threat from zoonotic enteric protozoa in wildlife? Internat J Parasitol Parasites Wildl 9:323–341
8. Bustin SA (2009) The PCR revolution: basic technologies and applications. Camb Univ Press
9. Kumar A, Chordia N (2015) *In silico* PCR primer designing and validation. PCR Primer Design, pp 143–151
10. Rogers SO, Bendich AJ (1994) Extraction of total cellular DNA from plants, algae and fungi. In: Plant mol biol manual. Spring, Dordrecht, pp 183–190
11. Gurusamy D (2014) Transcriptome analysis of cotton (*Gossypium hirsutum* L.) during boll development under drought condition and functional characterization of drought tolerance genes. Ph.D. thesis: https://hdl.handle.net/10603/215568
12. Girija S (2020) Practical manual on plant molecular biology and analytical techniques. Akin Pub
13. Armstrong JA, Schulz JR (2015) Agarose gel electrophoresis. Curr Protoc Ess Lab Tech 10(1):7–2
14. Borah P (2011) Primer designing for PCR. Sci Vis 11(3):134–136
15. Farrell RE Jr (2009) RNA methodologies: laboratory guide for isolation and characterization. Aca Press

Chapter 15

Characterization of Transgenic Cotton Plants by Southern Blotting

Janani Sree Sivakumar, Dhandapani Gurusamy, Selvi Chellamuthu, Sakthi Ambothi Rathnasamy, and Vasanth Krishnan

Abstract

Blotting is the key technique of current molecular biology, and it is an efficient and a sensitive way of determining if a sample contains a certain biomolecule. The array of technique known as "Southern blotting," in which a particular restriction fragment can be recognized against the background of more restriction fragments. The process comprises transferring DNA fragments from the electrophoresis gel onto a nitrocellulose membrane in a way that evoke the gel's DNA banding pattern. After that, hybridization probing takes place to detect the desired restriction fragment. To investigate gene targets, Southern blotting is frequently used as the benchmark for identifying transgenic integration and effective homologous recombination. This chapter enlightens the following various steps to perform the Southern blotting that includes genomic DNA isolation, restriction digestion of isolated DNA, agarose gel electrophoresis, depurination, denaturation, neutralization then transfer of DNA from agarose gel to nylon membrane (blotting), pre-hybridization, probe preparation and random labeling of DNA fragments, hybridization, and post-hybridization (washing) and exposed the x-ray film on the membrane and autoradiography or scanning to visualize the hybridized DNA fragments.

Key words Transgenic cotton, Genomic DNA, Restriction digestion, Random labeling, Hybridization, Autoradiography, Southern blotting

1 Introduction

Transgenic crops created revolutionary agriculture by establishing particular traits that increases the performance of crop and sustainability. Transgenic cotton grabs the attention compared with other transgenic crops because of it improvised fiber quality, enhanced insect pest resistance, and enhanced herbicide tolerance. The incorporation of the foreign genes within the cotton crop become a remarkable solution to agricultural challenges such as pest attack and excessive utilization of chemical pesticides [1]. However, the promising development as well as commercialization of transgenic

Nitish Kumar (ed.), *Transgenic Plants: Methods and Protocols*, Methods in Molecular Biology, vol. 2973, https://doi.org/10.1007/978-1-0716-4803-2_15,

cotton depends on the characterization in order to confirm the integration along with expression of transgenes.

There are numerous techniques for characterization of transgenic crops; among them Southern blotting is commonly employed. The technique was developed by "Edwin Southern" in 1970s that equips the researchers to identify specific gene sequence in the entire genome set [2]. Southern blotting is especially utilized for detecting transgene integration and copy number that is crucial for understanding its stability and inheritance of the transgenic crop traits. There are some of the steps involved in Southern blotting which includes extraction of genomic DNA isolation, restriction digestion of isolated DNA, agarose gel electrophoresis, depurination, denaturation, neutralization, transfer of DNA from agarose gel to nylon membrane (blotting), pre-hybridization, probe preparation and random labeling of DNA fragments, hybridization, post-hybridization (washing), exposing the X-ray film on the membrane, and autoradiography or scanning to visualize the hybridized DNA fragments. By visualizing DNA fragments that correspond to transgenes, researchers can ascertain whether the transgenes have been successfully integrated into the cotton genome and whether they are present in multiple copies or as single insert [3]. Furthermore, the application of Southern blotting in transgenic cotton research provides insights into the integration the transgene and copy number identification. While the presence of a transgene is essential, understanding its functionality and expression level is equally important. Variations in expression can affect plant phenotype and, ultimately, the success of the transgenic trait in field conditions [4]. Therefore, analyzing the integration and copy number of transgenes through Southern blotting is essential for assessing the overall performance and stability of transgenic cotton.

Other than its biological significance, the characterization of transgenic cotton has critical legal ramifications. This signifies that thorough evaluations of GMOs are conducted for their approval for human consumption and environmental risk [5]. In essence, comprehensive molecular characterization, which includes Southern blot analysis, lays a firm basis in the minds of the regulators as it substantiates genetic stability and possible risks through transgenic crops. The characterization of transgenic cotton plants through Southern blot analysis remains a principal point of modernity in plant biotechnology. This technique is promoting the detection and analysis of transgenes while fulfilling an essential function in the determination of the safety and performance by transgenic plants. The increasing pressure for sustainable agriculture will, no doubt, ensure that molecular techniques such as Southern blotting will prove relevant in the development of better crop varieties in the near future. This understanding of these methods will, in due course, allow researchers to decide the fate of transgenic crops for innovations that are mutually beneficial to farmers and consumers because they will provide for informed decision-making.

2 Materials (*see* Note 2)

1. *Genomic DNA Extraction Buffer* (Table 1) (*see* **Note 3**).
2. *1 M Tris (for 100 mL):* 100 mL of 1 M Tris to be prepared by dissolving 12 grams of Tris base in 60 mL of distilled water. After dissolving, adjust pH to 8.0 using concentrated HCl. Then, the final volume should be made up to 100 mL with distilled water and autoclaved at 15 psi pressure for 20 min, with denaturation solution, neutralization solution, pre-hybridization and hybridization solutions, restriction digestion analysis of plant genomic DNA buffer (Tables 2, 3, and 4).
3. *0.5 M EDTA (for 100 mL):* Prepare 100 mL of 0.5 M EDTA by dissolving 18.6 grams of EDTA in 75 mL of distilled water. NaOH pellets can be added to dissolve the EDTA as it taken time for dissolving completely. The pH should be adjusted to 8.0 using 1 N NaOH. Finally, the volume should be made up

Table 1
Genomic DNA extraction buffer (For 100 mL)

Sterile DDH$_2$0	38 mL
2% CTAB	20 mL (5% stock)
1.4 M NaCl	28 mL (5 M stock)
100 mM Tris.Hcl	10 mL (1 M stock)
20 mM EDTA	4 mL (0.5 M stock)

Table 2
Denaturation solution (500 mL)

NaCl	29.22 gm (1 M)
NaOH	10 gm (0.5 M)

Dissolve in 400 mL of MilliQ/double distilled water and made the final volume of 500 mL. Autoclave and stored at room temperature

Table 3
Neutralization solution (500 mL)

NaCl	43.83 gm (1.5 M)
Tris	30.28 gm (0.5 M)

Dissolve in 350 mL of MilliQ water and adjust the pH to 7.0 with concentrated HCl and made the final volume of 500 mL. Autoclave and store at room temperature

Table 4
Restriction digestion analysis of plant genomic DNA

Nuclease Free H_2O	(make up to volume 50 μL)
Genomic DNA	20 μg
Restriction buffer	5 μL (1×)
Restriction enzymes	5 μL (50 units)

Twenty micrograms of cotton genomic DNA is restricted with specific restriction enzymes and incubate in water bath at 37 °C for overnight (12–16 h)

to 100 mL with distilled water and autoclaved at 15 psi pressure for 20 min.

4. *5% CTAB (for 100 mL):* Prepare 100 mL of 5% CTAB by dissolving 5 grams of CTAB in 80 mL distilled water, and thereby 2 grams of NaCl should be added to it. Then the final volume should be made up to 100 mL and autoclaved at 15 psi pressure for 20 min.
5. *5 M NaCl (for 100 mL):* 29.2 g of NaCl to be added in to 80 mL of distilled water and dissolved completely. Finally, make up to 100 mL and autoclaved at 15 psi pressure for 20 min.
6. *0.1× TE (for 20 mL):* Prepare 20 mL of 0.1× TE by adding 0.02 mL (20 μL) of 1 M Tris (pH 8.0) and 0.004 mL (4 μL) of 0.5 M EDTA (pH 8.0) to 18 mL of distilled water. After dissolving properly, the final volume should be made up to 20 mL and autoclaved at 15 psi pressure for 20 min.
7. *Chloroform: isoamyl alcohol (24:1) (for 100 mL):* 100 ml of 24:1 Chloroform: Isoamyl alcohol was prepared by adding 4 ml of isoamyl alcohol in 96 ml of chloroform in the Amber reagent bottle and mix well. This mixture should be ice-cold before use.
8. *20×SSC (1 L):* Dissolve 175.3 gm of NaCl and 88.2 gm of sodium citrate in 800 mL of Milli-Q water. Adjust the pH to 7.0 with concentrated HCl. Make the final volume to 1 L. Autoclave and store at room temperature (*see* **Note 16**).

3 Methods

3.1 Extraction of Genomic DNA Through Modified CTAB Method (for Plants Containing High Polysaccharide and Polyphenol Components) [6]

1. 1.0 g of plant leaf tissue should be ground using with chilled mortar and pestle, and transfer the powder to polypropylene tube (30 mL) using chilled spatula (50 mg of PVP for 0.5 g tissue should be added when grinding the tissue) (*see* **Note 2**).
2. 10 mL of DNA extraction buffer (preheated at 60 °C) should be added, mix the tubes by inversion, keep at 60 °C for 45 min,

and mix once in 15 min. Take the tubes and bring to room temperature (RT).

3. Add 10 mL of chloroform: Isoamyl alcohol (24:1) and mix by inversion until it forms emulsion. Centrifuge for 10 min at 6000 rpm (25 °C), and repeat the steps one more time to remove the cloudiness from the samples.
4. Carefully transfer the supernatant to fresh tubes, add and mix 1/2 volume of 5 M NaCl, add 2.5 volumes of 95% ethanol (pre-chilled at 4 °C), mix the tubes, and then keep at −20 °C for an hour.
5. Spin at 6000 rpm for 10 min (25 °C), discard the aqueous solution, wash the pellet with 70% ethanol, and spin for 5 min. Air-dry the pellet for 1 h, add 300 μL of 0.1× TE to dissolve the pellet, and keep for 4 °C overnight.
6. Dissolved DNA should be transferred to 1.5 mL Eppendorf tube, and 5 μL of *RNase* A (10 mg/mL) should be added and incubated for 1 h at 37 °C water bath (*see* **Note 3**).
7. 5 μL of proteinase K (1 mg/mL) should be added and incubated for 30 min at 37 °C water bath.
8. Extract equal volume of neutral phenol and two times with ether solution to remove phenol solution completely.
9. Add 1/10 volume of 3 M sodium acetate pH 5.2 and 2.5 volume of 90% ethanol to the solution for precipitation of genomic DNA. Keep the tubes overnight at -70 °C.
10. Next day, spin the tubes for 12,000 rpm, 10 min at 4 °C, discard the supernatant, and wash the pellet twice with 70% ethanol (pre-chilled). Finally dry the pellet and dissolve with 100 μL of 0.1× TE buffer.

3.2 Southern Blotting

3.2.1 Restriction Digestion (see Notes 1 and 4) [8]

1. Electrophorese restricted DNA samples on 0.8% agarose gel (*see* **Notes 10**, **15**, and **16**).
2. After that, stain the gel in ethidium bromide (10 mg/mL) solution for half an hour, capture the image, and confirm the complete restriction before processing the next step.
3. Process the gel before blotting through three steps, viz., depurination, denaturation, and neutralization.

3.2.2 Depurination

1. Immerse the gel into depurination solution (0.125 M HCl) and agitate for 15–20 min.
2. During this time the bromophenol blue dye present in the sample will change to yellow.

3.2.3 Denaturation

1. Submerge the gel into denaturation buffer, and incubate for 1 h with gentle agitation.
2. The bromophenol blue dye will return to its original blue color.

3.2.4 Neutralization

1. Wash the gel with distilled water three to four times to remove denaturation solution completely, and submerge to sufficient neutralization buffer for 1 h with gentle agitation.
2. The processed gel is now ready for blotting in which a setup is made for capillary transfer of depurinated DNA fragments from agarose gel to the nylon membrane.

3.2.5 Blotting

1. Sheet of Nylon membrane to be cut as per the size of gel (*see* **Notes 1**, **5**, **6**, and **17**).
2. Half-fill a tray with the transfer buffer (20× SSC) and use for setting up of blot transfer.
3. A platform must be made and covered with a wick made from three sheets of 3MM Whatman filter paper saturated in transfer buffer.
4. The treated the gel is placed on the wick platform. Trapping any air bubbles between the gel and wick has to be avoided.
5. The gel surround with cling film to prevent the transfer buffer being absorbed directly into the paper towels.
6. The membrane is placed on top of the gel. Three sheets of 3MM Whatman paper cut to gel size saturated in transfer buffer are placed on top of the membrane.
7. A stack of absorbent towels is placed on top of the 3MM paper at least 6–7 cm high. Finally, a glass plate is placed, and a weight is put on top of glass plate (*see* **Note 9**).
8. The transfer is usually allowed to proceed for 16–18 h.
9. After blotting, the transfer apparatus is carefully dismantled.
10. Before separating the gel and membrane, the membrane is marked to allow identification of the tracks with a pencil. Remove the blot after overnight transfer and rinse briefly (30 s) in 2× SSC (*see* **Note 8**).
11. Air-dry and the nucleic acid is fixed to the membrane by using UV cross-linker UVP for 20 s. Furthermore, the blot is processed for pre-hybridization, hybridization, and post-hybridization.

3.2.6 Pre-hybridization

1. Keep the membrane inside the hybridization bottle by using a sterile forceps, and add the pre-hybridization solution (150 μL/cm^2 membrane).
2. Remove the air bubbles using a thin spatula. Incubate at 65 °C for 30 min in a hybridization oven (*see* **Note 7**).

3.2.7 Random Labeling of DNA Fragments

1. 100 ng of gel eluted DNA fragments which use as probe (gene of interest which needs to detect the integration and copy number) to be taken and 5 μL of random primer (Amersham Oligo labelling Kit) should be added in a microfuge tube. The volume of 33 μL should be made with sterile distilled water.
2. Denature the DNA by heating for 5 min over boiling water. Transfer the tube to a beaker containing hot water (~70 °C), and allow it to cool down to room temperature gradually.
3. Centrifuge for 5 s, and make the volume to 33 μL with sterile distilled water.
4. To denature DNA, 10 μL of labeling mix is provided in the Amersham Oligo Labeling Kit, 5 μL of α-^{32}P radioactivity, and dCTP, and 2 μL of Klenow fragment is added. Mix gently by tapping and centrifuge for 5 s in a microfuge.
5. Incubate at 37 °C for 20 min, and stop the reaction by adding 50 μL of nick translation dye mix or 0.5 M EDTA.

3.2.8 Hybridization

1. Denature the labeled probe in a boiling water bath for 5 min, and cool rapidly on ice for 10 min.
2. Remove pre-hybridization solution and fresh pre-hybridization solution (same volume) to be added to the bottle.
3. Add denature probe to the bottle, mix by gentle shaking, and incubate at 65 °C for 20 hrs in the hybridization oven (*see* **Note 18**).

3.2.9 Post-hybridization Washes (High Stringency Wash)

1. Discard hybridization solution, and rinse the membrane briefly with 20 mL of 3× SSC/0.1% SDS. Three washes are done with 40 mL each of 3× SSC/0.1% SDS, 0.5× SSC/0.1% SDS, and 0.1× SSC/0.1% SDS one after the other for 20–30 min each at 65 °C.
2. Finally, briefly rinse with 2× SSC at room temperature, then fill up the bottle with 2× SSC, and remove the membrane carefully.
3. Air-dry the membrane, and expose it to X-ray film for autoradiography or visualizing the images by phosphorimaging instrument.

3.2.10 X-Ray Film Exposing for Autoradiogram (see ***Notes 11, 12, 13****, and* ***14****)*

1. Completely dry the nylon membrane used for the Southern blotting process, properly affix to the cassette and expose with an X-ray film in the dark room.
2. Place the X-ray sheet properly, press lock it with the cover, label the cassette properly, and store it at −70 °C in a polyethylene cover until it is subjected to developing.

3.2.11 X-Ray Film Developing and Fixing (see ***Note 11*, *12*, *13***, *and* ***14****)*

1. The developer and fixer are to be kept ready before the X-ray film, which has been exposed for sufficient period of time, is subjected to developing.
2. The developer is prepared as specified on the box, as per manufacturer's instructions (KODAK). The developer is light-sensitive.
3. The developer and fixer are stored in amber (brown) bottles and stored in dark. They can be reused until no precipitation is observed. Care should be taken to avoid mixing of developer and fixer. Developing should be done in the dark room.
4. Small size trays are used into which the developer, fixer, and water are poured (separately) before use.
5. The X-ray film is removed from the cassette in the dark by holding it in one corner and placed into developer solution. It is gently pressed in four corners to ascertain that it is thoroughly dipped in the developer. The tray is then tilted to ~10–20° from the ground level and reverted back to its normal position. Repeat it several times so that the film is developed for ~2 min.
6. Then it should be transferred into a tray with tap water and washed for 1 min.
7. Transfer it into the fixer solution and wash again for 2 min. Once in a while, the film is inverted so as to obtain a better fixing.
8. Wash thoroughly in running tap water (at least 10 min) until the film appears non-sticky. (Till this step perform in dark) and finally hang to dry.

3.2.12 Autoradiography

1. The developed X-ray film shows the bands which represent the integration and copy number of the transgene (Fig. 1) (*see* **Notes 19**, **20**, and **21**).

4 Notes

1. Non-powdered gloves should be worn to avoid contamination.
2. Use a more reliable DNA extraction kit made for plant materials or a modified CTAB method, optimize the preparation of plant tissue and the DNA extraction protocol, and look for possible contamination with polysaccharides or other interfering chemicals to prevent low DNA yield.
3. Use fresh, high-quality reagents, reduce shearing pressures during lysis, and think about adding RNase to eliminate RNA contamination in order to prevent DNA degradation.

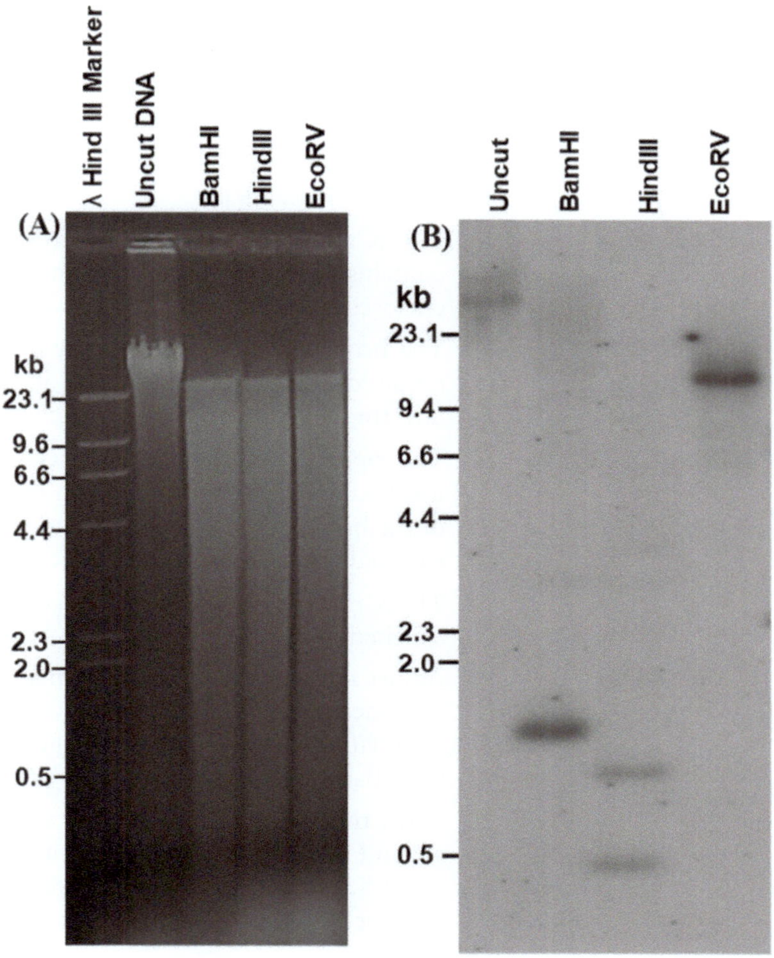

Fig. 1 Southern blot analysis for copy number identification in cotton. **a** Image of ethidium bromide stained gel restricted with *Bam*HI, *Hind*III and *Eco*RV enzymes and electrophoresis with 0.8% agarose. **b** Autoradiography of blot shows single band in *Bam*HI and *Eco*RV restriction represents single copy of the gene and two bands are observed in the *Hind*III restriction indicate that one specific restriction site present within the gene represent single copy of the gene [7, 8]

4. Check for any enzyme inhibitors in the plant extract, adjust digestion conditions (temperature, salt content), and confirm that the enzyme concentration and incubation period are suitable, to avoid incomplete digestion of genomic DNA.
5. The membrane used for the blotting should be always be slightly larger compared to gel size.
6. To prevent contamination, avoid handling the membrane directly even if the hands are gloved instead use a forceps (blunt-end).
7. Formation of air bubbles and heat may be due to unsaturated filters. Make sure the filters are fully saturated using buffer.

8. Rinsing the membrane with 2×SSC enables for leaching the excessive salt.
9. Make sure not to use excessive weight on glass plate because it may crush the gel and spoil transfer.
10. To avoid short circuit, the apparatus should be wrapped with plastic material [9].
11. The X-ray film box should "never" be opened in light. It contains X-ray sheets with inter-leaves which together are enclosed in a black polythene cover.
12. The box should be opened only in dark, and after using the required number of X-ray films, it "has to be" replaced back into the box "in dark."
13. The opening of the cover has to be folded properly and inserted back into the box in such a way that it faces the bottom of the box. The lid is put back and fastened with rubber bands to avoid the lid from releasing, accidentally.
14. The X-ray film should be handled on only one corner and never touched in any other portion.
15. Adjust agarose concentration to maximize fragment resolution (the ideal agarose gel percentage is typically 0.8%), and optimize running parameters (voltage, running time) to prevent poor band separation.
16. Verify the pH and content of the transfer buffer, make sure the gel and membrane make excellent contact during the transfer, and adjust the voltage and transfer time to prevent ineffective transfer.
17. Consider using a different type of membrane, and pre-treat the membrane with the proper blocking solution if the DNA is not adhering to it.
18. Design probes with high sequence specificity to target the transgene and steer clear of areas that have a lot of resemblance to other plant genomic sequences in order to prevent low probe specificity. Additionally, adjust labeling reaction conditions (enzyme activity, labeling time), and monitor for possible probe degradation to prevent inadequate probe labeling.
19. If no signal is found in the X-ray film (autoradiography), check for potential problems at each stage, such as insufficient DNA, incorrect enzyme selection, poor transfer, nonspecific probe binding, or excessively harsh washing conditions. Other important areas to look at include DNA quality and quantity, the effectiveness of restriction enzyme digestion, agarose gel electrophoresis, transfer of gel to the nylon membrane, probe design and labeling methods, pre-hybridization, hybridization, and post-hybridization conditions.

20. If there is no signal or only a weak signal is visible, the hybridization temperature should be adjusted to maximize probe binding. If there is a lot of background noise, the washing stringency should be increased (higher temperature, lower salt concentration), and the membrane should be pre-hybridized with blocking solution to minimize nonspecific binding.
21. To ensure the experiment is functioning properly, always include a positive control sample (known transgenic plant DNA) and perform the same experiment twice or thrice to ensure consistent results.

References

1. Bonny S (2016) Genetically modified herbicide-tolerant crops, weeds, and herbicides: overview and impact. Environ Manag 57(1):31–48
2. Klionsky DJ (2017) Blame it on southern, but it's a western blot. Auto 13(1):1–2
3. Harlow E, Lane D (1988) Antibodies a laboratory manual. Cold Spring Har Lab Pre 579:44
4. He Q, Shang X, Tian R, Zhu X, Guo W (2023) An efficient and accurate droplet digital PCR method for rapid transgene copy number detection and homozygous identification in cotton (*Gossypium hirsutum*). Indus Cro Prod 204: 117284
5. Balasubramani G, Raghavendra KP, Das J, Kumar R, Santosh HB, Amudha J, Kranthi KR (2021) Critical evaluation of GM cotton. In: Cotton preci breed. Springer, Cham, pp 351–410
6. Rogers SO, Bendich AJ (1994) Extraction of total cellular DNA from plants, algae and fungi. In: Plant molecular biology manual. Spring, Dordrecht, pp 183–190
7. Gurusamy D (2014) Transcriptome analysis of cotton (*Gossypium hirsutum* L.) during boll development under drought condition and functional characterization of drought tolerance genes. Ph.D. thesis: https://hdl.handle.net/10603/215568
8. Dhandapani G, Kanakachari M, Padmalatha KV, Phanindra MLV, Singh VK, Raghavendrarao S, Kumar PA (2015) A gene encoding cold-circadian rhythm-RNA binding-like protein (CCR-like) from upland cotton (*Gossypium hirsutum* L.) confers tolerance to abiotic stresses in transgenic tobacco. Plant Mol Biol Report 33: 22–42
9. Brown T (2001) Analysis of DNA sequences by blotting and hybridization. Curr Prot Mol Bio

Chapter 16

Characterization of Transgenic Cotton Plants by Northern Blotting

Janani Sree Sivakumar, Selvi Chellamuthu, Sakthi Ambothi Rathnasamy, Vasanth Krishnan, and Dhandapani Gurusamy

Abstract

The main intent of northern blotting is an essential strategy in molecular biology to quantify the particular messenger RNA (mRNA). The very initial step is to identify the cells that express a specific gene, therefore providing a better understanding regarding the physiological role of the protein that has been encoded. Understanding the nutritional, hormonal, or environmental factors which governed by a particular gene is the primary aim of measuring mRNA. In addition to identifying transcript presence, it also provides size and a comparative analysis for transcript abundance over the single membrane. It's been effectively adapted in recent years to investigate and validate the expression in addition to size of short noncoding RNA. The denaturing agarose/polyacrylamide gel should be used to segregate RNA based on their size before it can be transferred on the solid surface in a way that maintains the topology distribution of the gel for the purpose of preparing northern blot to hybridization. This chapter clearly emphasizes the step-by-step protocol for northern blotting.

Key words Transgenic cotton, mRNA, Hybridization, Autoradiography, Northern blotting

1 Introduction

Researchers may gain understanding regarding gene function as well as regulation through the examination for gene expression. The following approach may represent for distinguishing from the multitier strategies utilized to determine the pattern of gene expression. This techniques is employed for identifying the pattern of gene expression for one or few genes, entirety transcriptome-based techniques, and aimed on next-generation sequencing assessments for the particular gene sets or else genomic regions that have been chosen [1]. RNA molecules tend to be naturally unstable, thus handling them needs to be done with considerably greater caution. Supplies devoid of RNase must be used for all procedures [2]. The special benefit of offering information regarding transcript

Nitish Kumar (ed.), *Transgenic Plants: Methods and Protocols*, Methods in Molecular Biology, vol. 2973,
https://doi.org/10.1007/978-1-0716-4803-2_16,

size along with expression level, which could be utilized to recognize deletions or processing errors within transcripts, identify alternative splice products from the same genes or the repetitive sequence motifs, along with isolate unique transcripts from the heterogeneous pools of mRNA [3–5].

Instead of having the extrapolated value driven by the effectiveness of RT or even amplification similar to RT-PCR, the signal strength derived using northern blotting seems directly correlated with the number of copies within the gene inside original sample. The actual quantitation of the signal is potentially derived by simply comparing variations in RNA level among samples with a single membrane. As the benchmark for validating information gathered via high-throughput gene expression examination, northern blotting represents undoubtedly an approach typically provides highly reliable information on gene expression [6].

When particular proteins have overexpressed with neoplastic cells, it has been investigated whether it's because of gene amplification (the exponential rising the genes numbers), gene derepression (in which one gene implies increased mRNA), or the posttranscriptional processes including decreased cellular excretion [7]. This chapter aims to equip researchers with the knowledge and methodologies necessary to effectively utilize northern blotting in their studies of transgenic cotton.

2 Materials

Transgenic cotton leaf sample, liquid nitrogen, spatula, mortar and pestle, microfuge tubes, vortex, sterile microtips, pipettes, cooling centrifuge, diethyl pyrocarbonate (DEPC) RNase-free water/nuclease-free water, TRI reagent, chloroform, isopropanol, 75% ethanol, sterile filter paper/tissue paper, etc.

1. *10× MOPS buffer:* 0.2 M MOPS, 0.05 M NaAc, and 0.005 M EDTA; adjust to pH 7.0 with NaOH.
2. *Sample denaturation mix (100 μL):* 64.6 μL formamide, 22.6 μL formaldehyde, and 13 μL 10× MOPS (*see* **Note 11**).
3. *Loading dye mix:* 50% glycerol, 0.3% xylene cyanol, 0.3% bromophenol blue, and 1 mM EDTA.
4. *20× SSC:* 3 M NaCl, 0.3 M sodium citrate, and 0.02 M EDTA; adjust to pH 7.4; prepare by using DEPC-treated water.
5. *Pre-hybridization solution:* Ambion ULTRAhyb® Ultrasensitive Hybridization Buffer (Thermo Fisher Scientific).
6. *Wash-buffer:* (i) 2× SSC/0.1% SDS (2 × 5 min), (ii) 0.1×SSC/0.1% SDS (2 × 15 min).

3 Methods

3.1 Total RNA Isolation by TRIzol Method (see Notes 1, 2, 3, 8, and 9)

1. Using sterile pre-chilled mortar and pestle, grind 500 mg of leaf sample with liquid nitrogen (*see* **Notes 4** and **5**).
2. Transfer 500 mg of the ground powders into microfuge tubes using sterile pre-chilled spatula, add 2.5 mL of TRI reagent into the tubes, and vortex the sample immediately for 1 min. Keep the sample at room temperature for 5 min, and add 0.5 mL of chloroform (*see* **Note 5** and **9**).
3. Close the lid and shake vigorously for 15 s, and incubate at room temperature for 10 min.
4. Centrifuge the mixture at 14,000 × g for 15 min at 2–8 °C. The centrifuge sample has three phase upper aqueous phase contains RNA, middle phase contains DNA, and lower phase contains proteins (*see* **Note 11**).
5. To a fresh microfuge tube, transfer the upper aqueous phase, and 0.5 mL of isopropanol should be added per mL of TRI reagent used in the **step 2**.
6. Incubate the sample at room temperature for 5–10 min.
7. Centrifuge the sample at 2–8 °C for 12,000 × g.
8. The RNA is precipitated as pellet in side and bottom of the tube.
9. Discard the supernatant and then wash it with 700 μL of 75% ethanol.
10. Gently mix it by inversion, and then centrifuge at 2–8 °C for 7500 × g.
11. Repeat **step 10** and again wash the pellet using 75% ethanol 7500 × g at 2–8 °C.
12. Carefully decant the ethanol and air-dry the pellet in a tube on a sterile filter paper in an inverting position for 5 min, followed by keeping 37 °C incubator for about 5 min (*see* **Note 6**).
13. Add 10–20 μL of RNase-free water and incubate it overnight at 4 °C.
14. Analyze the quality and quantity of total RNA by denaturing MOPS gel electrophoresis and NanoDrop spectrophotometer (*see* **Notes 7**, **8**, **10**, and **12**).

3.2 Total RNA Isolation by Spectrum Plant Total RNA Extraction Kit (Sigma-Aldrich) (Refer Chapter 14) (see Notes 1, 2, and 3)

3.3 Analysis of Total RNA by Denaturing MOPS Agarose Gel Electrophoresis (Refer Chapter 14) (see Note 10)

3.4 Northern Blot Hybridization Requirement [8]

3.4.1 Transfer to Solid Support and Immobilization (see Note 13)

1. For northern transfer, the gel should be rinsed by several times with Milli-Q water to remove formaldehyde.
2. Cut a piece of nylon membrane according to the size of gel, rinse in distilled water for 5 min, and keep it in 20× SSC (DEPC) at least for 5 min before use.
3. Performed the capillary blotting in 20× SSC for 14–18 h. The gel should be lifted along with nitrocellulose paper and marks the position of each well on nylon membrane, using pencil. The membrane is rinsed with 2× SSC and air-dried by keeping on Whatman 3 MM filter paper (*see* **Note 14**).
4. The blot is baked at 80 °C in an oven for 2 h for immobilization/UV cross-linked.

3.4.2 Pre-hybridization and Hybridization with Probe

1. Pre-hybridize the nylon membrane containing RNA in hybridization bottle using ~10 mL of Prehybridization Buffer (Ambion ULTRAhyb® Ultrasensitive Hybridization Buffer (*see* **Note 17**).
2. Excise the gene-specific cloned DNA fragments by running the restricted DNA on low melting point agarose gel.
3. Excise the fragment of interest by taking minimum volume of agarose using sterile blade, and purify by gel elution (QIAGEN gel extraction kit).
4. Prepare the radiolabelled probes using labelling kits (e.g., Mega Prime DNA Labelling Kit (Amersham Biosciences) (*see* **Note 15**).

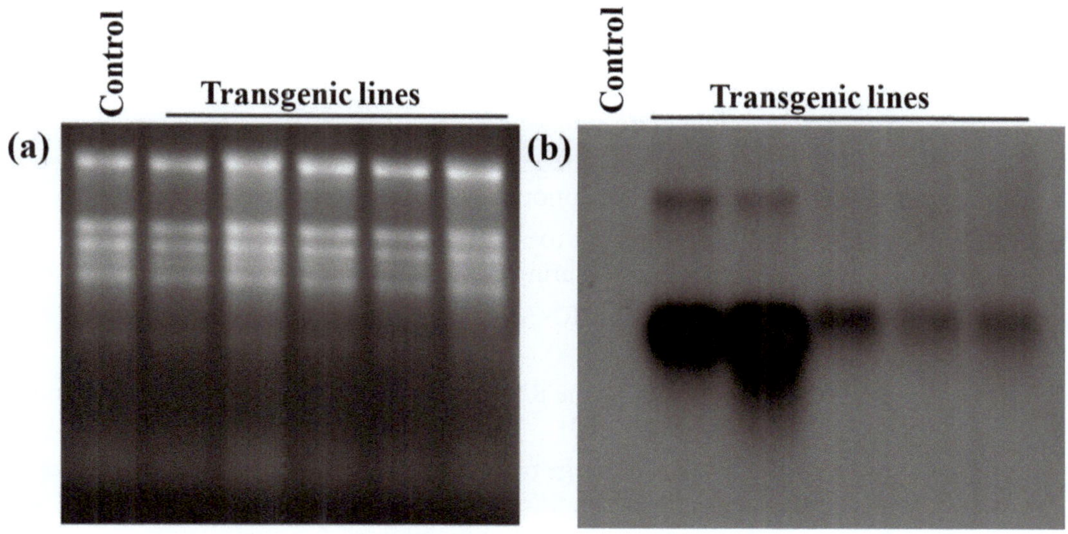

Fig. 1 Northern blot analysis of transgenic cotton. **a** Denature agarose gel (1.2%) electrophoresis of total RNA. **b** Autoradiography of blot shows intensity or quantity expression of the transcripts [9]

5. Required amount of DNA (100 ng per probe) was taken. The labeled DNA can be used by purified using Sephadex G-50 column/PCR purification column.
6. Perform the hybridization by adding probe and incubate it for 24 h at 42 °C (*see* **Notes 16** and **18**).

3.4.3 Washing of the Blots and Autoradiography

1. Wash the blots with the following solutions in the given order till background counts are reduced to two to five counts per second as detected by Mini-Monitor GM tube.
2. 2× SSC and 0.1% SDS for 5 min × 2 times at 42 °C 2; 0.1× SSC and 0.1% SDS for 15 min × 2 times at 42 °C.
3. Remove the membrane from hybridization bottle, and air-dry then wrapped with saran wrap.
4. Expose the blots to the X-ray films (e.g., Kodak Biomax MR) in cassettes with intensifying screen at −70 °C. Develop the autoradiogram after 1–2 days depending on the counts (Fig. 1) (*see* **Notes 17**, **19**, and **20**).

4 Notes

1. The centrifuge tubes, mortar and pestle, and spatula is treated with 0.1% diethyl pyrocarbonate (DEPC) water and autoclaved prior to use.
2. The working area should be cleaned with RNaseZap solution or 70% ethanol.

3. The DEPC is suspected to be carcinogenic and handle with care.
4. Use pre-chilled mortar and pestle to avoid instant thawing and denaturation of sample.
5. Improper homogenization of leaf sample results in incomplete cell wall disruption and leading to negative results.
6. Make sure not to air-dry the precipitated RNA after 5 min that leads to denaturing RNA.
7. High-quality RNA is the primary criteria for an efficient northern blot.
8. To maintain the RNA integrity, RNase-free conditions are most significant [10].
9. Use RNA integrity number (RIN) to assess RNA quality, optimize lysis conditions to minimize RNase contamination, and consider adding RNase inhibitors during isolation.
10. Optimize operating parameters (voltage, time) to achieve good separation, adjust gel concentration based on RNA size range, and make sure samples are loaded properly to prevent overloading if poor RNA separation is evident.
11. To prevent RNA smearing during RNA preparation, check for possible DNA or protein contamination in the sample, and adjust denaturing conditions (formaldehyde concentration).
12. Insufficient RNA while loading in NanoDrop results in false results. Load at least 2 μL of RNA sample for accurate reading.
13. It is necessary to wear gloves, sanitize all equipment with alcohol, and treat all solutions with DEPC before blotting [11].
14. After transferring the gel, it should be illuminated with UV to verify whether any remaining RNA is present in the gel.
15. Designing probes with high specificity to the target transcript, ensuring adequate labeling efficiency, and looking for potential probe secondary structures that can impede hybridization are all ways to prevent low hybridization signals.
16. Optimize hybridization stringency settings (temperature, salt concentration, etc.) to avoid nonspecific binding, and think about employing blocking agents to reduce background noise.
17. In hybridization and washing steps: To avoid excessive background noise, fully pre-hybridize the membrane to restrict nonspecific binding sites. Then, optimize the washing processes to efficiently remove the unbound probe.
18. To prevent a poor signal, improve probe binding by adjusting the hybridization temperature and, if needed, extending the hybridization period.

19. Confirm the presence and expression level of the transgene using other methods like qPCR to ensure a detectable signal with northern blotting.
20. Include a loading control (housekeeping gene) probe to normalize RNA levels across samples, and always include positive and negative controls to monitor experiment quality.

References

1. Bartlett JMS (2002) Approaches to the analysis of gene expression using mRNA. Mol Biotechnol 21:149–160
2. Schlickeiser S, Pleyer U (2007) Western, northern, and southern blotting. In: Basic science and technology in clinical practice. Springer, London, pp 48–57
3. Dvorak Z, Pascussi JM, Modrianský M (2003) Approaches to messenger RNA detection-comparison of methods. Biomed Pap 147: 131–135
4. Green MR, Sambrook J (2012) Molecular cloning. In: A laboratory manual, 4th edn, p 448
5. Bohm-Hofstatter H, Tschernutter M, Kunert R (2010) Comparison of hybridization methods and real-time PCR: their value in animal cell line characterization. Appl Micro Biotech 87:419–425
6. Yang T, Zhang M, Zhang N (2022) Modified northern blot protocol for easy detection of mRNAs in total RNA using radiolabeled probes. BMC Genomics 23(1):66
7. Hayes PC, Wolf CR, Hayes JD (1989) Blotting techniques for the study of DNA, RNA, and proteins. BMJ Br Med J 299(6705):965
8. Gurusamy D (2014) Transcriptome analysis of cotton (*Gossypium hirsutum* L.) during boll development under drought condition and functional characterization of drought tolerance genes. Ph.D. thesis: https://hdl.handle.net/10603/215568
9. Dhandapani G, Kanakachari M, Padmalatha KV, Phanindra MLV, Singh VK, Raghavendrarao S, Kumar PA (2015) A gene encoding cold-circadian rhythm-RNA binding-like protein (CCR-like) from upland cotton (*Gossypium hirsutum* L.) confers tolerance to abiotic stresses in transgenic tobacco. Plant Mol Biol Report 33:22–42
10. He SL (2013) Northern blot. Methods Enzymol 530:75
11. Hoopes LLM (2012) Nucleic acid blotting: southern and northern. Cur Pro Ess Lab Techn 6(1):8–2

Chapter 17

Development of Fungal-Resistant Transgenic Fennel Plant

Abhishek Sharma, Manali Singh, and Lovelesh Singh Saini

Abstract

Fennel (*Foeniculum vulgare*), a key spice crop in India, is valued for its flavor and medicinal properties but faces significant yield losses (15–30%) due to fungal diseases. Traditional breeding for fungal resistance in fennel is slow and limited, while genetic engineering offers a faster and more effective solution. The development of fungal-resistant transgenic fennel plants through genetic engineering is a key approach to minimize this yield loss. Key defense genes such as chitinase can be introduced into fennel to enhance its immune response to fungal pathogens. These genes, sourced from plants and fungi, are essential in degrading fungal cell walls and inhibiting growth. This approach offers a sustainable solution to increasing fennel production and reducing the impact of fungal diseases, ensuring higher yields and meeting rising market demands.

Key words Fungal resistance, Genetic engineering, Defense genes, Crop disease management

1 Introduction

India is well-known as "Land of Spices" across the world for a long time [1]. Fennel, a seed spice, native of Southern Europe and Mediterranean area, is one among them. Civilizations since long are using fennel due to its valuable aroma, taste, flavor, and medicinal properties. The aroma of fennel is due to the presence of a medicinally important compound "Anethole" in the seeds. It has digestive, stomachic, carminative, stimulant, and appetizer properties and used to treat diseases like biliousness, cholera, dysentery, diarrhea, cough, cold, constipation, and ailment of the chest, lungs, and kidney [2].

The main aim of agriculture practices to increase the production to meet the market demand increasing day by day. Major fennel crop varieties, viz., GF1, GF2, GF11, GF12, RF15, RF35, RF101, RF125, etc., are used for commercial agriculture in Gujarat and Rajasthan region and are under sever attack of several fungal pathogen [3]. Out of many, the major fungal-borne diseases responsible for 15–30% Fennel yield loss are damping off of

Nitish Kumar (ed.), *Transgenic Plants: Methods and Protocols*, Methods in Molecular Biology, vol. 2973,
https://doi.org/10.1007/978-1-0716-4803-2_17,

seedlings, Ramularia blight, Cercospora blight, Alternaria blight, Fusarium wilt, Downy mildew, and Powdery mildew caused by fungal pathogen *Pythium aphanidermatum*, *Ramularia foeniculi*, *Cercospora foeniculi*, *Alternaria alternata*, *Fusarium oxysporum*, *Plasmopara nivea*, *Leveillula taurica*, respectively.

Few commercial varieties of fennel demonstrated mildly resistant against fungal infection, viz., GF7/GF9 (leaf blight mildew), RF15 (Alternaria), RF35 (leaf-spot and leaf-blight), RF101, and RF 125 (blight and powdery mildew), however, still suffers a considerable damage to Fennel crop. Also, the exact mechanism behind this mild resistance is not known so far; however, the production of enzymes capable of degrading the cell walls of invading phytopathogenic fungi can be an important component of the defines response in plants. This natural mechanism of defense by fennel can be improved to obtain a superior fungal resistance through transgenic approaches [4].

To improve fungal disease-resistance genetically, plant breeding approaches can also be employed, however, its time-consuming and can take up to 15–20 years. Genetic engineering has the advantage of integrating genes which produce resistance proteins from any species to any crop for improving disease resistance [4]. The role of multiple genes associated with plant immune response after fungus infestation, and the various pathways intricated within were identified. These defense-responsive genes, viz., chitinase, β-glucanase, ribosome-inactivating protein, trichomania, antifungal proteins, along with synthetically prepared cationic peptide-msrA3 can be utilized to produce fungal-resistant transgenic Fennel varieties. These genes have both plant and fungal origin and play a crucial role in different enzymatic reactions leading to the elimination of fungal growth.

2 Materials

1. HEPA filters.
2. 70% ethanol solution,
3. Autoclave.
4. Scalpels and forceps.
5. MS culture media (*see* Table 1).
6. 1 N NaOH.
7. 1 N HCl.
8. *Agrobacterium tumefaciens* LBA4404.

Table 1
Composition of Murashige and Skoog medium

S. no.	Ingredients	Composition of stock solutions	Volume of stock in 1 L medium
1.	*Macronutrients (10×/100 mL)*		
	NH_4NO_3 KNO_3 $MgSO4.7H_2O$	16.5 gm/200 mL 38 gm/400 mL	20 mL 20 mL
		248 mg	5 mL
	KH_2PO_4	6.8 gm/200 mL	
	$CaCl_2$	33.2 gm	
2.	*Micronutrients (100×/100 mL)*		
	$MnSO_4.4H_2O$	892 mg/200 mL	5 mL
	H_3BO_3	248 mg	
	$ZnSO_4$ $CuSO_4.5H_2O$	344 mg	
		5 mg/200 mL	1 mL
	$COCl_2.2H_2O$	5 mg/200 mL	
	$NaMoO_4.2H_2O$	50 mg/200 mL	1 mL
	KI	33.2 mg	
3.	*FeEDTA (100×/100 mL)*		
	Na_2EDTA $FeSO_4.7H_2O$	1.49 gm 1.114 gm/200 mL	5 mL
4.	*Vitamins (100×/100 mL)*		
	Myoinositol	2 gm/200 mL	
	Thiamine HCl	10 mg	
	Nicotinic acid	50.0 mg/100 mL	11 mL
	Pyridoxine HCl	50.0	
	Glycine	200.0 mg	
5.	Sucrose (g)	30.0	–
6.	Agar (g)	0.7%	–
7.	pH	5.8 ± 0.2	–

3 Methods

3.1 In Vitro Regeneration

1. Basal MS medium supplemented with 3% (w/v) sucrose was used for culture initiation [5].
2. The pH of the medium was adjusted to 5.8 with 0.1 N NaOH or 0.1 N HCl before adding 0.7% (w/v) agar.

3. The medium was then sterilized using autoclave at 15–18 lb./inch2 at 121 °C for 15–18 mins. and dispensed into jam bottles (250 mL).
4. The seeds (*see* **Note 1**) were surface-sterilized with 0.1% mercuric chloride (HgCl2) solution for 3–4 min (*see* **Note 2**). and rinsed three to four times (for approx. 2–3 min) with distilled water. Water was removed completely, and seeds were dried by placing them on the sterile paper towel [6].
5. Using sterilized forceps, the surface-sterilized seeds were inoculated onto the half-strength MS media without any growth regulator in jam bottles.
6. The cotyledonary petioles and hypocotyls were excised from 5-day-old seedlings using sterilized scalpel.
7. Cotyledonary petioles were planted on the callus induction medium with their abaxial surface in contact, and hypocotyls were placed horizontally on the media. While plating the explants, it was ensured that the cut end of the petiole and hypocotyl remained either in contact or slightly immersed in the media.
8. MS medium here was augmented with different concentrations of (BAP 1–5 mg/L and NAA 0.1–0.5 mg/L) for callus induction (*see* **Note 3**).
9. Calli so induced were subcultured on MS basal medium for somatic embryogenesis in every 2 weeks (*see* **Note 4**).
10. Somatic embryos obtained from calli were transferred to MS medium supplemented with BAP and NAA (mg/L) for shoot induction.
11. Ethylene inhibitor AgNO3 (0.5 mg/L) was also added to the medium to induce multiple shoots.
12. The shoots obtained were excised from callus and transferred for shoot elongation to the MS basal medium supplemented with the same combinations of growth regulators used for shoot induction.
13. Shootlets of length 2–4 cm were then transferred for root induction to the MS basal medium and were supplemented with (0.25–1.0 mg/L) IBA for the root initiation in the regenerated shootlets.

3.2 Maintenance of Cultures

1. All the cultures obtained from tissue culture and transformation experiments were maintained at 25 ± 2 °C, 60% relative humidity.
2. These cultures were subcultured to fresh medium after every 2 weeks.

3. Agrobacterium strain harboring pCAMBAR.chi11 was maintained on yeast extract peptone (YEP) medium containing 50 mg/L kanamycin sulfate with continuous shaking at 28 °C and 180 rpm in a shaker for 2 days (*see* **Note 5**), while fresh YEP plates streaked from stock plates can be stored up to 1 month at 4 °C.

3.3 Agrobacterium-Mediated Transformation

1. The explants were co-cultivated with bacterial suspensions of different OD values at 600 nm (UV Spectrophotometer, Shimadzu) (0.6–1.2) [7].
2. Single colony of Agrobacterium strain LBA4404 harboring pCAMBAR.chi11 was taken from a freshly streaked plate and was inoculated into 50 mL YEP liquid medium having 50 mg/L kanamycin and was incubated on shaker for 48 h at 28 °C, and so obtained fresh culture was used for transformation.
3. The cotyledonary petioles and hypocotyls were excised from 5-day-old seedlings with help of sterilized scalpel and were used as explants here [8].
4. To determine the effect of preculture on transformation frequency, cotyledonary petioles and hypocotyls were cultured for 24–72 h on a regeneration medium before co-cultivation with Agrobacterium [8].
5. A single colony of *A. tumefaciens* strain LBA4404 containing pCAMBAR.chi11 was cultured in 10 mL liquid YEP medium containing 50 mg/L kanamycin sulfate at 28 °C overnight with continuous shaking at 150 rpm.
6. Optical density of the culture was adjusted to O.D.600 (0.6–0.8). Cultured bacteria were collected by centrifugation at 5000 rpm for 5 min and were suspended in 15 mL of *Agrobacterium*-infection medium (MS medium supplemented with 100 μM acetosyringone).
7. The explants were incubated in 15 mL *Agrobacterium*-infection medium (with the cut end of explants immersed in Agrobacterium infection medium) for 15 min at 28 °C with gentle shaking in dark conditions [6].
8. The infected cotyledonary petioles and hypocotyls were then blotted dry on sterile filter paper and afterward co-cultivated on solid regeneration medium supplemented with 100 μM acetosyringone (*see* **Note 8**) for 1–3 days and were maintained at 25 ± 2 °C in dark conditions [6].
9. The cotyledonary petioles and hypocotyls were washed two to three times for 5 min each in sterile water containing 250 mg/L cefotaxime and then blotted dry on sterile filter paper [6].

10. The explants were then placed on selection medium supplemented with 2 mg/L AgNO3, 250 mg/L cefotaxime, and 20 mg/L hygromycin (*see* **Notes 6**, **7**) and were maintained at 25 ± 2 °C under 16 h light photoperiod.
11. The explants were subcultured onto fresh selection medium every fortnight until shoot buds developed (*see* **Note 9**).
12. Selected hygromycin-resistant shoots of 2–4 cm in length have apparent leaves, nodes, internodes, and apical buds were then transferred to rooting medium (MS medium supplemented with indole-3-butyric acid) containing 20 mg/L hygromycin, 250 mg/L cefotaxime.
13. Regenerated plantlets with well-developed root system were transferred to plastic cups having autoclaved garden soil, farmyard soil, and sand in the ratio of 2:1:1 and grown for 15 days.
14. These plantlets were irrigated with 1/8 MS basal salt solution destitute of sucrose and myoinositol after every 2 days for 1 week.
15. The plantlets were covered with porous polyethylene sheets for maintaining high humidity and were kept inside the culture room conditions.
16. Acclimatized plantlets were then transferred to soil in earthen pots to allow development of the root system and cultured in the greenhouse until mature seeds were harvested.

4 Notes

1. Change the seed lot if it is an old one and take only intact healthy seeds.
2. Surface sterilization (with 70% ethanol) of seed should not be done for more than 90 s.
3. Use freshly prepared plant hormones during preparation of callus induction medium.
4. Maintain the inoculated seeds in the Petri plates in dark during callus induction period.
5. Do not grow the *Agrobacterium* beyond an O.D600 of 1.0 in YEP medium.
6. Use freshly prepared cefotaxime for selection medium preparation and during washing.
7. Add cefotaxime and hygromycin in the culture medium after autoclaving once its temperature reaches around 48 °C.
8. Always use freshly prepared acetosyringone.
9. Avoid shifting brown or black calli during subsequent transfer to selection medium.

References

1. Kumawat R, Singh D, Kumawat KR, Kumawat S, Choudhary M (2022) Genetic variability, heritability and genetic advance studies in coriander (Coriandrum sativum L.) under normal and staggered moisture regimes. Pharm Innov J 11(2):2033–2036
2. Noreen S, Tufail T, Ain HBU, Awuchi CG (2023) Pharmacological, nutraceutical, functional and therapeutic properties of fennel (foeniculum vulgare). Int J Food Prop 26:915. https://www.semanticscholar.org/paper/ff2a939eb9c5f3f45a2633b55715a2b151986a43
3. Филипюк ОМ, Вишневська Л (2022) Study of some pharmacotechnological, physicochemical and pharmacognostic properties of common fennel (Foeniculum vulgare) fruit. Farmatsevtychnyi Zhurnal:84. https://www.semanticscholar.org/paper/b844942c56651a3265aa6801390c78ceb6055d2f
4. Ceasar AS, Ignacimuthu S (2012) Genetic engineering of crop plants for fungal resistance: role of antifungal genes. Biotechnol Lett 34:995–1002. https://doi.org/10.1007/s10529-012-0871-1
5. Murashige T, Skoog F (1962) A revised medium for rapid growth and bio assays with tobacco tissue cultures. Physiol Plant 15(3):473–497
6. Singh VV, Verma V, Pareek AK, Mathur M, Yadav R, Goyal P, Kumar S (2009) Optimization and development of regeneration and transformation protocol in Indian mustard using lectin gene from chickpea [Cicer arietinum (L.)]. J Plant Breed Crop Sci 1(9):306–310
7. Sambrook J, Fritsch ER, Maniatis T (1989) Molecular cloning: a laboratory manual, 2nd edn. Cold Spring Harbor Laboratory Press, Cold Spring Harbor, NY
8. Gasic K, Korban SS (2006) Indian mustard [Brassica juncea (L.) Czern.]. Methods Mol Biol 343:281–289. https://doi.org/10.1385/1-59745-130-4:281

Chapter 18

Agrobacterium-Mediated Genetic Transformation in Ashwagandha (*Withania somnifera*)

Manali Singh, Shivani Bhutani, Lovelesh Singh Saini, Sumit Purohit, and Nisha Dinkar

Abstract

Withania somnifera (Ashwagandha) contains pharmacologically active metabolites such as withaferin A and withanolide A, and it has been considered one of the most important medicinal plants in the Indian Ayurvedic system. However, the commercial demand is too large, and the concentration of these bioactive metabolites in plants is too low. The goal of the current work is to produce these bioactive metabolites more efficiently by using a biotechnological approach. The present chapter focuses on optimizing the techniques for *Agrobacterium* rhizogenes-mediated transformation of *Withania somnifera* using different tissues, viz., root, stem, and leaves. *Agrobacterium rhizogenes* is a great resource for researching gene function, plant-microbe transformation, and increasing secondary metabolite production. In plant biotechnology, *A. rhizogenes* offers creative approaches to functional genomics and genetic manipulation, making it a promising field. Scientists work to understand the molecular processes underlying the change induced by *A. rhizogenes*, the bacterium of agricultural benefits.

Key words Withanolides, *Withania somnifera*, Ashwagandha, *Agrobacterium rhizogenes*

1 Introduction

A. rhizogenes is a member of the *Rhizobiaceae* family of soil-borne bacteria. *A. rhizogenes*, like its close sibling *Agrobacterium tumefaciens*, is well-known for its capacity to introduce genetic material into plant cells and induce the development of mutated tissues. Hairy roots are induced to grow by *A. rhizogenes* [1]. *A. rhizogenes* enters plant cells by transferring T-DNA, a particular section of its DNA. Hairy roots arise as a result of adventitious roots being formed at the infection site by the genes transmitted from the Ri plasmid. Hairy roots with uncontrollably large and branched roots are the outcome of *A. rhizogenes* transformation of plant cells. The genes that produce plant hormones are found on the Ri plasmid of *A. rhizogenes*. These genes produce plant hormones found on the

Nitish Kumar (ed.), *Transgenic Plants: Methods and Protocols*, Methods in Molecular Biology, vol. 2973,
https://doi.org/10.1007/978-1-0716-4803-2_18,

Ri plasmid of *A. rhizogenes* like auxins and cytokinins, which play crucial roles in root development [2]. The enzymes involved in the manufacture of opiates, which provide the bacteria with nourishment, are likewise encoded by the transferred genes. In plant biotechnology, *A. rhizogenes* has been used to create transgenic plants with enhanced root systems. *A. rhizogenes* hairy root cultures are used to produce secondary metabolites because they show higher bioactive compounds productivity [3].

We have used *A. rhizogenes* to optimize the transformation conditions of *Withania somnifera* in the current investigation. *A. rhizogenes* and its interactions with plants have been extremely helpful in understanding the molecular processes involved in plant genetic transformation. Researchers examine plant gene expression and regulation through the use of *A. rhizogenes*-mediated transformation as a tool for functional genomics. Plant genetic engineering and the creation of transgenic plants for a variety of uses, such as agriculture and the manufacture of pharmaceuticals, have greatly benefited from our growing understanding of the biology and applications of *A. rhizogenes* [4]. Withanolides are a class of naturally occurring chemicals mostly found in plants belonging to the Solanaceae family. They are classified as steroidal lactones. Known by many as Indian Ginseng, Ashwagandha or *Withania somnifera* is widely used medicinally. It has been found that *A. rhizogenes* is a helpful for increased production of withanolides. T-DNA is inserted into *Withania somnifera* cells using *A. rhizogenes.* The genes on the Ri plasmid cause the plant to develop hairy roots [5]. The bacterial genes are carried by these genetically altered roots. The plant's metabolism is altered as a result of *A. rhizogenes* genetic transformation, producing more withanolides. Compared to non-transformed roots, the hairy roots function as a bioreactor, generating higher concentrations of bioactive chemicals. Compared to conventional plant cultures, *A. rhizogenes* hairy root cultures offer a number of benefits. They can be cultivated in vitro, which offers a regulated setting for the synthesis of withanolides [6].

The biotechnological synthesis of withanolides for pharmacological and therapeutic uses can benefit from the utilization of *A. rhizogenes*-mediated transformation in *Withania somnifera* [7–10]. Withanolides are useful for medical applications because of their diverse pharmacological characteristics, which include anti-inflammatory, antioxidant, and anticancer effects. *A. rhizogenes* is frequently used by researchers as a tool to investigate the control of the genes involved in withanolides biosynthesis and to improve production conditions. High-yielding plant lines are developed through the use of *A. rhizogenes* in genetic engineering techniques to discover important genes involved in the withanolide biosynthesis pathway. To summarize, the process of using *A. rhizogenes* to produce withanolides entails transforming *Withania somnifera* hairy roots, which leads to a sustained and increased synthesis of these beneficial bioactive compounds. This biotechnological method may offer a more effective and regulated system for increased withanolide production, such as withaferin A, withanolide A, etc. a strong anticancer agent, with predominant therapeutic and rejuvenating qualities.

2 Materials

Yem broth, MS media, agarose, needles, spectrophotometer, orbital shaker, HPLC, cefotaxime, distilled water, thermo cycler, isopropanol, 70% ethanol, TE buffer, 1X TAE buffer.

3 Methods

Transgenic root cultures are created by co-cultivating *A. rhizogenes.* Two strains of *A. rhizogenes* strains (15834 and A4) were used for the transformation of 1-month-old seedlings of elite varieties (Jawahar-20 and Poshita) of *Withania somnifera* obtained from plant tissue culture in order to observe the effect of transformation on the biosynthesis of bioactive contents of withanolide A/withaferin A and root biomass. After 6 months, the transformed roots were harvested, their fresh and dry weights were recorded, and the contents of withanolide A and withaferin A were quantified using high-performance liquid chromatography (HPLC). The altered roots were then further subcultured on root proliferating media.

3.1 Preparation of YEM Broth

In order to prepare a fresh bacterial culture of *A. rhizogenes* (A4 and 15834).

1. A loop from the glycerol stock of bacterial culture was taken in inoculation loop and placed in 250 mL of YEM broth (*see* **Notes 1**, **6**, **7**, and **8**).
2. The infected broth was kept in an orbital shaker at 100 r.p.m. for 48 h at 28 °C in the dark.
3. At 600 nm, the broth's OD was measured.
4. When the broth's OD was less than 1, it was utilized to transform various in vitro-grown explants, such as the leaves, nodes, and roots of two promising varieties of *Withania somnifera*, namely, Jawahar and Poshita.

3.2 Procedure of Transformation of Withania somnifera Using A. rhizogenes

1. The leaves, nodes, and roots of elite cultivars of *Withania somnifera* (Poshita and Jawahar-20) that were grown in vitro for 1 month were carefully removed from the seedlings and injected all over with a needle before being placed in YEM

broth that included a culture of *Agrobacterium rhizogenes* (A4 and 15834) (*see* **Notes 2**, **3**, and **8**).

2. The explants were maintained in an orbital shaker at 28 °C and 100 r.p.m. for varying durations of 10, 15, 20, and 30 min.
3. The explants were removed from the broth at various intervals and placed on MS half media at 25 °C in the dark.
4. Following a 5-day period, the explants were moved to MS medium supplemented with 250 mg/L of cefotaxime to ensure that *Agrobacterium rhizogenes* did not continue to contaminate the media and the explants.
5. Additionally, the concentration of cefotaxime was progressively decreased to 150 mg/L, 100 mg/L, 50 mg/L, and so forth throughout subculturing.
6. Eventually, the explants were moved to MS media containing IAA 0.5 ppm and IBA 2.0 ppm once there was no longer any contamination in the subsequent subcultures.
7. Later, once the hairy roots had developed, the roots were carefully removed and put on an orbital shaker in MS (liquid) media containing IAA 0.5 ppm and IBA 2.0 ppm to encourage further root growth.
8. Following a 6-month period, the roots were removed, cleaned with double-distilled water to eliminate any remaining media residue, and their fresh and dry weights were determined.
9. HPLC was then used to quantify the amount of withanolide and withaferin. Additionally, the transformed roots for molecular characterization were examined to validate.

4 Molecular Characterization of *Withania somnifera* Plants Transformed by *Agrobacterium rhizogenes*

Genomic DNA was extracted from *A. rhizogenes* transformed and untransformed lines, and PCR analysis was used to confirm the transformation at the molecular level.

4.1 Isolation of Genomic DNA

1. With a few modifications, the CTAB technique was used to isolate genomic DNA from both transformed and untransformed lines.
2. After washing the roots and leaves in 90% ethanol, the water was removed by drying them with autoclaved Whatman filter paper.
3. The samples were then ground in a sterile, pre-cooled mortar and pestle using liquid nitrogen.

4. 1 g tissue sample was placed into centrifuge tubes along with 2 mL of CTAB extraction buffer that had been gently warmed for 10–15 min at 65 °C and thoroughly vortexed.
5. After that, the tubes were incubated for an hour at 65 °C in a water bath with vortexing.
6. After 5 min of room temperature cooling, the tubes were centrifuged for 15 min at 1200 rpm.
7. One milliliter of the supernatant was transferred to a wide-bore pipette-equipped Eppendorf tube, and 500 μL of phenol, chloroform, and isoamyl alcohol were added in a 25:24:1 ratio.
8. For 5 min, the mixture was gently shook with inversions.
9. After that, the mixture was centrifuged for 15 min at 1200 rpm.
10. After removing the supernatant, 24:1 chloroform and isoamyl alcohol were added.
11. After that, the mixture was centrifuged once again for 10 min at 1200 rpm.
12. After adding equal amounts of pre-chilled isopropanol to the supernatant, the combination was stored at −20 °C for 2 h.
13. The solution was centrifuged once again for 8 min at 8000 rpm after 2 h.
14. After discarding the supernatant, 300 μL of 70% ethanol was added.
15. After that, the mixture was centrifuged for 8 min at 8000 rpm.
16. The ethanol was disposed of after centrifugation, and the pellet that was left at the bottom of the tube was dissolved in 100 μL of TE buffer and kept for later use at −20 °C.

4.2 Genomic DNA Quantification

1. The direct technique of quantifying genomic DNA involved measuring the absorbance at 260 nm (OD 260).
2. Using a UV-visible spectrophotometer, the absorbance of DNA samples was measured against TE buffer as a blank.
3. The following formula was used to determine the concentration of DNA based on the optical density:

   A_{260}X50XDilution factor/1000 = Conc. in μg/μL.
4. The amount of RNA or protein in the DNA sample is shown by the ratio A 260/A 280.
5. The ideal ratio for DNA is 1.8; a result of 1.8 or higher indicates the presence of RNA in the sample, while a value of 1.8 or less suggests the presence of protein (*see* **Notes 9** and **10**).

4.3 Primer Preparation

1. All of the sequences for the rol C, rol B, TR DNA, TLT DNA, vir C, and vir D genes were examined for multiple sequence alignment using Clustal W in order to confirm the primer sequence.
2. Primer 3 was used to design the primer after the conserved region was determined. The hair pin loop and self-complementarity of the primers were further examined with the use of the Oligo calc tool.

4.4 Amplification by PCR

1. By examining the amplified DNA fragment for the presence of the rol C, rol B, TR DNA, TLT DNA, vir C, and vir D genes, the transformation was confirmed.
2. A separate set of primers was employed for amplification. Untransformed plant genomic DNA was utilized as the negative control [11].
3. A 50 μL reaction volume was used for the PCR amplification (Tables 1 and 2).

4.5 Agarose Gel Preparation

1. 1.5 g of agarose was dissolved in 100 mL of either 1X TAE or 0.5X TBE solution to create an agarose gel 1.5% [12] (*see* **Notes 4** and **5**).
2. EtBr was added to the solution at a rate of 1.5 μg/μL. The DNA loading dye and sample were combined in a 25:2 v/v ratio for each well.
3. A micropipette was used to load the DNA sample. In 1X TAE / 0.5X TBE electrophoresis buffer, the electrophoresis was carried out at 50 V for 4 h.
4. The gel's image was examined and saved in the gel Doc system.

Table 1
The composition of reaction mix for the PCR analysis

Components	Concentration	Volume taken in a single tube (μl)
DNA template	200 μg/μL	1.0
dNTPs	10 mM each	1
Taq polymerase(10 μL)	0.5 U	1
$MgCl_2$	25 mM	2.5 μL
Reaction buffer	10X	2.5
Forward primer		1.0
Reverse primer		1.0
ddH_2O		40
Total volume		50 μL

Table 2
The details of the PCR cycle conditions

	Denaturation		Annealing		Extension	
Cycle	**Temp.**	**Time**	**Temp.**	**Time**	**Temp.**	**Time**
1st cycle	94 °C	5 min	–	–	–	–
35 cycles	94 °C	30 s	37 °C	30 s	72 °C	30 s
Final cycle	–	–	–	–	72 °C	7 min

5 Notes

1. Properly measure the reagents, and add accordingly as per the calculated value.
2. Always use fresh and sterilized needles for injecting.
3. While inoculating the seedlings into the YEM broth, make sure to handle it carefully.
4. Make sure to use only freshly prepared buffers to avoid any sort of contamination and errors.
5. While doing agarose gel electrophoresis, be careful with the handling EtBr as it is highly carcinogenic. So handle it by wearing gloves.
6. While preparing yem broth, please avoid freezing and overheating.
7. Make sure to check the expiry dates for all the chemicals before use.
8. Always check for the physical signs of deterioration; do not use the reagents if it shows the signs of discoloration, drying, microbial contamination, or any other sign of deterioration.
9. Key points to consider while preparing sample for genomic DNA quantification are:
   - *Homogeneity*: Thoroughly mix your DNA samples to ensure even distribution before measurement.
   - *Dilution*: if the concentration of DNA is too high, dilute it appropriately to fall within the detection range of your spectrophotometer.
   - *Blank measurement*: always use a blank containing the same buffer which you are using to dissolve your DNA to account for background absorbance.
10. Be careful of potential contaminants like proteins or residual chemicals which can affect your readings.

Acknowledgments

We gratefully acknowledge the Centre for Research and Development of Parul University, for providing the financial support in carrying out this research work.

References

1. Yamazaki Y, Kitajima M, Arita M, Takayama H, Sudo H, Yamazaki M (2004) Biosynthesis of camptothecin: In silico and in vivo tracer study from [1 13C] glucose. Plant Physiol 134:161–170
2. Keeling PJ, Palmer JD (2008) Horizontal gene transfer in eukaryotic evolution. Nat Rev Genet 9:605–618
3. De Boer KD, Lye JC, Aitken CD, Su AK, Hamill JD (2009) The A622 genein *Nicotiana glauca* (tree tobacco): evidence for a functional role inpyridine alkaloid synthesis. Plant Mol Biol 69:299–312
4. Ono NN, Tian L (2011) The multiplicity of hairy root cultures: prolific possibilities. Plant Sci 180:439–446
5. Ozyigit II, Dogan I, Tarhan EA (2013) *Agrobacterium rhizogenes*-mediated transformation and its biotechnological applications in cropsof rhizodeposits in structuring rhizosphere bacterial communities. FEMS Microbiol Ecol 1903: 75–84
6. John J (2014) Therapeutic potential of *Withania somnifera*: a report on phyto pharmacological properties. Int J Pharm Sci Res 5:2131–2148
7. Singh M, Poddar NK, Singh D, Agrawal S (2020) Foliar application of elicitors enhanced the yield of withanolide contents in *Withania somnifera* (L.) Dunal (variety, Poshita). 3 Biotech 10:157
8. Singh M, Shah P, Punetha H, Agrawal S (2018) Varietal comparison of withanolide contents in different tissues of *Withania somnifera* (L.) Dunal (ashwagandha). Int J Life Sci Sci Res 4:1752–1758
9. Singh M, Shah P, Punetha H, Gaur AK, Kumar A, Agrawal S (2017) Isolation and quantification of a potent anti cancerous compound, Withaferin A from the aerial parts of *Withania somnifera* (Ashwagandha). Ad In Plant Sci 30:231–235
10. Singh M, Suyal DC, Dinkar N, Joshi S, Srivastava N, Maurya VK, Agnihotri A, Agrawal S (2022) *Agrobacterium rhizogenes* as molecular tool for the production of hairy roots in *Withania somnifera*. J App Biol Biotech:1–10
11. Bandyopadhyay M, Jha S, Tepfer D (2007) Changes in morphological phenotypes and withanolide composition of Ri-transformed roots of *Withania somnifera*. Plant Cell Rep 26:599–609
12. Sambrook J, Fritsch EF, Maniatis T (1989) Molecular cloning: a laboratory manual. Cold Spring Harbour Laboratory Press, Plainview

Chapter 19

Agrobacterium Infection in Cotton for the Development of Transgenic Cotton Plants

Masoud Tohidfar

Abstract

Cotton (*Gossypium hirsutum* L.) is an important fiber crop in the world. Abiotic and abiotic stresses are the main challenges of yield reduction in cotton. Traditionally, four to six times of spraying per growing season of environmentally harmful chemical insecticides is used for pest control. There are various methods for producing transgenic cotton, among which the *Agrobacterium tumefaciens* method is widely used. In order to produce transgenic cotton tolerance to stresses, explants are transformed with *Agrobacterium tumefaciens*. Also, selectable marker gene is used as a marker. Inoculated tissue sections are placed onto co-cultivation medium. Transformed tissues are selected on culture medium containing suitable antibiotic. Plantlets are subsequently regenerated from putative transgenic explants. In this chapter, the steps of genetic transformation of cotton by *Agrobacterium* method using embryogenic calli have been completely explained.

Key words *Agrobacterium tumefaciens*, Transgenic, Cotton, Explants, Transformation, Embryogenic calli

1 Introduction

Genetic transformation is a technology used for improving traits in plants. Among the available methods, *Agrobacterium tumefaciens* is widely used for genetic transformation and gene transfer in plants, especially in dicotyledonous [1]. *Agrobacterium* is a gram-negative, soil-borne bacterium that causes crown gall disease in plants [2]. *Agrobacterium* contains Ti plasmid on which T-DNA and *vir* genes are located. The T-DNA, which means transfer DNA, are transferred to the genome of the wounded plant cell with the help of *vir* genes. The T-DNA contains genes that cause crown gall disease in plants. *Agrobacterium* attaches to wounded plant cells and finds an appropriate signal [3], and then *vir* genes are activated with a plant phenolic metabolite [4, 5]. Subsequently, T-DNA is excised by nucleases and enters into the plant genome [6, 7]. *Agrobacterium* ability to transfer DNA to the plant allows it to introduce

Nitish Kumar (ed.), *Transgenic Plants: Methods and Protocols*, Methods in Molecular Biology, vol. 2973,
https://doi.org/10.1007/978-1-0716-4803-2_19,

useful genes into the plant without causing tumors [8]. One of the main limitations of *Agrobacterium*-mediated gene transfer in cotton is that only a limited number of cotton genotypes have regeneration ability [9]. In this method, the integration of several copies of the desired gene is relatively common, although this method is the most efficient method to create a single copy integration [7]. Another challenge of *Agrobacterium*-mediated transformation is the presence of *Agrobacterium* in the tissue, which leads to systemic contamination [10]. Nevertheless, the *Agrobacterium* method is an appropriate method for genetic transformation and manipulation in cotton. So far, many genes have been transferred to cotton for resistance and tolerance to biotic and abiotic stresses [1, 12–17]. The cultivated area of transgenic cotton commercialized in the world reaches more than 20 million hectares. All these were produced by *Agrobacterium*. [11]. Most of *Agrobacterium* gene transfer protocols included explant infection, induction of embryogenic callus, screening of transformed calli, shooting, and plant regeneration [16].

2 Materials

2.1 Plant Material

1. Seeds of upland cotton (*Gossypium hirsutum* L.) cultivar Coker 312 (*see* **Note 1**).
2. Sodium hypochlorite.
3. Sterile distilled water.
4. Murashige and Skoog [18].
5. Ethanol.
6. Whatman No. 1 filter paper.
7. Sterile double-distilled water.
8. Seed germination medium (SGM): sucrose 30 g/L, MS medium, 100 mg/L of myoinositol, 0.5 mg/L of thiamin-HCl, 0.5 mg/L of nicotinic acid, 0.5 mg/L of pyridoxine HC (*see* **Note 2**).

2.2 Agrobacterium tumefaciens *Strain*

1. *Agrobacterium tumefaciens* strain LBA4404 harboring the recombinant binary vector pBI121-chi containing the *chi* gene insert and neomycin phosphotransferase (*nptII*) gene, as selectable marker (*see* **Note 3**).

2.3 Agrobacterium *Culture Media*

1. Luria-Bertani (LB): 20–25 g/L medium.
2. Agar: 12–15 g/L.
3. Kanamycin sulfate: 50 mg/L (*see* **Note 4**).
4. Rifampicin: 75 mg/L (*see* **Note 5**).

2.4 Infection of Explants with Agrobacterium

1. Whatman No. 1 filter paper.
2. Petri dish.
3. Scalpel.
4. Forceps.

2.5 Co-cultivation Medium (CCM)

1. CCM: Murashige and Skoog (MS) salts, B5 vitamin, 0.75 g/L $MgCl_2$, 0.46 μM kinetin, 0.45 μM (2,4Dichlorophenoxyacetic), 30 g/L glucose, pH 5.8, 0.2% (w/v) phytagel.
2. Acetosyringone: 30–40 mg/L (*see* **Note 6**).
3. DMSO.
4. 0.4 NaoH.
5. 1N HCl.

2.6 Selection of Transformants

1. Kanamycin sulfate: 25–50 mg/L.
2. Cefotaxime: 250–200 mg/L (*see* **Note 7**).
3. Callus induction medium: MS salts, vitamin B5, 0.75 g/L $MgCl_2$, 30 g/L glucose, 0.46 μM kinetin, 0.45 μM. (2,4Dichlorophenoxyacetic), pH 5.9, 0.2% (w/v) phytagel, 50 mg/L kanamycin, and 250 mg/L cefotaxime (*see* **Note 8**).
4. Embryo maturation medium: MS salts, vitamin B5, 0.75 g/L $MgCl_2$, 1.9 g/L KNO3, 0.25% (w/v), 30 g/L glucose, pH 5.8, 0.2% (w/v) phytagel, 50 mg/L kanamycin, and 250 mg/L cefotaxime.
5. Shooting medium: MS salts, 100 mg/L of myoinositol, 0.5 mg/L of thiamin-HCl, 0.5 mg/L of nicotinic acid, 0.5 mg/L of pyridoxine HCl, 3% (w/v) sucrose, pH 5.7, 0.15% (w/v) Gelrite, 25 mg/L of kanamycin, and 200 mg/L cefotaxime.
6. Rooting Medium: MS salts, 100 mg/L myoinositol, 0.5 mg/L thiamin-HCl, 0.5 mg/L nicotinic acid, 0.5 mg/L pyridoxine HCl, 3% sucrose, pH 5.8, 0.15% (w/v) Gelrite, and 200 mg/L cefotaxime (*see* **Note 9**).

2.7 Tissue Culture Media

1. B5 vitamins [19].
2. 0.2% (w/v) phytagel (*see* **Note 10**).
3. 3% (w/v) sucrose.
4. 30 g/L glucose.
5. 2, 4-D. Store at 4 °C.
6. Kin. Store at −20 °C.
7. ZT. Store at −20 °C.
8. $MgCl_2$.

9. KNO_3.
10. 0.15% (w/v) Gelrite.

2.8 Regeneration

1. MS salts including vitamins.
2. Pots.

2.9 Other Materials

1. Calcium chloride (*see* **Note 11**).

2.10 Equipment

1. 50 mL centrifuge Falcon.
2. Plant growth chambers.
3. Laminar flow hood Class II.
4. Shaker incubator.
5. Boiling water bath.
6. pH and Mv meter.
7. Microcentrifuge.
8. Laminar airflow Class 1 hood.
9. Phytotron or growth chambers.

3 Methods

3.1 Prepare the Medium of SGM

1. Adjust the pH of the SGM to 5.8 with 0.4 NaOH or 1N HCl (*see* **Note 12**).
2. Add 8% agar to medium.
3. Autoclave medium under121°C for 15 min.
4. Aliquot medium to test tubes, after the medium temperature reaches 50–60 °C.

3.2 Seed Disinfection

1. Place the seeds in sterile water for 24 h.
2. Remove the skin of the seed or make a scratch on it.
3. Soak seeds in 70% ethanol for 1 min.
4. Wash seeds with sterile distilled water for three times.
5. The seeds are surface-sterilized with 0.1% v/v sodium hypochlorite for 10 min.
6. Wash the seeds three times with sterile distilled water (*see* **Note 13**).
7. Place the seeds on a sterile paper towel to dry.
8. Place seeds in the glass jar containing the SGM, and keep in growth chambers under 16 h day/8 h night cycle with a light intensity of 2500 lux at 26 ± 2 °C. (*see* **Note 14**).

3.3 Growth of Agrobacterium LBA4404

1. Prepare culture medium LB.
2. Place 5 mL LB medium to Falcon 50 mL.
3. Autoclave under 121 °C for 15 min.
4. Add 50 mg/L kanamycin and 75 mg/L rifampicin to the cooled medium (50–60 °C) (*see* **Note 15**).
5. Inoculate a fresh single colony from the plate (*Agrobacterium tumefaciens* strain harboring the recombinant binary vector pBI121-chi) with 5 mL of LB liquid medium containing appropriate antibiotics under shaking (180 rpm) at 28 °C for 24–48 h (*see* **Note 16**).
6. Next day, dilute the overnight culture 1:100 with LB broth along with antibiotic and acetosyringone and shake at 28 C until the OD_{600} reaches 0.6–0.8 (*see* **Note 17**).

3.4 Prepare CCM

1. Prepare CCM medium that contains MS salts and B5 vitamin.
2. Add 0.75 g/L $MgCl_2$, 0.46 μM kinetin, and 0.45 μM (2,4Dichlorophenoxyacetic) into the medium.
3. Add 30 g/L glucose into the medium.
4. Adjust the pH of the medium to 5.9 with 0.4 NaoH or 1 N HCl.
5. Solidify medium with 0.2% (w/v) phytagel.
6. Autoclave under 121 °C for 15 min.
7. Aliquot the cooled medium (50–60 °C) to Petri dish 10 cm. Each Petri dish contains 25 mL medium.
8. Keep Petri dish in 4 °C.

3.5 Infection of Explants with Agrobacterium

1. Prepare hypocotyl pieces (0.5 cm^2 surface area) from 7 to 10 day old sterile seedlings (*see* **Note 18**).
2. Immerse hypocotyl pieces (explants) in *Agrobacterium* suspension ($OD_{600} = 0.6$–0.8).
3. Explants are gently shaken in the bacterial suspension for 5–10 s.
4. Ensure contact of all hypocotyl edges with bacterial cells.
5. Place the hypocotyl pieces on a Whatman No. 1 filter paper to dry (*see* **Note 19**).
6. Transfer the hypocotyl pieces to CCM without antibiotics.
7. Incubate for 2 days at 26 ± 2 °C in the dark.

3.6 Selection of Transformants and Regeneration of Plantlets

1. After co-culture, explants are transferred to callus induction medium and are incubated at 26 ± 2 °C with a 16 h photoperiod (90 μmol/m^2/s).
2. Subculture is done every 15 days.

3. After 3–4 weeks, calli are placed onto embryo maturation medium.
4. Mature embryos are picked up and transferred to embryo germination medium (*see* **Note 20**).
5. Germinated somatic embryos are placed in shooting medium in 500 mL jars.
6. Subculture is done every 15 days.

3.7 Rooting

1. Germinated somatic embryos are placed in 500 mL jars containing rooting medium.
2. After 3 weeks, roots are observed.
3. Prepare pots for rooted plantlets.
4. Prepare pots containing a (1:1:1 v/v/v) mixture of garden soil, compost and perlite (*see* **Note 21**).
5. Sterile pots.
6. The roots of the plants are washed with tap water to remove the remains of the culture medium.
7. Individual rooted plantlets are transferred to pots.
8. Pots are covered with polyethylene bags.
9. Spray the plants three times a day with sterile water.

4 Notes

1. Tissue culture and regeneration in cotton cultivars is difficult and depends on the genotype. The Coker is variety that easily responds to tissue culture. The plants resulting from the hybrid with Coker also respond to tissue culture.
2. Sometimes, instead of MS medium, 1/2 MS medium is used for seed germination. Adding vitamins is optional.
3. Different strains of *Agrobacterium* (LBA4404, AGL1, and EHA105) are used for transformation of cotton. Among them, LBA4404 has high efficiency and is commonly used in cotton. It is difficult to control strain EHA105 in the medium, and it is less used in transformation of cotton.
4. One gram of kanamycin is added to 10 cc of sterile H_2O and dissolved completely. Sterilize kanamycin stock through the prepared 0.22 μm syringe filter. Stock kept at −20 °C.
5. A stock solution is made by dissolving the rifampicin in the solvent of metanol. 100 mg rifampicin is added to 5 mL methanol; thus vigorously vortex for 1 min to dissolve completely. Stock kept at −20 °C.

6. Acetosyringone is soluble in organic solvents such as ethanol, DMSO, and dimethyl formamide (DMF). The solubility of acetosyringone in ethanol is approximately 2 mg/mL and approximately 20 mg/mL in DMSO and DMF. The acetosyringone stimulates *vir* genes and increases gene transfer efficiency in *Agrobacterium.*
7. The cefotaxime via affinity for penicillin-binding proteins in cell wall of bacteria prevents from the growth of *Agrobacterium* and has no effect on plant cells.
8. Embryogenic calli have a different structure from non-embryogenic, exhibiting a granular appearance with milky white color. Initial embryogenic callus are cream in color, and during subsequent sub-culturing, most of them turned to golden yellow and green.
9. Cotton is rich in phenolic compounds that sometimes inhibit rooting. In this case, 0.15% of activated charcoal is used into the medium to eliminate the effects of phenolic compounds. Also, this medium is not supplemented with kanamycin.
10. Phytagel is used to solidify the culture medium. To prevent clumping, phytagel should be added to rapidly stirring culture medium which is at room temperature.
11. Calcium chloride is used for the production of fairly high efficiency competent cell of bacteria for transformation vector into bacteria.
12. Make sure that the pH does not fluctuate too much.
13. Make sure the rinse is done thoroughly so that no sodium hypochlorite remains. Residues of sodium hypochlorite prevent germination.
14. It is better to use white light.
15. Add antibiotics under sterile conditions.
16. Adding antibiotics depends on the type of vector.
17. Sometimes acetosyringone is added in the co-culture step.
18. Sometimes, in order to better secrete phenolic compounds from plant cells, wounds or small scratches are created in the explants. Phenolic compounds cause the stimulation of *vri* genes.
19. This step causes excess bacteria to be removed from the explants. The time of this step should not be too long, it may cause drying of the explant.
20. Somatic embryo is obtained from embryogenic calli and have a globular, heart-shaped, and torpedo-shaped structure. Mature embryos have a bipolar structure with a size 5 mm or longer.
21. The composition of the pots may vary.

References

1. Srivastava A, Shukla A, Rama SD, Singh PK, Verma PC (2023) *Agrobacterium*-mediated genetic transformation of cotton and regeneration via somatic embryogenesis. Bio Protoc 13(10):e4677. https://doi.org/10.21769/BioProtoc.4677
2. Meyer T, Renoud S, Vigouroux A, Miomandre A, Gaillard V, Kerzaon I, Prigent-Combaret C, Comte G, Moréra S, Vial L (2018) Regulation of hydroxycinnamic acid degradation drives *Agrobacterium* fabrum lifestyles. Mol Plant Microbe Interact 31(8): 814–822
3. Guo M, Huang Z, Yang J (2017) Is there any crosstalk between the chemotaxis and virulence induction signaling in *Agrobacterium tumefaciens*? Biotechnol Adv 35(4):505–511
4. Wetzel ME, Olsen GJ, Chakravartty V, Farrand SK (2015) The repABC plasmids with quorum-regulated transfer systems in members of the Rhizobiales divide into two structurally and separately evolving groups. Genome Biol Evol 7(12):3337–3357
5. Baker CJ, Smith J, Rice C (2020) Apoplast redox metabolism: Effect of acetovanillone (apocynin) and acetosyringone, on their co-oxidation and redox properties. Physiol Mol Plant Pathol. https://doi.org/10.1016/j.pmpp.2020.101481
6. Nester EW (2015) *Agrobacterium*: nature's genetic engineer. Front Plant Sci 5:730
7. Gelvin SB (2017) Integration of *Agrobacterium* T-DNA into the plant genome. Annu Rev Genet 51:195–217
8. Caplan A, Herrera-Estrella L, Inzé D, Van Haute E, Van Montagu M, Schell J, Zambryski P (2019) Introduction of genetic material into plant cells. In: Biotechnology and biological frontiers. Routledge, Milton, pp 480–493
9. Tzfira T, Citovsky V (2006) *Agrobacterium*-mediated genetic transformation of plants: biology and biotechnology. Curr Opin Biotechnol. 17(2):147–154
10. Torres M, Jiquel A, Jeanne E, Naquin D, Dessaux Y, Faure D (2022) *Agrobacterium tumefaciens* fitness genes involved in the colonization of plant tumors and roots. New Phytol233:905–918.
11. James G (2019) Global status of commercialized biotech/GM Crops: 2019. N:55.
12. Wu J, Luo X, Guo H, Xiao J, Tian Y (2006) Transgenic cotton, expressing Amaranthus caudatus agglutinin, confers enhanced resistance to aphids. Plant Breed 125:390–394
13. Guo HN, Wu JH, Chen XY, Luo XL, Lur R, Shi YJ, Qin HM, Xiao JL, Tian YC (2003) Cotton plants transformed with the activated chimeric *cry1*Ac and AP1-B genes. Acta Botanica Sinica 45:108–113
14. Chen WX, Xiao GF, Zhu Z (2002) Obtaining high pest-resistant transgenic upland cotton cultivars carrying *cry1*Ac3 gene driven by chimeric OM promoter. Acta Botanical Sinica 44: 963–970
15. Tohidfar M, Mohammadi M, Ghareyazie B (2005) *Agrobacterium*-mediated transformation of cotton (*Gossypium hirsutum*) using a heterologous bean chitinase gene. Plant Cell Tissue Organ Culture 83:83–96
16. Tohidfar M, Ghareyazie B, Mosavi M, Yazdani S, Golabchian R (2008) *Agrobacterium*-mediated transformation of cotton (*Gossypium hirsutum*) using a synthetic *cry1*Ab gene for enhanced resistance against *Heliothis armigera*. Iran J Biotechnol 3:164–172
17. Tohidfar M, Rassouli H, Haghnazari A, Ghareyazie B, Najafi J (2009) Evaluation of stability of chitinase gene in transgenic offspring of cotton (*Gossypium hirsutum*). Iran J Biotechnol 1:45–50
18. Murashige T, Skoog F (1962) A revised medium for rapid growth and bioassays with tobacco tissue cultures. Physiol Plant 15:473–497
19. Gamborg OL, Miller RA, Ojima K (1968) Nutrient requirements of suspension cultures of soybean root cells. Exp Cell Res 50:151–158

Chapter 20

Characterization of Transgenic Cotton Plant to Confirm Gene Integration

Masoud Tohidfar

Abstract

Production of transgenic cotton from *Agrobacterium tumefaciens* strain LBA4404 harboring the recombinant binary vector pBI121 containing the *chi* gene (pBI121-chi) and neomycin phosphotransferase (*nptII*) gene has been used. In order to confirm transgenic cotton, molecular analysis methods are used. Polymerase chain reaction (PCR) and Southern blot analysis are used to confirm the integration transgenes into the plant genome. Western immunoblot analysis of proteins are extracted from leaves of transgenic cotton, which reveals the presence of an immunoreactive band with MW of approximately 31 kDa in transgenic cotton lines using anti-chitinase-I polyclonal anti-serum. Although there are other molecular methods, PCR, Southern blot, and Western blot are the most important of them, which will be discussed in this chapter.

Key words Molecular analysis, Transgenic, Cotton, Expression, Integration, PCR, Southern and Western blot hybridization

1 Introduction

In gene transfer with *Agrobacterium*, a large number of putative transgenic plants (events) are produced that differ in the level of transgene expression [1]. These differences have been generally attributed to copy number, intact and rearrangements, and position effect of transgene, also methods employed to generate transgenic and the presence of DNA parts of the vector in the events. In addition to these, the process of tissue culture can also cause variation (somaclonal variation). Rearrangements are more common in biolistic gun (gene gun) [2]. Therefore, identifying a limited number of plants with stable expression from among large number of plants is an important challenge in transgenic plants [3]. While such limitations do not exist in new methods such as genome editing [4], there are molecular methods such as PCR, Southern blot, and Western blot to identify transgenic cotton with stable expression

Nitish Kumar (ed.), *Transgenic Plants: Methods and Protocols*, Methods in Molecular Biology, vol. 2973,
https://doi.org/10.1007/978-1-0716-4803-2_20,

[5, 13]. The first step, using PCR the presence of the marker gene and transgene in the plants is confirmed. The Southern blots can provide useful information about events, the integration of the transgene into genome, intact or rearrangements, and single or multicopy of transgene, to determine the type of event [6–8]. The presence and accumulation of recombinant protein in the cell is evaluated using western blot analysis [9]. Furthermore, other methods such as sequencing, RT-PCR, and qPCR are used to detect transgenic plants [10–12].

2 Materials

2.1 Genomic DNA Extraction

For to extract genome DNA from ExgeneTM Plant SV mini Kit is used.

1. Transgenic cotton leaves (*Gossypium hirsutum* L.) cultivar Coker 312.
2. ExgeneTM Plant SV mini Kit (*see* **Note 1**).
3. SV Column Type G.
4. Collection tube.
5. EzSepTM Filter Column (with collection tube).
6. Buffer PL (*see* **Note 2**).
7. Buffer BD (*see* **Note 3**).
8. Buffer CW (*see* **Note 4**).
9. Buffer AE: 10 mM Tris-HCl, pH 9.0, and 0.5 mM EDTA (*see* **Note 5**).
10. RNase A (100 mg/mL) (*see* **Note 6**).
11. Liquid nitrogen.
12. Microcentrifuge.
13. 65 °C water bath or heating block.

2.2 PCR Master Mix 2X

The PCR Master Mix offers convenient reagents for PCR amplifications. The reagent of Master Mix is an optimized 2X PCR mixture of Taq DNA polymerase (recombinant), PCR buffer, $MgCl_2$, and dNTPs. Master Mix contains all components for PCR, except DNA template and primers. Additionally, sterile and PCR grade water is supplied. The kit is stable at −20 °C until expiration date. All solutions should be thawed on ice, gently vortexed, and briefly centrifuged.

1. PCR Master Mix (2X): 0.2 units/μL Taq DNA polymerase, 3 mM $MgCl_2$, 0.4 mM dATP, 0.4 mM dCTP, 0.4 mM dGTP, 0.4 mM dTTP, and PCR buffer.

2. The sequences of the primer pairs for *nptII* gene: F: 5′-GAA CAA GAT GGA TTG CAC GC-3′ R: 5′-GAA GAA CTC GTC AAG AAG GC-3′.
3. The sequences of the primer pairs for *chi* gene: F: 5′-GAG TGG TGT GGA TGC TGT TG-3′ R: 5′-GCC ATA ACC GAC TCC AAG CA-3′.
4. DNase-free water.

2.3 Southern Blot Hybridization Analysis

In this protocol from DIG DNA Labeling and Detection Kit is used for preparation of probe and detection.

1. Plasmid (pBI121-chi) (Fig. 1).
2. Appropriate restriction enzymes (*see* **Note** 7 and Fig. 1).
3. Whatman 3 MM paper.
4. Nylon membranes, positively charged.
5. Tissue paper.
6. Glass.
7. SSC 20X: 3 M NaCl, 0.3 M Na3Citrate, pH 7.
8. Shaker with adjustable temperature.
9. Prehybridization buffer: 50 mg sterile N-laurylsarcosine, 10 mg sterile SDS, 5 cc sterile blocking reagent 10x, 12.5 cc SSC 20X. Make up 100 cc with sterile ddH2O.
10. Hybridization buffer: This solution is the same as prehybridization buffer plus the single strain probe.
11. Wash buffer: 0.1 M malic acid, 0.15 M NaCl (pH 7.5), %0.3 (v/v) Tween 20.

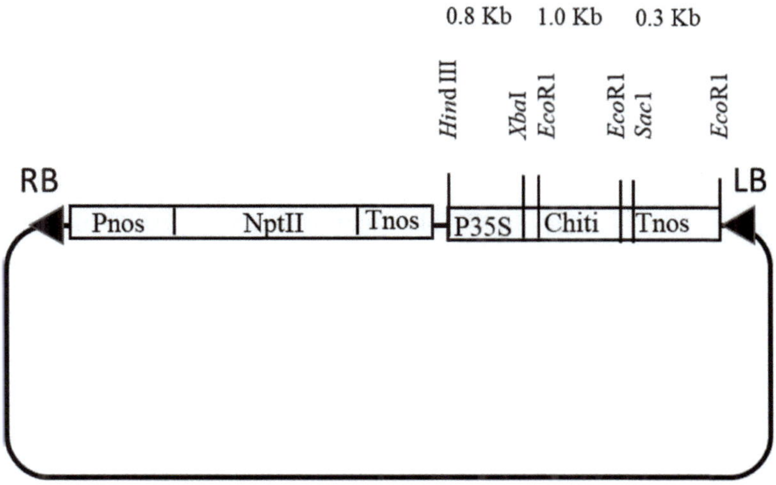

Fig. 1 The map of the recombinant binary vector pBI121-chi carrying the chitinase (*chi*) gene and *nptII* gene as a selectable

12. Malic acid buffer: 0.1 M malic, 0.15 M NaCl, pH 7.5.
13. DIG-11-dUTP (provided by kit).
14. Blocking solution: blocking reagent (provided by kit) %10 (w/v), maleic acid buffer under 65 °C.
15. Anti-digoxigenin-AP (provided by kit) (*see* **Note 8**).
16. Detection buffer: 0.1 M Tris-HCl, 0.1 M NaCl, pH 9.5.
17. NBT and BCIP (provided by kit).
18. TE buffer: 10 mM Tris-HCl, pH 8.0, 1 mM EDTA, pH 8.0 with NaOH.
19. Post-hybridization buffer 1: 2x SSC, 0.1% SDS.
20. Post-hybridization buffer 2: 0.2x SSC, 0.1% SDS.
21. Double-distilled water (ddH_2O).
22. Agarose.
23. Liquid nitrogen.
24. TAE buffer: 40 mM Tris-HCl base, 20 mM acetic acid, 1 mM EDTA, pH 8.0.
25. Depurination solution: 250 mM HCL.
26. Neutralizing solution: 0.5 M Tris-HCl, pH 7.5, 1.5 M NaCl.
27. DIG/dNTP mixture: 100 M dATP, 100 M dGTP, 100 M dCTP, 0.65 M dTTP, 0.35 DIG-11-Dutp

Prepare all solutions with Nuclease-free and sterile ddwater (ddH_2O).

2.4 Western Blot Analysis

1. Leaf tissue.
2. Liquid nitrogen.
3. Protein extraction buffer: 4% (w/v) SDS, 5% 2-mercaptoethanol, 20% glycerol, 68 mM Tris-HCl, pH 6.8.
4. Hybond-C membrane.
5. The Electrotrans blot.
6. The anti-chit-I anti-serum as primary antibody (1:2000).
7. The goat-anti-rabbit IgG alkaline phosphatase conjugate (1: 2000) as secondary antibody.
8. Microcentrifuge.
9. TEMED.
10. 10% ammonium persulfate.
11. Sample buffer: 500 mM Tris-HCl, pH 6.8, 500 μL glycerol, 10% SDS, 10 mg bromophenol blue, 2% 2-mercapthoethanol.
12. Coomassie brilliant blue G-250.

13. Destaining solution: 10 mL glacial acetic acid, 40 mL methanol, 50 mL ddH_2O.
14. Staining solution: 50% methanol, 10% glacial acetic acid, 0.1% Coomassie brilliant blue G-250.
15. Transfer buffer: 48 mM Tris-HCl, 39 mM glycine, pH 9.2, 20% methanol.
16. Whatman 3 MM paper.
17. Blocking solution: 20 mM Tris-HCl, 500 mM NaCl, pH 7.6, gelatin 3%.
18. TBS: 20 mM Tris-HCl, 500 mM NaCl, pH 7.6.
19. TTBS: 20 mM Tris-HCl, 500 mM NaCl, pH 7.6, 0.05 % Tween 20.
20. Antibody buffer: gelatin 1%, 100 mL TTBS.
21. AP buffer: 100 mM Tris-HCl, pH 9.5, 100 mM NaCl, 5 mM $MgCl_2$.
22. NBT and BCIP (provided by kit).
23. Color development solution: 45 μL NBT, 35 μL BCIP, 25 mL AP buffer.
24. 5x SDS electrophoresis buffer: 125 mM Tris-HCl base, 960 mM glycine, 0.5% SDS, pH 8.3.
25. Resolving gel %12.5: 1.875 mL 1.5 M Tris-HCl, pH 8.8, 3 mL acrylamide %30, 105 μL SDS %10, 27 μL ammonium persulfate %10, 7 μL Temed, 2.2 mL ddH_2O [14].
26. Stacking gel %6: 1.25 mL 0.5 mM Tris-HCl, pH 6.8, 830 μL acrylamide %30, 50 μL SDS %10, 27 μL ammonium persulfate %10, 7 μL Temed, 2.47 mL ddH_2O [14].
27. Hybond-C membrane.
28. Acrylamide %30.
29. TE buffer: 10 mM Tris-HCl, pH 8.0, 1 mM EDTA, pH 8.0.

Prepare all solutions with nuclease-free and sterile ddwater (ddH_2O).

3 Methods

3.1 PCR Analysis

1. Genomic DNA is extracted and purified from immature leaves of putative transgenic cotton based on the ExgeneTM Plant SV mini Kit.
2. PCR is performed in a total volume of 12.5 μL reaction mixture consisting of 6.25 μL PCR Master Mix 2X, 10x reaction buffer, 10 PM of each primer, 100 ng DNA template and make up to 12.5 μL with nuclease-free and sterile water. PCR is carried out

using the following conditions: initial denaturation at 94 °C for 5 min followed by 35 cycles of denaturing at 94 °C for 1 min, annealing at 55 °C for *nptII*, annealing at 60 °C for *chi* gene for 1 min, extension at 72 °C for 1 min, and final extension at 72 °C for 5 min.

3.2 Southern Blot Hybridization

1. Prepare a probe using PCR and *chi*-specific primer. PCR is performed in a total volume of 25 μL reaction mixture consisting of 10x reaction buffer, 0.7 μL 25 mM $MgCl_2$, 2.5 μL DIG/dNTP mixture, 10 PM of each primer, 5 ng pBI121-*chi* plasmid as template, 1U Taq DNA polymerase, and make up to 25 μL with nuclease-free and sterile ddwater. PCR is carried out using the following conditions: initial denaturation at 94 °C for 5 min followed by 35 cycles of denaturing at 94 °C for 1 min, annealing at 60 °C for 1 min (for *chi* gene), extension at 72 °C for 1 min, and final extension at 72 °C for 5 min.
2. Run 5 μL PCR product (*see* **Note 9**, Fig. 3).
3. Digest 5 μg of genomic DNA and 5 ng plasmid (as positive control) with appropriate restriction enzymes (*see* **Note 7**).
4. Use 1U of restriction enzyme for digest 1 μg of genomic DNA in appropriate volume.
5. Incubate tubes in appropriate temperature (for *Eco*R1 and *Hin*dIII 37 °C) for 12 h.
6. Run 5 μL of digested product at an 0.8% agarose gel to ensure complete digestion (*see* **Note 10**, Fig. 4).
7. Add 2 volume of cold %96 EtOH to the digestion product; thus incubate in −70 °C for 1 h.
8. Centrifuge at 10,000x *g* for 5 min at 4 °C; thus remove the supernatant.
9. Add 70% ethanol to remove salts; thus remove the supernatant.
10. Resuspend pellet in 10 μL nuclease-free and sterile ddH_2O.
11. Prepare an agarose gel %0.8 and load digested products and DIG-labeled DNA molecular weight marker.
12. Run the gel at voltage 50 for 30 min thus reduce voltage to 30.
13. After electrophoresis, wash gel slowly with ddH_2O for 2 min.
14. Submerge the gel in depurination solution for 15 min with gently shake (*see* **Note 11**).
15. Remove the HCl and wash the gel with sterile ddH_2O for two times.
16. Submerge the gel in NaOH = 0.4 N for 15 min with gently shake for two times.
17. Remove the NaOH and wash the gel with sterile ddH_2O for two times.

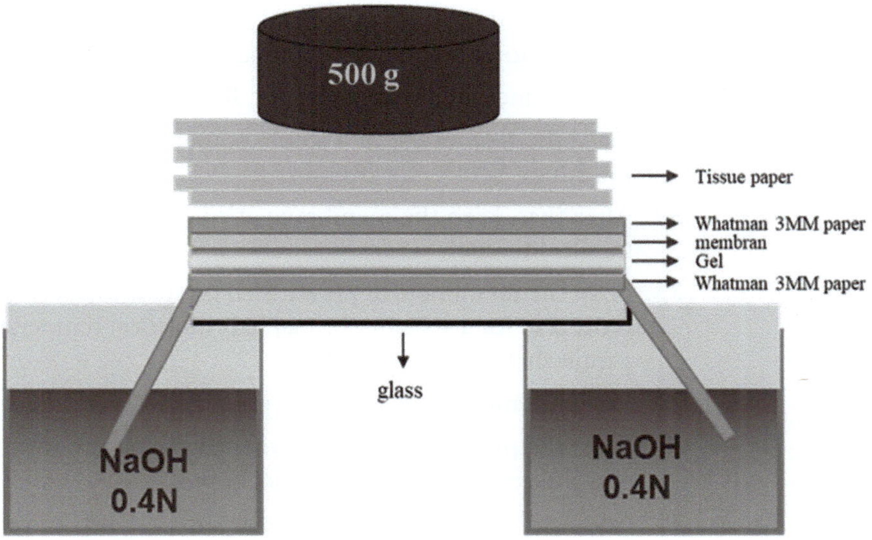

Fig. 2 Structure and components of blotting in Southern blot hybridization

18. Submerge the gel in neutralizing solution for 15 min for two times.
19. Place gel in the 2XSSC for 10 min.
20. Place two containers containing NaOH = 0.4 N and connect them with glass (Fig. 2).
21. Cut two pieces of Whatman paper 3 MM the same size as the glass so that the edge fits into the NaOH container (Fig. 2). Presoak the Whatman paper in NaoH 0.4 for 10 min.
22. Cut two Whatman papers the same size as the glass and two Whatman papers the same size as the gel and one Whatman paper exactly to the size of the membrane, and place it on previous Whatman paper. Roll a sterile pipette over the Whatman 3 MM paper to remove air bubbles. These papers should be presoaked in NaOH solution for 10 min.
23. Turn the gel upside down and place on the Whatman 3 MM paper (Fig. 2).
24. Roll a sterile pipette over the gel gently to remove air bubbles.
25. Cut a piece of membrane exactly the size of the gel and place it on the gel.
26. Remove air bubbles as above.
27. The four sides of the membrane should be covered in such a way that the buffer is the only way to pass through the gel.
28. Cut two Whatman papers exactly to the size of the membrane, and place them on the membrane. Remove air bubbles as above. Thus, place paper towels and a 300 g weight on it overnight.

29. The next day, take the membrane, and place it in 2xSSC solution for a few seconds with gentle shaking.
30. Dry the membrane on paper for 10 min.
31. Place the membrane in a sterile container, and store at 80 °C for 2 h or 120 °C for 30 min.
32. After that, keep the membrane at room temperature for a few minutes to cool.
33. Place the membrane in prewarmed prehybridization buffer (*see* **Note 12**) for 1 h at an optimal hybridization temperature with gentle shaking (*see* **Note 13**).
34. Place the membrane into container with sealed lid.
35. Place the probe in the 0.5 cc vial, and denature through boiling; thus quickly place on ice for 5 min.
36. Add 1 μL of the denatured probe into a 1 mL of prewarmed prehybridization buffer, and make hybridization buffer.
37. Immediately, add hybridization buffer to the container (with sealed lid) containing the membrane. Incubate in optimal hybridization temperature (*see* **Note 13**) for overnight with gentle shaking.
38. Next day, harvest hybridization buffer and store in −20 °C.
39. Transfer the membrane to a new and sterile the container.
40. Wash the membrane with post-hybridization buffer 1 for 5 min at RT with gentle shaking. Repeat.
41. Transfer the membrane to a new and sterile the container. Wash the membrane with prewarmed post-hybridization buffer 2 for 15 min at 60 °C (*see* **Note 14**) with gentle shaking.
42. Wash the membrane in wash buffer for 1–5 min with gentle shaking.
43. Transfer the membrane to a new and sterile the container. Incubate in blocking solution for 30 min.
44. Centrifuge anti-DIG in 11000 rpm for 5 min in 4 °C. Take 5 μL of the surface solution, and add to 25 mL (Appropriate and sufficient volume) of blocking solution. Incubate for 30 min at RT.
45. Transfer the membrane to a new and sterile the container. Add wash buffer and gently shake for 15 min at RT for two times.
46. Transfer the membrane to a new and sterile the container. Add detection buffer, and gently shake for 2–5 min at RT.
47. Transfer the membrane to a new and sterile the container. Add detection buffer (enough to cover the membrane), 45 μL NBT, and 35 μL BCIP for 10 mL detection buffer (or 200 μL mix NBT/BCIP for 10 mL detection buffer). Incubate for 30 min–16h in dark.

48. After the band appears, wash the membrane in sterile ddH_2O, and for better resolution of the bands, place the membrane in TE solution for 1 h, and take a photo. Thus, place it in a plastic bag with TE buffer and store at 4 °C.

3.3 Protein-Based Detection

Western immunoblotting is a very useful and effective method in determining the present of a specific protein in a complex protein mixture. In this method, the proteins are separated from each other using electrophoresis and then transferred to the nitrocellulose membrane. Next, they identify a target protein by using specific antibodies. In the western test, the antibody reacts with a specific antigen [15]. For protein extraction, segments of leaves of transgenic and control plants are ground to a fine powder with a mortar and pestle in liquid nitrogen [16].

1. Quickly, transfer 100 of powdered leaves into 1.5 cc vial on ice.
2. Add 1 mL of extraction buffer into the vial.
3. Vortex for 30–60 s.
4. Centrifuge 12000 rpm for 5 min at 4 °C.
5. Remove the supernatant slowly and transfer to a new vial.
6. Store at −20 °C.
7. Quantify the protein [17].
8. Prepare %12.5 resolving gel, pour the gel into the cassette, add 1 mL saturated isobutanol to the top gel, and keep for 1 h.
9. Prepare %6 stacking gel, and pour the gel on top of the separating gel. Place the comb in the gel and keep for 45 min.
10. Next, remove slowly the comb and wash the wells with distilled water.
11. Mix 10 μg of protein solution with one volume of 2x sample buffer; thus allow boil for 2 min. Immediately, put them on ice.
12. Load sample per well. Adjust the voltage to 20 mA.
13. Cut 5 pieces of the Whatman 3 MM paper and a membrane exactly to the size of the gel.
14. Place the Whatman 3 MM paper, membrane, and sponges in transfer buffer in a clean plastic container for 1 h.
15. After electrophoresis, remove the stacking gel, and mark the top and bottom of the gel.
16. Place the gel in transfer buffer for 15–30 min to remove SDS.
17. Prepare a sandwich including bottom sponge, two Whatman 3 MM paper, membrane, gel, three Whatman 3 MM paper, and top sponge, respectively. Put the sandwich in the Electrotrans blot (*see* **Note 15**) and run. An electric current 60 V is established from the gel side to the membrane side for 12 h.

18. After transfer, place the membrane in the blocking solution for 2 h with gentle shaking.
19. Wash the membrane three times and each time for 5 min with TTBS solution.
20. Transfer the membranes to a new sterile container, and add the diluted primary antibody (the anti-chit-I anti-serum). Allow gentle shaking for 2 h at room temperature. After 2 h, harvest the diluted primary antibody and store in −20 °C.
21. Wash the membrane three times and each time for 5 min with TTBS solution at RT.
22. Add the goat-anti-rabbit IgG alkaline phosphatase conjugate as secondary antibody to membrane. Shake slowly for 2 h at RT.
23. Wash membrane in TTBS for 5 min at RT.
24. Wash membrane in AP buffer for 15 min at RT.
25. Transfer the membrane to a sterile container, and add the color development solution. Incubate for 5 min in dark.
26. After the band appears (Fig. 5), wash the membrane in sterile ddH_2O, and for better resolution of the bands, place the membrane in TE solution for 1 h, and take photo. Thus, place it in a plastic bag with TE solution. Store at 4 °C.

4 Notes

1. All components of the kit should be stored at room temperature (15–25 °C). The buffers included in kit contain irritant which is harmful when in contact with skin or eyes, or when inhaled or swallowed. Care should be taken during handling. Always wear gloves and eye protector, and follow standard safety precautions. Buffer BD contains chaotropes. It can form highly reactive compounds when combined with bleach. DO NOT add bleach or acidic solutions directly to the sample preparation waste.
2. During delivery or storage under cold ambient condition, a precipitate may be formed in Buffer PL. Heat the bottle to dissolve completely before use.
3. Before using for the first time, add absolute ethanol (ACS grade or better) into Buffer BD as indicated on the bottle.
4. Before using for the first time, add absolute ethanol (ACS grade or better) into Buffer CW as indicated on the bottle, which contains sodium azide as a preservative.
5. To prepare solution 0.5 mM EDTA, refer to Sambrook J and Russell DW (2001).

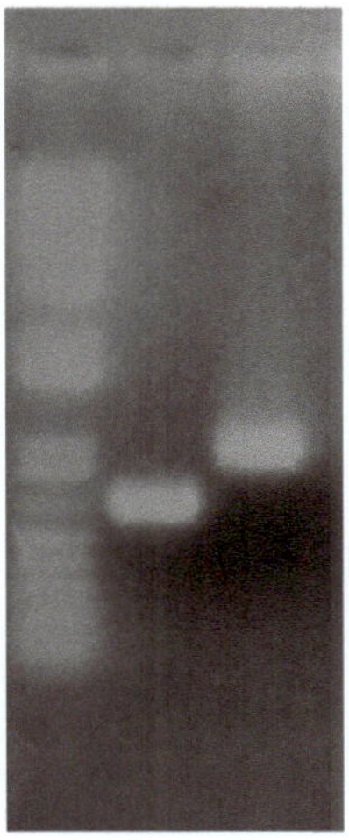

Fig. 3 PCR product for preparation probe using primer pairs specific for amplification of 900 bp chitinase (*chi*) gene in agarose gel. Lane M, 1.0 kb plus DNA ladder; Lane 1, unlabeled PCR product; Lane 2, PCR product labeled (probe)

6. RNase A is delivered under ambient conditions and can be stored at room temperature for 6 months without significant decrease in activity. But for prolonged conservation of activity, storing at storing at −20~8 °C is recommended.
7. In this experiment, *Hind*III and *Eco*R1 enzymes are used for copy number and intact of the gene, respectively. An enzyme that has one cutting site in the gene cassette is used to determine the copy number (Fig. 1).
8. Before use, anti-digoxigenin-AP should be centrifuged, then removed from its surface, and diluted 1:5000 in blocking solution.
9. Labeled DNA is heavier than unlabeled. Therefore, the probe is placed higher in the electrophoresis than the unlabeled (Fig. 3).
10. If complete digestion is done, a homogeneous smear is observed. If digestion is incomplete, continue digestion with more enzyme (Fig. 4).
11. For the plant genome, the time increases to 20 min. Depurination is required only when the size of our target fragment is greater than 5 kb.
12. Use 1 mL of prehybridization buffer for every 10 cm^2 of membrane.
13. For calculation of optimal hybridization temperature, see Roche (2003) DIG application manual for filter hybridization. Roche Diagnostic GmbH.

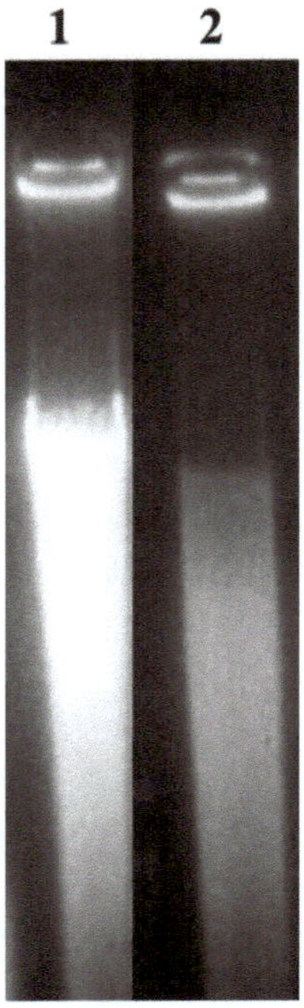

Fig. 4 *Hind*III digested DNA from transgenic cotton. Lane 1, partial digestion. Lane 2, complete digestion

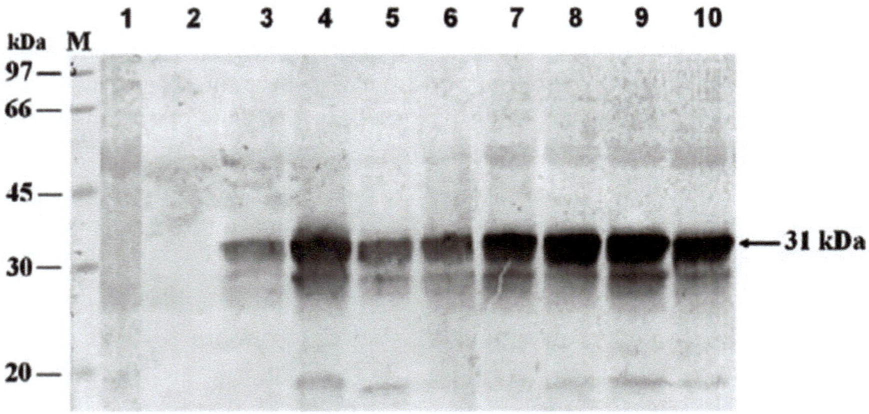

Fig. 5 Western immunoblot analysis of chitinase (chi) gene expression in leaf extract of transgenic cotton lines. Lanes 1 and 2, extract from untransformed cotton; Lanes 3–10, leaf extracts from transgenic lines. M, molecular size markers in kDa

14. For calculation of optimal hybridization temperature, see: Roche (2003) DIG application manual for filter hybridization. Roche Diagnostic GmbH (p: 95).
15. Pour enough transfer buffer into the Electrotrans blot.

References

1. Kohli A, Melendi PG, Abranches R (2006) The quest to understand the basis and mechanisms that control expression of introduced transgenes in crop plants. Plant Signal Behav 1: 185–195
2. Bhat SR, Srinivasan S (2002) Molecular and genetic analyses of transgenic plants: considerations and approaches. Plant Sci 163:673–681
3. Nagaya S, Kato K, Ninomiya Y (2005) Expression of randomly integrated single complete copy transgenes does not vary in *Arabidopsis thaliana*. Plant Cell Physiol 46:438–444
4. Delight H, Yasmina R, Ke Y, Chen J (2024) CRISPR/Cas genome editing in plants: mechanisms, applications, and overcoming bottlenecks. March 2024 functional & integrative. Genomics 24(2)
5. Southern E (1975) Detection of specific sequences among DNA fragments separated by gel electrophoresis. J Mol Biol 98:503–511
6. Neuhaus-Url G, Neuhaus G (1993) The use of the nonradioactive digoxigenin chemiluminescent technology for plant genomic Southern blot hybridization: a comparison with radioactivity. Transgenic Res 2:115–120
7. Engler-Blum G, Meier M, Frank J, Muller GA (1993) Reduction of background problems in nonradioactive northern and southern blot analysis enables higher sensitivity than 32P-based hybridizations. Anal Biochem 210: 235–244
8. Dietzgen RG, Abedina M, Higgins CM et al (1999) Nonradioactive detection of transgenes in plant genomic southern blots: a method using DIG chemiluminescent technology. Biochemica 1:19–20
9. Renart J, Stark GR (1979) Transfer of proteins from gels to diazobenzyloxymethyl-paper and detection with antisera: a method for studying antibody specificity and antigen structure. Proc Natl Acad Sci 76:3116–3120
10. Udvardi MK, Czechowski T, Scheible WR (2008) Eleven golden rules of quantitative RT-PCR. Plant Cell 20:1736–1737
11. Pfaffl MW (2001) A new mathematical model for relative quantification in real-time RT–PCR. Nucl Acids Res 29:e45
12. Valasek MA, Repa JJ (2005) The power of real-time PCR. Adv Physiol Educ 29:151–159
13. Tohidfar M, Mohammadi M, Ghareyazie B (2005) *Agrobacterium*-mediated transformation of cotton (*Gossypium hirsutum*) using a heterologous bean chitinase gene. Plant Cell Tissue Organ Cult 83:83–96
14. Sambrook J, Russell DW (2001) Molecular cloning: a laboratory manual. Cold Spring Harbor Laboratory Press, Cold Spring Harbor
15. Kurien BT, Scofield RH (2006) Western blotting. Methods 38:283–293
16. Tohidfar M, Rassouli H, Haghnazari A, Ghareyazie B, Najafi J (2009) Evaluation of stability of chitinase gene in transgenic offspring of cotton (*Gossypium hirsutum*). Iran J Biotechnol 1:45–50
17. Bradford MM (1976) A rapid and sensitive method for the quantification of microgram quantities of protein utilizing the principle of protein–dye binding. Anal Biochem 72:248–254

Chapter 21

Characterization of Transgenic Rice Plants by Non-radioactive Southern Hybridization

Rajasekaran Sripriya, Mahitha Koonjangadan Padinjhareveedu, Ajith Thaliyakkattil Girijan, and Jasmine M. Shah

Abstract

Southern hybridization is a method for detection of specific DNA sequences and gene analysis. It is performed by electrophoresis of the DNA samples after digesting with an appropriate restriction enzyme and transferring it to a nylon membrane and hybridizing it with a labeled probe. In this chapter we describe the protocol for transgene detection in rice plants by capillary transfer of DNA to a nylon membrane and hybridizing it with a Digoxin-labeled oligonucleotide probe and detection using alkaline phosphatase conjugated anti DIG antibody.

Key words Capillary transfer, Depurination, Denaturation, Neutralization, Alkaline phosphatase, Anti-DIG antibody, Prehybridization, Blocking, Chemiluminescence, CDP-star substrate

1 Introduction

Southern hybridization is a hybridization-based method for the detection of specific DNA fragment in a given sample. It is done by immobilizing the DNA on a membrane and hybridizing it with a labeled probe. This method was first developed by Edwin Southern in 1975 [1]. Southern blotting is the most reliable method for detecting transgenes in a genetically modified plant. Southern aids in the detection of copy number of the transgene [2], identification of truncated events [3], zygosity analysis [4], and differentiating a transgene from its endogenous copy [5]. The basic steps of Southern hybridization analysis are digestion of the DNA samples using restriction enzymes, electrophoretic separation of the DNA fragments, transfer of the DNA fragments and their immobilization onto a membrane, hybridization with a labeled probe, and detection of the hybridized probe using radioactive or non-radioactive methods.

Nitish Kumar (ed.), *Transgenic Plants: Methods and Protocols*, Methods in Molecular Biology, vol. 2973,
https://doi.org/10.1007/978-1-0716-4803-2_21,

Hybridization with radioactively labeled probes is a robust and reliable method because of its sensitivity to detect weak signals at high resolution. However, the risk of exposure to radioactivity and disposal cost of the radioactive wastes are the principle concern of radioactive methods [6].

Non-radioactive Southern hybridization using principles of chemiluminescence or fluorescence for detection of probe hybridization is an alternative and safer method for analysis of transgenic rice plants. Labeling the oligonucleotide probes with fluorescein or digoxigenin and detection using the respective antibodies conjugated with alkaline phosphatase has been successfully used for plant genomic DNA [7]. Transgene detection and copy number analysis in transgenic rice plants using non-radioactive Southern have been reported earlier [8]. The major limitations of non-radioactive Southern are low sensitivity and high signal to background ratio [9]. Reducing the background noise can improve the quality of non-radioactive Southern [10]. High-throughput detection methodologies and state-of-the-art imaging systems have significantly improved the sensitivity and clarity of non-radioactive Southern [11]. Improved sensitivity and rapid detection of transgene in plants using non-radioactive Southern can be achieved by using chemiluminescent substrates like CDP-Star that produces light that can be captured on X-ray film [12].

In this chapter, the detection of transgenes in rice plants using non-radioactive Southern hybridization is described. The Southern blotting methodology was adapted from Sambrook et al. [13] and Sambrook and Brown [14]. The rice DNA digested with the appropriate enzyme was fractionated by electrophoresis in a 0.8% agarose gel. The rice DNA in the agarose gel was hydrolyzed using strong base and neutralized prior to transfer to a nylon membrane. Upward capillary transfer of the DNA fragments to the nylon membrane as described by Southern (1975) [1] was used. The DNA was permanently cross-linked to the membrane by UV exposure. The nylon membrane was blocked to prevent nonspecific attachment of the probe followed by hybridization with a DIG-labeled probe. Detection with anti-DIG antibody conjugated with alkaline phosphatase and its substrate CDP-Star^{TM}.

2 Materials

Prepare all solutions using double-distilled water (*see* **Note 1**). Autoclave the solutions wherever recommended, and store in the appropriate temperature. Use analytical grade chemicals for preparation of solution.

2.1 DNA Transfer

1. 10X TBE buffer (pH 8.2): Dissolve 107.78 g of Trizma base, 8.41 g of Na_2 EDTA. 2 H_2O and 55 g of boric acid in 600 mL of distilled water. Make up the volume to 1 L and autoclave. (Check and confirm if the pH is about 8.2.)
2. 10 mg/ml ethidium bromide stock: Prepare by dissolving 10 mg of ethidium bromide in 1 mL sterile double-distilled water. Use at a working concentration of 1 μg/mL.
3. Denaturation solution: Dissolve 29.22 g of NaCl (1 M) and 10 g of NaOH (0.5 M) and in 400 mL of double-distilled water. Make up the volume to 500 mL. Autoclave and store at room temperature.
4. Neutralization solution (pH 7.0): Dissolve 43.83 g of NaCl (1.5 M) and 30.28 g of Tris base (0.5 M) and in 350 mL of double-distilled water. Adjust the pH to 7.0 with concentrated HCl, and make up the volume to 500 mL. Autoclave and store at room temperature.
5. 20 X SSC: Dissolve 175.3 g of NaCl and 88.2 g of sodium citrate in 800 mL of double-distilled water. Adjust the pH to 7.0 with con HCl. Make up the volume to 1 L. Autoclave and store at room temperature.

2.2 Probe Labeling

1. Probe DNA: 150 ng of DNA in a total volume of 15 μL (*see* **Note 15**).
2. Random hexanucleotides: As provided by the supplier.
3. dNTP mix: As provided by the supplier.
4. 1 U/μL Klenow fragment: As provided by the supplier.
5. 0.2 M EDTA, pH 8.0.
6. 4 M $LiCl_2$.
7. 95% and 70% ethanol: Use molecular biology grade, or double-distilled.

2.3 Hybridization and Detection

1. Nylon membrane.
2. 10% blocking reagent: Dissolve 5 g of blocking powder (skimmed milk powder) in 50 mL of maleic acid buffer. Heat gently to dissolve completely. Autoclave and store at 4 °C
3. Maleic acid buffer (stock): Dissolve 11.607 g of maleic acid (0.1 M) and 8.76 g of NaCl (0.15 M) in 800 ml of double-distilled water. Adjust the pH to 7.5 with NaOH pellets (around 80). Make up the volume to 1 L. Autoclave and store at room temperature.
4. 10% N-lauryl sarcosine Na salt stock: Dissolve 10 g of N-lauryl sarcosine Na salt in 90 mL double-distilled water, and make up to a final volume of 100 mL. Autoclave and store at room temperature.

5. 2 M $MgCl_2$ stock: Dissolve 10.165 g of $MgCl_2$ in 20 mL of double-distilled water, and make up the volume to 25 mL. Autoclave and store at room temperature.
6. 10 % SDS stock: Dissolve 10 g sodium dodecyl sulfate in 100 mL of double-distilled water. Autoclave and store at room temperature.
7. Hybridization solution: Mix 5X SSC, 0.1 % w/v N-Lauryl sarcosine Na salt, 0.02 % SDS, and 1 % blocking reagent (*see* **Note 2**).
8. Wash buffer: Add 0.3% v/v Tween 20 in maleic acid buffer (add 1.5 mL of Tween 20 in 500 mL maleic acid buffer). Note to add Tween 20 after autoclaving the maleic acid buffer.
9. 1% Blocking reagent: Add 20 mL of 10% Blocking reagent to autoclaved 180 mL of maleic acid buffer (*see* **Note 3**).
10. Detection buffer: Dissolve 1.21 g of Trizma base (0.1 M) and 0.58 g of NaCl (0.1 M) in 80 ml of double-distilled water. Adjust the pH to 9.5 with 4 N HCl. Make up the volume to 97.5 mL autoclave it. After autoclaving add 2.5 mL of 2.0 M $MgCl_2$ (50 mM).

2.4 X-ray Film Developer and Fixer

Prepare according to the manufacturer's protocol

3 Methods

3.1 Agarose Gel Electrophoresis

1. Digest 15 μg of rice DNA with an appropriate restriction enzyme. Make up the reaction volume up to 20 μL with sterile double-distilled water (*see* **Note 4**). Allow overnight digestion at 37 °C (*see* **Note 5**).
2. Separate the DNA in 0.8% agarose gel (*see* **Note 6**) in 1X TBE buffer at 80 V (*see* **Note 7**). Run the gel till the dye front reaches three-fourth of the gel.
3. The controls that should be loaded for the Southern blot include:
   (a) For the positive control, use the probe DNA fragment (1 pg) or the linearized plasmid from which the probe was obtained.
   (b) Negative control DNA from a non-transgenic rice plant digested with the same enzyme that was used for digesting the transgenic rice lines.
   (c) For the digestion control, use undigested rice DNA (transgenic or non-transgenic) to check proper digestion of the DNA.

4. Place the gel in a glass tray containing ethidium bromide (1 μg/μL) for 30 min for staining the DNA and photograph the gel (*see* **Note 8**).

3.2 Depurination and Denaturation

1. Transfer the ethidium bromide stained gel to a glass tray, and add 250 mL of 0.15 N HCl till the dye color changes to yellow to depurinate the DNA (approx. 20 min) (*see* **Note 9**).
2. Transfer the gel to a glass tray, add 250 mL of denaturation solution, and continuously shake the tray in a rocker platform for 45 min. Remove excess denaturation solution by washing the gel four times (with 250 mL of sterile double-distilled water each time) to remove any excess of denaturation solution.
3. After completely removing the water, add 250 mL of neutralization solution, and gently shake the gel in a rocker platform for 45 min.

3.3 DNA Transfer

1. Use the same platform that was used for casting the gel for DNA transfer. Keep the platform upside down in a glass tray containing 200 mL of 20 X SSC.
2. Make a wick using Whatman No. 3 paper. Fold the end of the papers so that they are dipped in the 20 X SSC solution in the tray (*see* **Note 10**).
3. Place the gel on the wick such that the bottom side of the gel (where the DNA is present) facing upward, in contact with the nylon membrane. Avoid air bubbles.
4. Cut a Zeta-Probe nylon membrane (Bio-Rad) exactly to the size of the gel. Soak in sterile double-distilled water for a few seconds and then in 20 X SSC for 2 min, and place it on the gel using a forceps without any air bubble (*see* **Note 11**). Mark the right top corner of the gel (*see* **Note 12**).
5. On top of the gel, carefully lay two Whatman No. 3 sheets cut to the size of the gel, wetted in 20 X SSC.
6. Place one dry Whatman No. 3 sheet cut to the size of the gel to initiate transfer process.
7. Stack crude filter papers cut exactly to the size of the gel on top of the dry Whatman No. 3 paper (to about 8–10 cm height) (*see* **Note 13**). On the stack of crude filter paper place a glass tray and a weight (200 gm) (Fig. 1).
8. Remove the membrane after overnight transfer and rinse briefly (20 s) in 2 X SSC. Air-dry on Whatman No 1 paper and perform UV cross-linking (*see* **Note 14**).

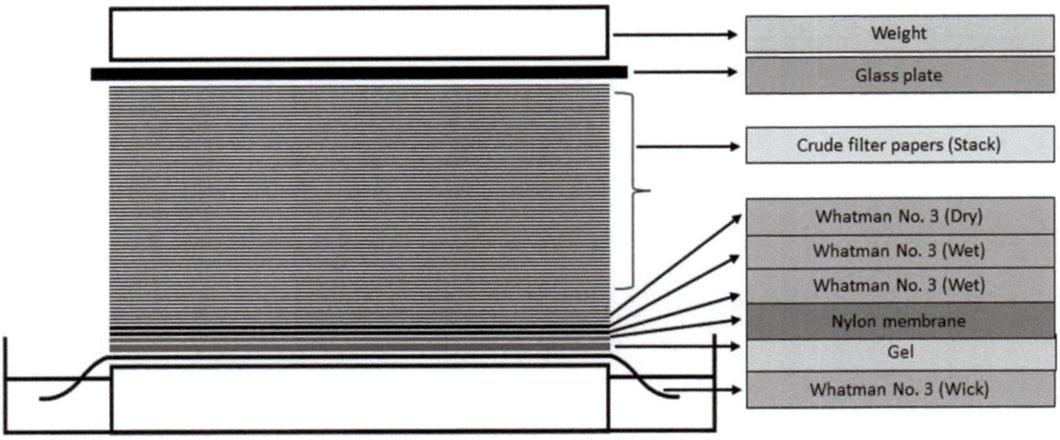

Fig. 1 Set up for the upward capillary rise method of Southern blot. (Adapted from [1])

3.4 Probe Preparation for DIG Labeling

1. Denature the probe DNA (150 ng in 15 μL) by placing it in a boiling water bath for 5 min, and snap cool on ice for 10 min (*see* **Note 16**). This will keep the DNA in a denatured condition.
2. Make up the volume to 15 μL if there is any loss due to evaporation during boiling. Collect the solution to the bottom of the tube by a short spin. Then to the tube add 2 μL of hexanucleotide, 2 μL of dNTP mix, and 1 μL of Klenow fragment.
3. Seal the tube with Parafilm and keep it in 37 °C water bath for 20 h.
4. Stop the reaction with 2.0 μL of 0.2 M EDTA, pH 8.0.
5. Precipitate the DNA with 2.5 μL of 4 M $LiCl_2$ and 75 μL of 95% ethanol (double-distilled), and keep it at −20 °C overnight. Centrifuge for 10 min at 12,000 rpm at 4 °C. Wash the pellet with 100 μL of 70% ethanol for 5 min at 12,000 rpm. Dry the pellet and dissolve in 50 μL of 0.1 X TE pH 8.0 (*see* **Note 17**).
6. Probe (50 μL) is denatured by boiling it in a water bath for 5 min and snap cooling it on ice for 20 min. This is used for hybridization.

3.5 Prehybridization

Keep the membrane inside the hybridization bottle by using a sterile flat forceps and add pre-warmed (65 °C) hybridization solution (20 mL/100 cm^2) at the bottom of the bottle. Slowly add the hybridization solution through the walls of the bottle and wet the membrane by rotating the bottle to avoid formation of air bubbles between the bottle and the membrane. Place the bottle in a hybridization oven, and maintain the oven temperature at 65 °C all through the reaction. Perform prehybridization for 1 h (*see* **Note 18**).

3.6 Hybridization

1. Remove the hybridization solution and add fresh pre-warmed (65 °C) hybridization solution (10 mL) to the bottle.
2. Add the denatured probe DNA (*obtained from Subheading 3.4, Step 6*) to the bottle, mix by gentle shaking, and incubate at 65 °C for 12–20 h in the hybridization oven.

3.7 Post-Hybridization Washes (High Stringency Wash)

Three post-hybridization washes are done by decreasing the SSC concentration step by step with 40 mL of wash solution for each step.

1. Discard the hybridization solution, and rinse the membrane briefly with 20 mL of pre-warmed (65 °C) 2 X SSC/0.1 % SDS solution. For the first wash, add 2 X SSC and 0.1% SDS at 65 ° C for 30 mins.
2. Repeat step 1 with 0.5 X SSC/0.1% SDS at 65 °C.
3. Repeat step 1 with 0.5% SSC/0.1% SDS at 65 °C (*see* **Note 19**).

3.8 Immunological Detection

1. Remove the wash solution and wash the membrane with 40 mL of wash buffer (maleic acid buffer + 0.3% v/v Tween 20) for 5 min at room temperature.
2. Incubate the membrane in 1% blocking solution in maleic acid buffer for 1 h at 37 °C (70–80 mL/bottle).
3. Discard the solution and add anti-DIG-AP conjugate diluted to 1:10,000 times (i.e., 1 μL in 10 mL- calculated for a 100 cm^2 blot) in a 1% blocking solution. Incubate for 30 min at 37 °C.
4. Wash two times with wash buffer 15 min each time (80 mL/bottle) at 37 °C.
5. Equilibrate the membrane with detection buffer (40 mL/100 cm^2) for 5 min at room temperature.
6. Remove the membrane, and incubate in CDP-STAR substrate diluted to 1:100 (10 μL in 1 mL detection buffer) for 5 min in dark (0.5 mL/100 cm^2). Keep the wet membrane on a transparent sheet, add 0.5 mL substrate for 100 cm^2 membrane along one edge of the membrane, and slowly lay over a transparent sheet on the membrane without any air bubble. Keep it in dark for 5 min.
7. Remove the membrane and let liquid drip off. Air-dry the membrane on Whatman No.1 paper for 2 min, fix it on black sheet, wrap it with saran wrap, and expose it to an X- ray film in an exposure cassette for 15 min to 1 h. The chemiluminescence (Fig. 2) can be captured on X-ray films by autoradiography.
8. Develop the X-ray film by dipping in the developer solution for 1 min, wash it with water, and dip in fixer solution for 1 min in a dark room.

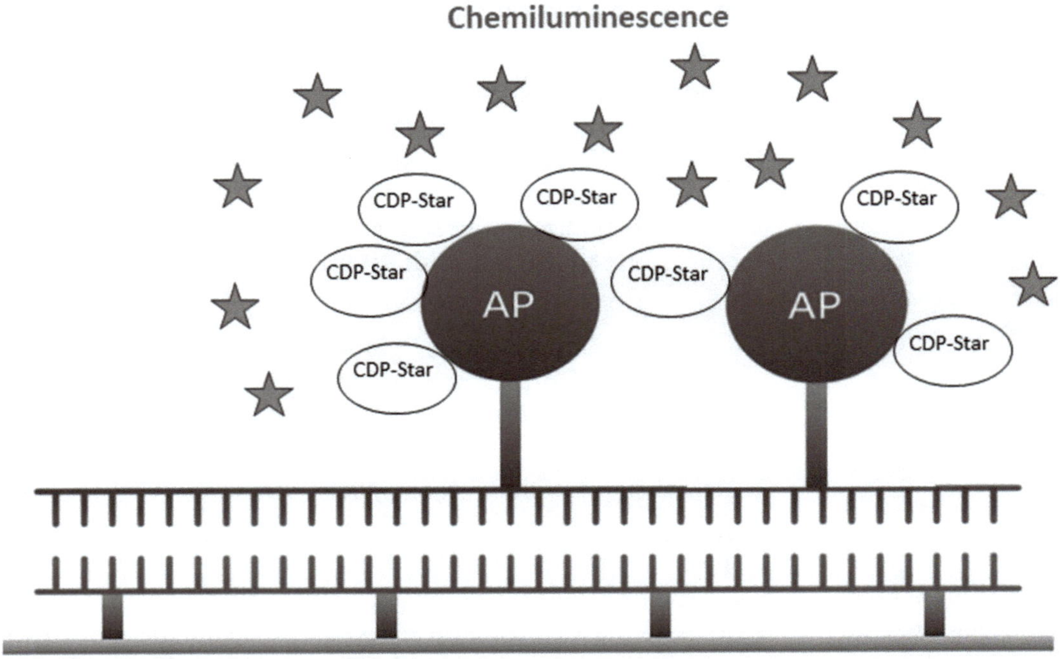

Fig. 2 Outline of DIG labeling and detection. Probes labeled with DIG are detected by anti-DIG antibody conjugated with alkaline phosphatase. Alkaline phosphatase reacts with CDP-Star substrate, and produces chemiluminescent signals that are detected by autoradiography

4 Notes

1. For preparation of solution the solutes should be added one at a time, ensuring that each is fully dissolved before adding the next.
2. A total volume of 100 mL of hybridization solution is required for one blot.
3. Blocking solution has to be prepared freshly during each hybridization.
4. Diluted reaction mix will aid in better digestion as the contaminants will be diluted.
5. Use water bath for maintaining the temperature as it will ensure even maintenance of temperature.
6. Add 0.8 g of agarose to 100 mL of 1 X TBE solution, boil it till it melts. Pour the agarose once it cools to bearable warmth, and add to a clean platform. Ensure no air bubbles are formed while pouring the gel. Allow the gel to solidify at least for 20 min. Transfer the gel to the electrophoresis tank filled with the 1 X TBE tank buffer. The level of the tank buffer should be above the level of the gel.

7. Running the gel at higher voltage will result in heating of the gel that may result in fuzzy signals.
8. Wear gloves while handling ethidium bromide stained gel.
9. Depurination improves the rate of transfer of large fragments of DNA [15].
10. Care should be taken to avoid air bubbles trapped between the platform and the Whatman paper. Gently roll a glass pipette over the Whatman paper to remove any air bubble trapped under it. Air bubbles hinder DNA transfer.
11. Soaking the membrane in water and buffer improves binding efficiency, transfers uniformly, avoids background signals, and improves signal quality.
12. Marking the corner of the membranes is useful in marking the gel boundaries and facilitates in measuring the size of the DNA fragments in the photographic sheet.
13. Oversized papers or small-sized papers will hinder even transfer of DNA.
14. Baking or UV cross-linking is done to permanently fixes the DNA to the membrane which prevents DNA loss during hybridization.
15. The probe fragment is generally obtained from the plasmid (generally the vector that has been used for transformation). The plasmid should be digested with an appropriate enzyme/s to release the probe fragment, and it should be electrophoresed in an agarose gel and eluted (the fragments should be well separated to avoid contamination of the probe with the vector backbone).
16. Poke a hole on the lid of the centrifuge tube during boing as it may snap open during the denaturation process.
17. Precipitation of the probe is done to remove other chemical contaminants that are used during labeling process.
18. The temperature of the prehybridization solution should be maintained at 65 °C. Lower temperatures will result in back ground signals.
19. The high stringency washes are used to identify DNA sequences with high similarity to the probe sequences (>95% similarity). For identification of sequences that have lesser percentage of similarity with the probe sequence, less stringent washing conditions can be used (higher SSC concentration and lower temperature).

Acknowledgments

The authors Sripriya R and Shah JM gratefully acknowledge Prof. K. Veluthambi and Mrs. Balamani Veluthambi for training them in non-radioactive Southern hybridization technique.

References

1. Southern EM (1975) Detection of specific sequences among DNA fragments separated by gel electrophoresis. J Mol Biol 98:503–517
2. Shah JM, Singh R, Veluthambi K (2013) Transgenic rice lines constitutively co-expressing *tlp-D34* and *chi11* display enhancement of sheath blight resistance. Biol Plant 57:351–358
3. Shah JM, Veluthambi K (2010) *DIANTHIN*, a negative selection marker in tobacco, is non-toxic in transgenic rice and confers sheath blight resistance. Biol Plant 54:443–450
4. Parameswari C, Sripriya R, Veluthambi K (2010) Establishment of homozygous transgenic rice lines after elimination of unlinked albino traits from transgenic events through segregation. In Vitro Cell Dev Biol 46:395–402
5. Sripriya R, Parameswari C, Veluthambi K (2017) Enhancement of sheath blight tolerance in transgenic rice by combined expression of tobacco osmotin (*ap24*) and rice chitinase (*chi11*) genes. In Vitro Cell Dev Biol Plant 53:12–21
6. Mansfield ES, Worley JM, McKenzie SE, Surrey S, Rappaport E, Fortina P (1995) Nucleic acid detection using non-radioactive labelling methods. Mol and Cell Probes 9:145–156
7. Neuhaus-Url G, Neuhaus G (1993) The use of the nonradioactive digoxigenin hemiluminescent technology for plant genomic Southern blot hybridization: a comparison with radioactivity. Transgenic Res 2:115–120
8. Chen P-Y, Tsai Y-T, Ng C-Y, Ku MS-B, To K-Y (2017) Transformation and characterization of transgenic rice and *Cleome spinosa* plants carrying the maize phospho*enol*pyruvate carboxylase genomic DNA. Plant Cell Tiss Organ Cult 128:509–519
9. Accotto GP, Vaira AM, Noris E, Vecchiati M (1998) Using non-radioactive probes on plants: a few examples. J Bioluminescence Chemiluminescence 13:295–301
10. Englerblum G, Meier M, Frank J, Muller GA (1993) Reduction of background problems in nonradioactive northern and southern blot analyses enables higher sensitivity than 32P-based hybridizations. Anal Biochem 210:235–244
11. Osborn J (2000) A review of radioactive and non-radioactive-based techniques used in life science application. Life Sci News 6
12. Liu G, Yuan Y, Jiang H, Bao Y, Ning G, Zhao L, Zhou X, Zhou H, Gao J, Ma N (2021) *Agrobacterium tumefaciens*-mediated transformation of modern rose (*Rosa hybrida*) using leaf-derived embryogenic callus. Hort Plant J 7:359–366
13. Sambrook J, Fritsch EF, Maniatis T (1989) Molecular cloning a laboratory manual. Cold Spring Harbor Laboratory Press, Cold Spring Harbor
14. Brown T (2001) Southern blotting. Curr Protoc Mol Biol Unit 2.9A
15. Reed KC, Mann DA (1985) Rapid transfer of DNA from agarose gels to nylon membranes. Nucleic Acids Res 13:7207–7221

Chapter 22

Characterization of Transgenic Rice Plants by Radioactive Northern Hybridization

Rajasekaran Sripriya, Ajith Thaliyakkattil Girijan, Mahitha Koonjangadan Padinjhareveedu, and Jasmine M. Shah

Abstract

Northern hybridization analysis of transgenic plants is done to confirm the presence of the transgene-derived transcript. Additionally, information on the size of the complete transcript and its relative expression can also be obtained by this technique. Presence of multiple forms (if any) of a transcript can also be visualized using this method. Radioactive labelling of DNA probe for Northern hybridization results in sensitive detection of the transcript. An elaborate protocol of Northern blotting and hybridization for detecting desired RNA from transgenic rice plants is described in this chapter.

Key words Radioactive Northern hybridization, Northern blotting, Rice, RNA

1 Introduction

Northern hybridization, as the name indicates, is a hybridization-based technique used to identify the presence and size of a desired RNA transcript that is separated in a matrix (generally an agarose gel) and blot-transferred to a membrane [1]. The complementary DNA fragment of the desired gene of interest is generally chosen for probing a Northern blot [1]. The probe DNA can either be labeled with radioisotopes, or chemical tags, or fluorescent molecules. The former one is therefore used in radioactive Northern hybridization method and latter two in non-radioactive methods. Northern hybridization has been routinely used to detect the expression of transgenes in transgenic crop plants, including rice. The main advantage of Northern hybridization is that the transgene-derived transcript size can be visualized and measured [2]. For example, in the transgenic rice plants transformed with *Trichoderma*-derived *Cht42* gene, the desired transgene-transcript of size 1.5 kb was obtained [3]. On the other hand, in rice plants transformed with tobacco osmotin-encoding *ap24* gene, apart

Nitish Kumar (ed.), *Transgenic Plants: Methods and Protocols*, Methods in Molecular Biology, vol. 2973,
https://doi.org/10.1007/978-1-0716-4803-2_22,

from the desired 1.1 kb band, another band of lower size, probably due to some post-transcriptional modification, was seen in all transgenic T_0 as well as T_2 plants [4]. Also, Northern blots can be used to understand the relative expression of the transcript if compared with their respective stained-gels. For example, in the *Cht42*-transformed rice plants, the transgene expression was evidently lower in one of the transgenic plants, despite having RNA concentration comparable to others [3]. In transgenic rice plants expressing the *Rice Stripe Virus* coat protein-encoding gene (*CP*), the relative expression of the transgene-derived transcript was measured using densitometric method [5]. Such a visualization of the transcript's complete size, along with intensity, is not possible in any other transcript analysis technique.

The protocol for Northern hybridization of RNA from transgenic rice plants, using radioactive labeling method, is described in this chapter. The RNA from rice plants can be extracted either by the method involving an acid-guanidinium-phenol-based reagent [3] or LiCl-involving method [6]. The procedure for RNA gel electrophoresis, involving formaldehyde and deionized formamide, was adapted from Sambrook et al. [7]. The capillary transfer method is same as that of Southern hybridization [8], except for the difference in the membrane and transfer buffer composition. Radioactive labeling of the DNA probe is as described in [9]. The hybridization protocol was adapted from [7, 9]. The described protocol has been successfully used several times for the Northern hybridization analysis of transgenic rice of two *indica* varieties, Pusa Basmati 1 [3, 9, 10] and White Ponni [6].

2 Materials

Use well-washed, clean bottles/measuring cylinders/beakers/trays for storing/preparing the reagents/transfer. Rinse all them with DEPC-treated water (unless specified otherwise) and leave them wetted with DEPC-treated water for at least 1 h prior to autoclaving/reagent preparation. Store all reagents at room temperature, unless specified otherwise.

2.1 Common Reagents

1. 0.1% v/v diethyl pyrocarbonate (DEPC)-treated water (1 L): Add 1 mL of DEPC to 1 L of autoclaved double-distilled water kept in an autoclaved bottle. Keep the solution for 2 h stirring to ensure proper mixing of the DEPC.
2. 2 M Tris stock, pH 7.5 (100 mL): Dissolve 24.22 g Tris base in 70 mL of nuclease-free water or autoclaved double-distilled water. Set the pH to 7.5 by using concentrated hydrochloric acid (HCL). Make up the volume to 100 mL and autoclave. (Note "not" to use DEPC-treated water for preparing Tris.

Also, do not rinse the reagent bottle/measuring cylinder/beaker with DEPC-treated water.)

3. 10X 3-(Morpholin-4-yl)propane-1-sulfonic acid (MOPS) buffer stock (500 mL): Dissolve 20.9 g MOPS (0.2 M), 1.86 g ethylenediaminetetraacetic acid (EDTA) (10 mM), and 3.4 g sodium acetate (50 mM) in 400 mL DEPC-treated water. Adjust the pH to 7.0 using 0.2 N NaOH. Make up the volume to 500 mL and autoclave. The solution turns light yellow upon autoclaving.
4. Deionized formamide (100 mL): Add 5 g of mixed bed resin to 100 mL of formamide, and stir gently on a stirrer for 30 min. Filter using a nuclease-free filter or a funnel with Whatman 1 filter paper and store at −20 °C. Note that this should crystalize at −20 °C.

2.2 RNA Gel Electrophoresis and Transfer

1. 5X gel loading dye stock (5 mL): Dissolve 0.625 g of Ficoll 400 (12.5%) to 4.98 mL double-autoclaved DEPC-treated water or nuclease-free water. To this, add 2 μL EDTA (from 0.5 M stock at pH 8.0, to get a final concentration of 2 mM), and dissolve 12.5 mg of Bromophenol blue (0.25%) and 12.5 mg of xylene cyanol (0.25%). Store the dye at 4 °C in a RNAse-free or autoclaved DEPC-treated glass screw cap bottle.
2. 1.2% agarose-formaldehyde gel (50 mL): Melt 0.6 g of agarose in 44 mL of autoclaved-DEPC-treated water. Simultaneously, warm 5 mL of 10X mops buffer to 60 °C. Cool the melted agarose solution to 60 °C. To this, add the warm 10X Mops buffer and 1.5 mL of 35% formaldehyde. Mix thoroughly, yet gently to avoid air bubbles. Pour this solution into the RNase-free gel casting platform containing the comb. The gel casting platform and comb can be made RNAse-free by soaking them in DEPC water for 1 h. The tank of the electrophoresis apparatus can be used for soaking these so that it also becomes RNAse-free. Let the gel to solidify for 1 h.
3. 1X tank buffer (500 mL): To 455 mL of autoclaved DEPC-treated water, add 5 mL of 10 X MOPS buffer.
4. Transfer buffer, pH 6.5 (1000 mL): Dissolve 6 g of NaH2PO4 (50 mM) and 1.86 g of EDTA (5 mM) in 800 mL DEPC-treated water. Adjust the pH to 6.5 using 2 N NaOH. Make up the volume to 1000 mL with DEPC-treated water, and autoclave.

2.3 Northern Prehybridization

1. 20X saline-sodium phosphate-EDTA (SSPE) stock (1000 mL): Dissolve 31.202 g NaH_2PO_4 (0.2 M), 210.38 g NaCl (3.6 M), and 7.44 g EDTA (320 mM) in 800 mL of DEPC-treated water. Adjust the pH to 6.5 using 2 N NaOH. Make up the volume to 1000 mL using DEPC-treated water and autoclave.

2. 10% sodium dodecyl sulfate (SDS) stock (100 mL): Gently dissolve 10 g SDS in 100 mL of autoclaved DEPC-treated water or nuclease-free water. Allow the froth (if any) to settle.
3. 100X Denhart's solution stock (20 mL): In 15 mL of sterile DEPC-treated water, dissolve 0.4 g of Ficoll 400 (0.02%), 0.4 g of PVP (0.02%), and 0.4 g of BSA (0.02%). Make it up to 20 mL using sterile DEPC water, and filter it using nuclease-free syringe filter. Divide the solution into aliquots of 5 mL and store in −20 °C.
4. 1 mg/mL denatured calf thymus DNA solution (10 mL): Dissolve 10 mg of sheared calf thymus DNA in 10 mL of nuclease-free water by gentle inversion, overnight at 4 °C. Store at −20 °C.
5. Prehybridization mix (100 mL): Mix 50 mL of deionized formamide (50%), 25 mL of 20X SSPE (5X), 10 mL of 10% SDS (1%), 5 mL of 100X Denhart's solution (5X), 5 mL of 1 mg/mL denatured calf thymus DNA (50 μg/mL), and 5 mL of double autoclaved DEPC-treated water or nuclease-free water.

2.4 Random Primer Labeling of DNA

1. Nick translation dye mix (1 mL): Add 6 mg of Blue Dextran and 1 mg of Orange-G to 1 mL of 0.5 M EDTA (pH 7.0), and mix gently.
2. 1 U/μL Klenow fragment: Directly used as provided by the supplier.
3. Random primer labelling mix containing dATP, dGTP, dTTP: Directly used as provided by the supplier.
4. (α^{32}P)dCTP: Directly used as provided by the supplier (*see* **Notes 8** and **9**)
5. Column buffer (100 mL): To 88.92 mL of distilled water, add 10 mL of 1 M NaCl, 0.6 mL of 2 M Tris (pH 7.0), and 0.48 mL of 0.5 M EDTA (pH 7.0), and autoclave.
6. Sephadex G-50: Add 0.4 g of Sephadex G-50 to 6 mL of column buffer. Autoclave and store at 4 °C.
7. Sephadex G-50 column: Plug some glass wool into the neck area of a 2 mL Pasteur pipette (*see* **Note 11**). Fill 70% of this column with the autoclaved Sephadex G-50 (*see* **Note 12**).

2.5 X-ray Film Developer and Fixer

1. Developer (500 mL): Prepare according to the manufacturer's protocol, and store in brown (amber) bottle. Do not mix with the fixer.
2. Fixer (500 mL): Prepare according to the manufacturer's protocol, and store in brown (amber) bottle. Do not mix with the developer.

3 Methods

3.1 RNA Gel Electrophoresis

1. To a 15 μL premix made up of 10X MOPS (2 μL), 35 % formaldehyde (3 μL), and deionized formamide (10 μL), add 10 μg of DNase-treated RNA (up to 5 μL), bringing the total volume to 20 μL with double-autoclaved DEPC-treated water or nuclease-free water. Immediately, warm the samples to 60 ° C for 10 min. To this, mix 5 μL of 5X gel loading dye and load it immediately. Follow the same treatment for RNA ladder preparation as well.
2. Carefully remove the comb of the agarose-formaldehyde gel. Transfer the gel to the RNAse-free tank of the electrophoresis apparatus (*see* **Note 1**). Pour 1X Mops buffer into the tank to sufficiently submerge the gel. Load the RNA samples carefully into the wells. Run at a constant voltage of 100 V until the bromophenol blue dye has migrated sufficiently across the gel (*see* **Note 2**). Stain the gel in 0.1 mg/mL ethidium bromide for 20 min (*see* **Note 3**).
3. Transfer the gel to a tray containing 300 mL of transfer buffer. Incubate for 1 h to remove the formaldehyde. View in a UV gel documentation system, and take a photograph keeping the actual gel size (*see* **Note 4**).

3.2 Northern Transfer

1. The RNA is transferred from the gel to the nylon membrane (Hybond N^+) (*see* **Note 5**) by capillary transfer method (Fig. 1; similar to Southern blotting) using the transfer buffer.
2. After a transfer of 24 h, the transfer setup is carefully dismantled using a flat-tipped forceps, to reach the membrane. The wet nylon membrane is carefully lifted and placed with the gel-exposed region facing upward, on a piece of Whatman no. 1 paper (larger than the nylon membrane). This is air-dried, UV cross-linked, and finally baked for 2 h at 80 °C in a vacuum oven to fix the RNA to the membrane and to reverse the formaldehyde reaction.

3.3 Prehybridization

1. Carefully place the membrane into the hybridization bottle, with the gel-exposed side facing inward.
2. Warm the prehybridization solution to 60 °C, and pour about 50 $\mu L/cm^2$ of this into the hybridization bottle (*see* **Note 6**). Incubate this bottle at 42 °C at 10 rpm for 3 h in the hybridization oven.

3.4 Random Primer Labeling of the Probe DNA

1. To 50 ng of the DNA chosen for probe preparation (*see* **Note 7**) in a microfuge tube, add 5 μL of random primers. Make up the volume to 33 μL with sterile distilled water or nuclease-free water.

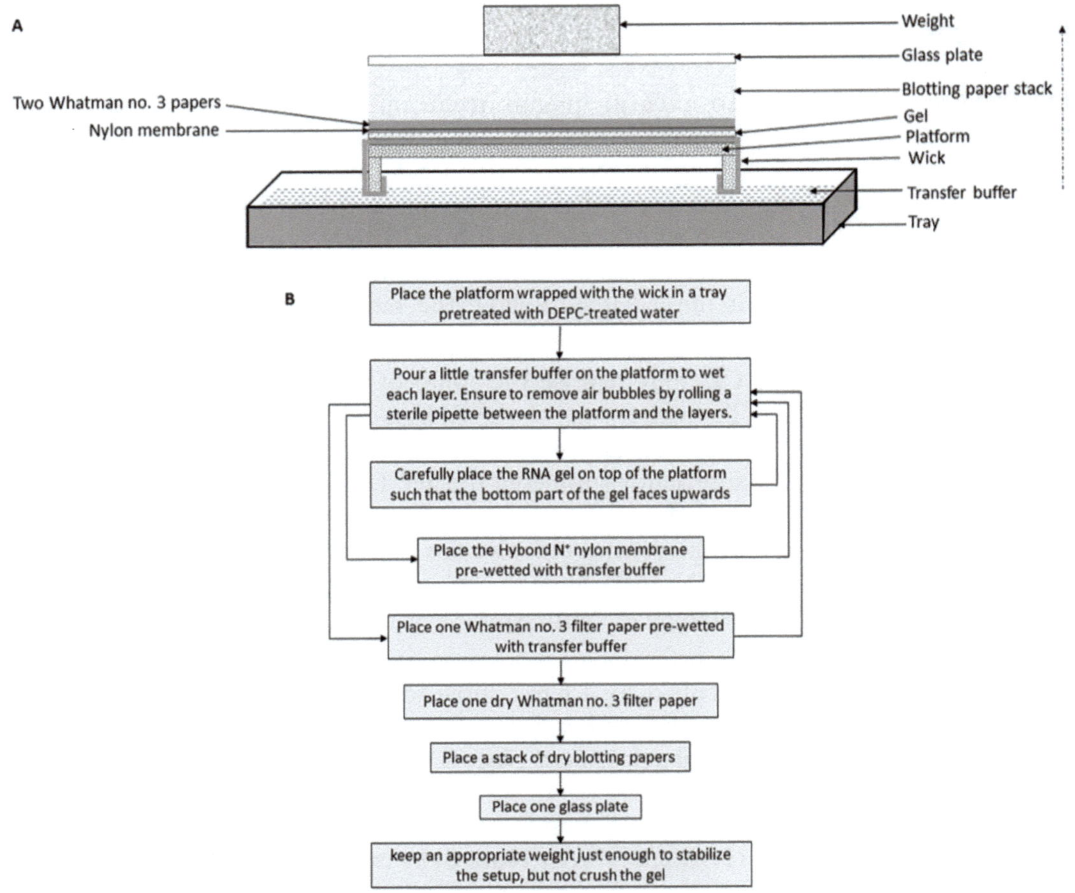

Fig. 1 Capillary transfer method of Northern blotting. (**a**) Northern blotting setup adapted from Southern (1975). (**b**) Steps for arranging the capillary transfer setup. The Hybond N^+ nylon membrane, Whatman no. 3 filter papers, and blotting papers are cut to the same size of the gel. The wick is made by cutting a Whatman filter paper no. 3 with length exceeding the length of the electrophoresis gel casting platform (the same one that was used for casting this gel) by 10–15 inches and breadth same as that of the platform. This paper is placed on the platform, and the extended paper from both sides is tucked in so as to form a wick, which would transfer the buffer from the tray toward the dry blotting paper stack. The dotted arrow indicates the direction of the capillary transfer

2. Denature the DNA by heating for 5 min in a boiling water bath. Allow it to cool down gradually to room temperature (for primer annealing).
3. Centrifuge for 5 s, and make up the volume to 33 μL with sterile distilled water or nuclease-free water.
4. To this DNA, add 10 μL of random primer labelling mix containing dATP, dGTP, and dTTP.
5. Add 5 μL (50 μCi) of (α^{32}P)dCTP (*see* **Notes 8** and **9**) and 2 μL (2 U) of Klenow fragment. Mix gently by tapping. Centrifuge for 5 s in a microfuge (*see* **Note 10**).

6. Incubate at 37 °C for 20 min in a water bath (*see* **Note 10**).
7. Stop the reaction by adding 50 μL of nick translation dye mix.

3.5 Purification of Labeled DNA

1. Keep the column on a stand (*see* **Note 10**) behind the protective shield. Keep a small container to collect the waste flow through.
2. Wash the column by allowing about 10 mL of column buffer to gradually flow out through the column.
3. Gently add more column buffer into the column, and carefully load the radio-labeled DNA mix into the column buffer. Allow it to enter completely into the Sephadex G-50 matrix. Gently add some more column buffer.
4. Collect the blue fraction separately in a 1.5 mL nuclease-free microfuge tube (*see* **Note 13**).

3.6 Hybridization

1. Denature the probe (blue fraction) by adding equal volume of deionized formamide and heating for 10 min at 90 °C. Snap cool on ice for 10 min (*see* **Note 10**).
2. Meanwhile, replace the prehybridization solution from the hybridization bottle with 10 mL of freshly added prehybridization solution warmed to 42 °C (*see* **Note 14**).
3. Add the denatured probe to this solution (*see* **Note 14**) and hybridize in the hybridization oven at 10 rpm for 24 h at 42 °C.

3.7 Washing

1. Carefully discard the hybridization solution. Add 40 mL of 0.1 X SSPE/0.5% SDS at room temperature, and briefly wash the membrane without removing it from the bottle. Replace this solution with 40 mL of same buffer preheated to 42 °C, and rotate at 10 rpm for 30 min at 42 °C in the hybridization oven. Once again repeat the last step.
2. Wash the membrane once with 0.1X SSPE at room temperature to remove the SDS.
3. Carefully remove the membrane from the bottle using a flat-tipped forceps, and air-dry it on a Whatman filter paper piece (slightly larger than the membrane) (*see* **Note 15**).
4. Wrap and seal the membrane in a plastic sheet (*see* **Note 16**).
5. Expose the wrapped membrane to an X-ray film (with labeled side facing the film), in a cassette with intensifier screen (*see* **Note 17**).
6. Keep it in a −70 °C freezer for 24 h (*see* **Note 18**).

3.8 X-ray Film Developing

1. Bring the X-ray cassette to room temperature before developing.
2. Inside the dark room, pour the developer and fixer in two different trays. In between these two trays, keep a tray of tap water.

3. Inside the dark room (*see* **Note 19**), remove the X-ray film from the cassette by holding one of its corners, and place it in the tray with developer solution. Ensure to thoroughly dip the film into the solution, gently rocking the tray for approximately 2 min.
4. Lift the film and hold it vertically, allowing the developer to drip. Soon transfer it to the tap water-containing tray, and wash by rocking the tray for 1 min.
5. Lift the film and hold it vertically, allowing the water to drip. Soon transfer it to the next tray containing the fixer. Ensure that the film is thoroughly dipped into the fixer solution. Gently rock the tray for approximately 2 min.
6. Lift the film and hold it vertically, allowing the fixer to drip. Wash the film thoroughly in running tap water.
7. Rinse with single-distilled water and air-dry the film by hanging it.

3.9 Identification of the Lanes

The lanes and the RNA ladder sizes can be identified and annotated (Fig. 2) by superimposing the developed X-ray film onto the gel image.

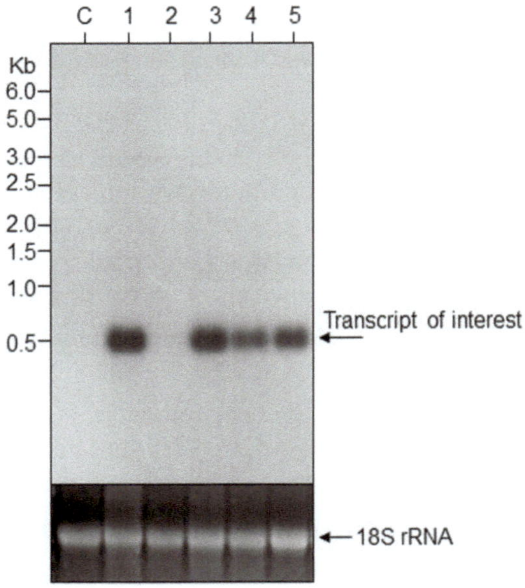

Fig. 2 Northern blot of transgenic rice lines. The panel below the Northern blot is the cropped ethidium bromide-stained gel image, representing the relative intensities of RNA samples. C denotes RNA from a non-transformed control plant. The numbers 1, 2, 3, 4, and 5 denote RNA from transformed rice plants. The absence of band in lane 2 indicates lack of transgene expression in this plant

4 Notes

1. Use a clean tank. To make it RNAse-free, rinse it with DEPC-treated water, and leave it wetted for at least 1 h. Air-dry it and use for electrophoresis.
2. Though about 70–80% migration across the gel by the dye is generally sufficient, this distance depends on the size of the expected target transcript. Therefore, the separation and sizes of ladder bands should be considered for deciding the dye migration distance.
3. Do not to touch ethidium bromide with bare hands/other body parts. Wear gloves while handling ethidium bromide. Do not contaminate any place/equipment with ethidium bromide-contaminated gloves. Remove the gloves carefully so that ethidium bromide does not touch your hands. Wash your hands well with soap. Commercially available alternatives for ethidium bromide can be also used, for which the instructions of the respective manufacturer need to be followed.
4. Take a printout of this image because it will be required later for superimposing the exposed X-ray sheet onto it and marking the ladder, wells, and the gel boundary, to correctly interpret the Northern hybridization result. Alternatively, photograph with a scale alongside so that it can be correlated with the image of the exposed X-ray sheet with the same scale. Alternatively, in the absence of a gel documentation system, trace of the gel covered with a transparent plastic sheet also can be taken under a shielded UV trans-illuminator, by wearing goggles and not getting any body parts exposed to the UV radiation.
5. Mark an “R” symbol with a pencil on the top right corner of the membrane. This indication is required identify the corners, and thus the orientation of the gel, after hybridization.
6. Prehybridization is done to reduce the background.
7. Electroelute or column-purify the DNA fragment desired for probe preparation (preferably 300–1000 bp long), derived after restriction digestion or PCR or reverse transcription. This is required to reduce the chances of background due to nonspecific hybridization.
8. Note to wear gloves and coat and take all precautionary measures to handle and discard the radioactive material. Perform radiolabeling reactions behind the radioactive shield. Read and follow the instructions provided by the radioactive material supplier. All work involving radioactivity should be done in the separate room exclusively allotted for such work.
9. Note that the specific radioactivity of the probe should be at least 1 × 108 CPM, which can be measured by scintillation

counting. Therefore, it is better to use the hot (α^{32}P)dCTP within 1 week of arrival to get good X-ray signals. As the half-life of radioactive ^{32}P is 14.3 days, the signals become fainter after that.

10. Keep aside a separate centrifuge, a water bath, a pipette stand, and ice bucket for handling radioactive material-containing microfuge tubes or columns. Do not use these for common purpose.
11. Wear faceshield and gloves while handling glass wool. Do not accidentally inhale or touch the glass wool with bare hands as it is toxic. Using pointed forceps, transfer some glass wool through the wider mouth of the pipette, and stuff it into the neck part. The purpose of glass wool is to prevent Sephadex G-50 to flow out of the column. Therefore, fill accordingly (about 10% of the pipette length).
12. Sephadex G-50 is insoluble in the column buffer. Hence, shake the bottle before pipetting out the Sephadex G-50 particles. For convenience, fix the glass wool-containing column to a pipette stand. First fill it with column buffer. Ensure that drops of the buffer flow out of the column. Keep a container to collect the wasted drops. Start adding the suspended Sephadex G-50 into the buffer in the column. While it is settling, gradually add some more to pack about 70% of the column. The filling should be such that it permits drop-by-drop slow falling of the buffer. If required, the wider end of the column can be wrapped with Parafilm to prevent evaporation.
13. All drops other than the blue fraction should be collected separately. Note that this waste flow through is also radioactive. Keep aside a few drops of orange fraction to mark the corners of the washed membrane.
14. Gradually decant the solution, not letting the membrane to fall. Do not touch the membrane directly with the probe.
15. While the membrane is left for drying, mark two tiny dots of the radioactive orange flow through using the sharp end of a toothpick, on the "R" marked on the membrane. This indicates top-right corner of the gel. Mark one tiny dot on the top-left corner of the membrane. These dots when exposed to the X-ray sheet would be leave an indication of the orientation of the original gel.
16. Take a transparent plastic cover that is slightly larger than the membrane. Cut it open from two adjoining sides. Place the membrane within this, and seal the cover with a sticky tape. Note to keep bits of the tape ready beforehand, behind the safety shield.

17. Perform this is a dark room. Open the fresh X-ray film only in dark. Exposing the X-ray film to light will completely mask the hybridization result. Note not to expose the intensifier screen directly to membrane; always have an X-ray sheet in between.
18. This time depends on the intensity of the radioactivity of the label. If required, the membrane can be re-exposed to a new X-ray film for more/less time depending on the band intensity obtained after developing the first film. The older the label, longer is the required exposure time.
19. Before opening the X-ray cassette, keep the positions of the trays in mind because their locations have to be identified in the dark. Faint red light can be used in the dark room if required.

Acknowledgments

The authors Sripriya R and Shah JM gratefully acknowledge Prof. K. Veluthambi and Mrs. Balamani Veluthambi for training them in radioactive Northern hybridization technique.

References

1. Trayhurn P (1996) Northern blotting. Proc Nutr Soc 55:583–589
2. He SL, Green R (2013) Northern blotting. Methods Enzymol 530:75–87
3. Shah JM, Raghupathy V, Veluthambi K (2009) Enhanced sheath blight resistance in transgenic rice expressing an endochitinase gene from *Trichoderma virens*. Biotechnol Lett 31:239–244
4. Sripriya R, Parameswari C, Veluthambi K (2017) Enhancement of sheath blight tolerance in transgenic rice by combined expression of tobacco osmotin (*ap24*) and rice chitinase (*chi11*) genes. In Vitro Cell Dev Biol Plant 53: 12–21
5. Hayakawa T, Zhu Y, Itoh K, Kimura Y, Izawa T, Shimamoto K, Toriyama S (1992) Genetically engineered rice resistant to rice stripe virus, an insect-transmitted virus. Proc Natl Acad Sci U S A 89:9865–9869
6. Shah JM, Singh R, Veluthambi K (2013) Transgenic rice lines constitutively co-expressing *tlp-D34* and *chi11* display enhancement of sheath blight resistance. Biol Plant 57:351–358
7. Sambrook J, Fritsch EF, Maniatis T (1989) Molecular cloning a laboratory manual. Cold Spring Harbor Laboratory Press, Cold Spring Harbor
8. Southern EM (1975) Detection of specific sequences among DNA fragments separated by gel electrophoresis. J Mol Biol 98:503–517
9. Sripriya R, Raghupathy V, Veluthambi K (2008) Generation of selectable marker-free sheath blight resistant transgenic rice plants by efficient co-transformation of a cointegrate vector T-DNA and a binary vector T-DNA in one *Agrobacterium tumefaciens* strain. Plant Cell Rep 27:1635–1644
10. Shah JM, Veluthambi K (2010) *DIANTHIN*, a negative selection marker in tobacco, is non-toxic in transgenic rice and confers sheath blight resistance. Biol Plant 54:443–450

Chapter 23

Isolation and Transfection of *Nicotiana benthamiana* Mesophyll Protoplasts

Julius Jeyaprakasam and Jebasingh Tennyson

Abstract

Protoplasts, being plant cells without cell walls, can efficiently uptake large macromolecules such as DNA, RNA, and proteins. This feature makes them exceptionally effective for transient gene expression experiments. This chapter provides a detailed protocol for the efficient isolation of mesophyll protoplasts from the model plant *Nicotiana benthamiana.* It also covers the subsequent transfection of these isolated protoplasts using PEG-$CaCl_2$ method with plasmid DNA, offering a comprehensive guide to ensure high-quality protoplast preparation and transfection for a successful genetic manipulation.

Key words Protoplasts, *Nicotiana benthamiana*, Transfection, PEG, $CaCl_2$

1 Introduction

Nicotiana benthamiana is indigenous to Australia, and it is extensively cultivated in research centers worldwide [1]. It is an important host for numerous plant pathogens, especially viruses, which makes it a key model for investigating plant-pathogen interactions [2]. It has widespread use due to its outstanding suitability for studying plant immunity and defense mechanisms, facilitated by its effectiveness in virus-induced gene silencing and transient gene expression studies [3–5]. *N. benthamiana* is equipped with a variety of functional genomics tools, including viral vectors, RNA interference (RNAi), ethyl methane sulfonate (EMS) mutagenesis, agroinfiltration, protoplast transfection, and *Agrobacterium*-mediated stable transformation. These methodologies support a wide range of research areas, including genomics, biochemistry, metabolomics, cell biology, pathology, and agricultural applications [6]. Protoplasts are increasingly utilized in various genome editing techniques, such as evaluating gRNA (guide RNA) efficiency [7], enabling transgene-free editing [8], performing base editing

Nitish Kumar (ed.), *Transgenic Plants: Methods and Protocols*, Methods in Molecular Biology, vol. 2973, https://doi.org/10.1007/978-1-0716-4803-2_23,

[9, 10], conducting prime editing [11], and facilitating CRISPR activation [12, 13].

In particular, *N. benthamiana*, with its large allotetraploid genome of 3.1 Gb [14], offers significant challenges for genome editing and functional gene studies. The CRISPR-Cas (clustered regularly interspaced short palindromic repeats-CRISPR-associated protein) genome-editing system will overcome these challenges by facilitating the precise DNA cleavage through the Cas protein, which is directed to specific DNA sites by a guide RNA (gRNA) [15]. The adoption of CRISPR-Cas has notably accelerated advancements in plant research and crop breeding in recent years [12, 15–17].

Transfer of genetic material can be accomplished through various methods, depending on the species involved. A key challenge in gene transfer is the cell membrane's hydrophobic layer, which acts as a barrier to genetic modification. Additionally, the highly charged nature of DNA makes its diffusion across the membrane particularly difficult [18].

Gene transfer methods fall into two main categories: indirect and direct [19]. Indirect methods use biological techniques, such as *Agrobacterium tumefaciens*-mediated transformation. In contrast, direct methods involve physical or chemical techniques to penetrate the cell wall. The direct chemical techniques include calcium phosphate co-precipitation, Lipofection, PEG-mediated transfection, and DEAE-dextran- mediated transfection. Direct physical methods include biolistic transformation, microinjection, macroinjection, electroporation, agitation with glass beads, ultrasound waves, shock wave-mediated techniques, vacuum infiltration, silicon carbide whiskers, and laser microbeams.

These methods each have their limitations. For instance, *Agrobacterium*-mediated transformation is restricted by plant genotype and explant type, while particle bombardment can lead to random DNA integration and significant tissue damage, which impairs regeneration efficiency [20, 21].

Protoplast transient transfection stands out as an excellent alternative. This method involves introducing plasmids encoding Cas and gRNA or pre-assembled Cas- ribonucleoprotein complexes (RNPs) into protoplasts [22]. Since protoplasts are single cells, any gene editing performed will be uniformly present in all cells of the regenerated plants [16, 23, 24]. PEG-mediated protoplast transfection is particularly effective due to its high efficiency and viability. It directly introduces genetic material into cells without complex vectors or carriers, thereby minimizing the risk of incorporating extra foreign DNA [23–25].

Here we provide the protocol addresses the isolation of protoplasts from *Nicotiana benthamiana* leaf mesophyll and effective transfection of plasmid DNA using PEG- $CaCl_2$-mediated methods.

2 Materials

2.1 Solutions and Reagents

1. 0.2 M MES KOH, pH 5.7: Dissolve appropriate amount of 4-morpholineethanesulfonic acid (MES) in molecular-grade water; adjust pH to 5.7 with KOH, and sterilize by passing through a 0.45 μm filter.
2. 0.8 M Mannitol: Dissolve appropriate amount of mannitol in molecular-grade water, and sterilize by passing through a 0.45 μm filter.
3. 1 M $CaCl_2$: Dissolve appropriate amount of calcium chloride in molecular-grade water, and sterilize by passing through a 0.45 μm filter.
4. 2 M KCl: Dissolve appropriate amount of potassium chloride in molecular-grade water, and sterilize by passing through a 0.45 μm filter.
5. 2 M $MgCl_2$: Dissolve appropriate amount of magnesium chloride in molecular-grade water, and sterilize by passing through a 0.45 μm filter.
6. 10% (wt/vol) BSA: Dissolve appropriate amount of bovine serum albumin in molecular-grade water, and sterilize by passing through a 0.45 μm filter.
7. Cellulase R10: Use as supplied, prepare working solution according to protocol if required.
8. Macerozyme R10: Use as supplied, prepare working solution according to protocol if required.
9. Enzyme solution: 20 mM MES (pH 5.7), 1.5% (wt/vol) cellulase R10, 0.4% (wt/vol) Macerozyme R10, 0.4 M mannitol, and 20 mM KCl; heat at 55 °C for 10 mins to inactivate DNases and proteases, which enhance enzyme solubility, and then cool to room temperature (25 °C), and then add 10 mM $CaCl_2$, 1–5 mM β-mercaptoethanol (optional), and 0.1% BSA.
10. 40% PEG 4000: Dissolve appropriate amount of polyethylene glycol (PEG 4000) in molecular-grade water, and sterilize by passing through a 0.45 μm filter.
11. Washing and incubation (WI) solution: 4 mM MES (pH 5.7), 0.5 M mannitol, 20 mM KCl; store at room temperature (22–25 °C).
12. W5 solution: 2 mM MES (pH 5.7), 154 mM NaCl, 125 mM $CaCl_2$, and 5 mM KCl; store at room temperature.
13. MMG solution: 4 mM MES (pH 5.7), 0.4 M mannitol, and 15 mM $MgCl_2$; store at room temperature.

14. PEG–calcium chloride transfection solution: Composition and preparation details to be provided in the relevant section.
15. Molecular-grade water or sterile deionized water: Use for all reagent preparation.

2.2 Equipment

- Fluorescence microscope.
- Luminometer.
- Fluorometer.
- Bench-top centrifuge.
- NanoDrop or spectrophotometer.
- Laminar flow hood.
- Rockymax.
- 0.45 mm syringe sterilization filter.
- Generic razor blade (single-edged blade).
- Nylon mesh (75 mm).
- Improved Neubauer 0.1-mm-deep Reichert hemacytometer.
- 30 mL round-bottomed tube.
- 2 mL round-bottomed natural microcentrifuge tube.
- 96-well plate (depending on the microscopic imaging systems).

3 . Methods

Maintain the laboratory temperature at 22–26 °C during all procedures.

3.1 Growing of Nicotiana benthamiana (~Timing 3–4 Weeks)

1. Sterilize *N. benthamiana* seeds by soaking in 75% ethanol for 1 min and then in a bleach solution for 20 min, and rinse thoroughly with autoclaved water (five times). Soak the seeds in sterilized water for 30 min to boost germination (*see* **Note 1**).
2. Place the sterilized seeds into a wide-mouth glass bottle with vermiculite in a greenhouse or a controlled environment chamber under in vitro conditions (*see* **Note 2**). Maintain a short photoperiod of 12 h light and 12 h dark at 20–22 °C, with low light intensity (75 μmol m^2 s^{-1}) and 40–65% relative humidity (*see* **Note 3**).

3.2 Protoplast Isolation (~Timing 5–8 h)

1. Choose tender, fully expanded leaves of 3–4 week-old *N. benthamiana* plants (five to seven leaves) for optimal protoplast isolation. Collect approximately 0.2–0.25 g of leaves before flowering and macerate them with 10–20 mL of enzyme solution within 5–6 h for the best results (*see* **Note 4**).

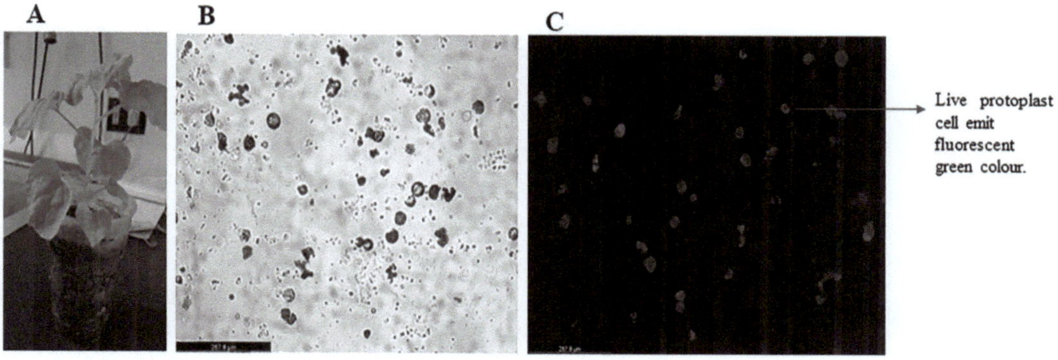

Fig. 1 Protoplast isolated from *N. benthamiana* leaf mesophyll cells. (**a**) *Nicotiana benthamiana* plant. (**b**) Protoplast under Brightfield. (**c**) Protoplasts stained with FDA. Scale bars: 100 μm

2. Use a fresh, sharp razor blade to cut 0.5–1 mm leaf strips from the central part of each leaf, avoiding crushing the tissue. A successful preparation usually produces approximately 10^7 protoplasts per gram of fresh weight. Digesting about 40–50 leaves in 80–100 mL of enzyme solution typically achieves this yield (*see* **Note 5**).
3. Using flat-tip forceps, quickly immerse the leaf strips (from five to ten leaves) into the enzyme solution (10–20 mL), ensuring that both sides of each strip are thoroughly covered (*see* **Note 6**).
4. Use a desiccator to vacuum infiltrate the leaf strips with the enzyme solution for 30 min in the dark to ensure complete solution absorption into the tissue.
5. Digest the protoplasts in enzyme solution at 22–23 °C with low shaking in a Rockymax at 10–12 rpm for 30 min. Then, continue the digestion in the dark at 22–23 °C for a minimum of 3 h without shaking. The enzyme solution should turn green upon gentle swirling, indicating successful release of protoplasts (*see* **Note 7**).
6. Examine the solution under the microscope to check for the release of protoplasts (see Fig. 1b, c). Mesophyll protoplasts typically measure approximately 30–50 μm in size.
7. Slowly add an equal volume of W5 solution to stop the enzymatic digestion and to remove undigested leaf tissues, then proceed with filtration (*see* **Note 8**).
8. Rinse a 75 mm nylon mesh with water to remove ethanol, and then remove excess water. Wet the mesh with W5 solution before filtering the enzyme solution containing protoplasts.
9. Spin the flow-through at 100 g for 1–2 min in a 50 mL round-bottom tube (20 mL enzyme solution + 20 mL of W5 solution)

to collect the protoplasts into a pellet. Carefully pour off the supernatant, and gently swirl to resuspend the protoplast pellet (*see* **Note 9**).

10. Take 10 μL of well-mixed protoplast suspension and load onto a hemocytometer to determine the protoplast density. Intact and healthy protoplasts should have a cell density of 2×10^5/mL in W5 solution. Incubate the solution on ice for 30 min.
11. Remove the W5 solution thoroughly by pipetting it out with large orifice tips to eliminate residual enzymes and prevent dilution of the transfection reagents. After 30 min of incubation, protoplasts should start settling at the bottom of the tube by gravity (*see* **Note 10**).

3.3 Significance and Risk Assessment

Excessive sucrose concentration or prolonged exposure may lead to osmotic stress, causing protoplast lysis or shrinkage. The osmotic properties of the sucrose solution are carefully controlled to minimize the risk of protoplast damage.

1. Add 3 mL of 20% (w/v) sucrose solution and resuspend the protoplast pellet. Carefully mix the contents by gentle swirling and flicking the tube. Centrifuge at 200× *g* for 3 min. Carefully transfer the dark green, intact, and healthy protoplasts collected at the surface of the sucrose solution into a new 50 mL Oakridge tube (*see* **Note 11**).
2. Resuspend protoplasts at 2×10^5 mL^{-1} (usually 20,000 cells per mL) in MMG solution, which is used to maintain the protoplasts and facilitate the uptake of genetic material.

3.4 Checking the Viability of Protoplasts

Protoplast isolation and transfection conditions can induce stress or damage to the cells (*see* **Note 12**) [26]. Fluorochromes are critical for evaluating protoplast viability (see Fig. 1c) and detecting any damage incurred during isolation and culture. Commonly used dyes include fluorescein diacetate (FDA) [27], for labeling living cells and Evan's blue for dead cells [28]. Additional nuclear fluorochromes such as 4′,6-diamidino-2-phenylindole (DAPI) [29] and Hoechst 33342 can also be used [30].

3.5 DNA-PEG–Calcium Transfection (~Timing Less Than 1 Hr)

1. Add 10 μL plasmid DNA (10–20 μg of plasmid DNA of 5–10 kb in size) to a 2 mL microfuge tube (*see* **Note 13**).
2. Add 100 μL of protoplast suspension (approximately 2×10^4 protoplasts), and mix gently. Next, add 110 μL of 40% PEG solution and thoroughly mix by gently tapping the tube.
3. Incubate the transfection mixture at room temperature for 10–15 min for efficient transfection (*see* **Notes 14** and **15**). Optimize the transfection time empirically [31].

4. Stop the transfection process by diluting the mixture with 400–440 μL of W5 solution at room temperature. Mix thoroughly by gently rocking or inverting the tube.
5. Centrifuge at 100 g for 2 min at room temperature using a bench-top centrifuge. Remove the supernatant and resuspend protoplasts gently with 1 mL WI solution in each well of a 6 well tissue culture plate (*see* **Note 16**).
6. Examine live protoplasts for any fluorescent protein-fused proteins (e.g., GFP, YFP, or CFP fusion proteins) using fluorescence microscopy (400–600 nm) without further sample preparation.
7. Transfer the protoplast solution to 2 mL round-bottom microcentrifuge tubes. Rinse the wells with 200 μL of W5, and add this rinse to the tubes. Centrifuge the tubes at 50 g for 1–2 min using a swing-out rotor with no braking. Carefully remove the supernatant without disturbing the protoplast pellet. The pellet is now ready for further analysis, such as Western blotting, RNA extraction, or qRT-PCR [32, 33].

3.6 Timing

Step 1: 3–4 weeks.

Steps 2–5: approximately 1 h (depending on numbers of leaves being collected).

Steps 5–14: approximately 5–6 h (depending on the digestion of leaf strips).

Steps 15–21: approximately 1 h (depending on numbers of samples being transfected).

Step 22–23: approximately 1–2 h (depending on the assay).

4 Notes

1. Enzyme Solution Preparation: Preheat the MES solution at 70 °C for 3–5 min before adding the enzymes to achieve a clear light brown solution. Filter the solution through a 0.45 μm syringe filter to ensure clarity. Prepare the enzyme solution fresh for optimal results.
2. PEG Solution Preparation: Prepare the PEG solution at least 1 h prior to transfection to ensure it is fully dissolved. Store at room temperature and use within 5 days for best results. Freshly prepared PEG solution is recommended for optimal protoplast transfection efficiency. Do not autoclave the PEG solution.
3. *N. benthamiana* Growth Conditions: Maintain a stable environment to prevent environmental fluctuations such as

drought, flooding, extreme temperatures, and mechanical disturbances that can affect plant health.

4. Leaf Selection for Protoplast Isolation: Use young, immature leaves. Mature leaves with thicker cell walls and higher secondary metabolite levels may require longer digestion times and yield fewer protoplasts. Avoid yellowed or unhealthy leaves.
5. Leaf Preparation: Ensure the enzyme solution is ready before cutting the leaves to streamline the process. Replace the blade after cutting four to five leaves, and use a clean white paper for precision and contamination prevention.
6. Submerging Leaf Strips: Quickly and completely submerge the leaf strips in the enzyme solution to maximize protoplast yield.
7. Digestion Time: Protoplasts are typically released within 3 h. Extending digestion to 16–18 h in the dark may stress the leaves and alter physiological responses. Avoid over-digestion, which can damage protoplasts.
8. Protoplast Yield: A good preparation yields approximately 10^6 protoplasts per 0.2–0.25 g of fresh leaf weight (from five to seven leaves digested in 10 mL of enzyme solution).
9. Centrifugation Speed: For protoplast recovery, use a centrifugation speed of 100–200 g (RCF) to avoid damaging protoplasts.
10. Gentle Handling: Handle protoplasts delicately throughout the procedure to avoid damage. Use wide-bore pipette tips or cut automatic pipette tips to minimize shear forces. Avoid touching the protoplast pellet directly [34].
11. Sucrose Solution Density: Note that a 0.55 M sucrose solution is denser than the W5-protoplast solution, so the latter will remain on top after pipetting [33].
12. W5 Solution: Minimize residual W5 solution as it can reduce transfection efficiency. Replace the W5 supernatant with MMG solution before PEG-$CaCl_2$-mediated transfection.
13. DNA Purity and Quantity: Ensure plasmid DNA purity with an A260/A280 ratio of 1.80–2.00. Typically, 10–30 μg of plasmid DNA is used for transfection, isolated using a Midiprep kit from 50 mL *Escherichia coli* overnight cultures [31].
14. Optimal Mannitol Concentration: Use 0.6 M mannitol for the transfection process for optimal results.
15. Transfection Buffer: Replace W5 with MMG solution before transfection for better results.
16. Protoplast Viability: To enhance viability, use sterile calf serum (5% vol/vol) to coat the culture plate briefly (1–2 s) to prevent protoplasts from sticking to the surface.

References

1. Knapp S, Bohs L, Nee M, Spooner DM (2004) Solanaceae—a model for linking genomics with biodiversity. Comp Funct Genomics 5(3): 285–291
2. Goodin MM, Zaitlin D, Naidu RA, Lommel SA (2008) *Nicotiana benthamiana*: its history and future as a model for plant–pathogen interactions. Mol Plant-Microbe Interact 21(8): 1015–1026
3. Abrahamian P, Hammond RW, Hammond J (2020) Plant virus–derived vectors: applications in agricultural and medical biotechnology. Annu Rev Virol 7(1):513–535
4. Bally J, Jung H, Mortimer C, Naim F, Philips JG, Hellens R, Waterhouse PM (2018) The rise and rise of *Nicotiana benthamiana*: a plant for all reasons. Annu Rev Phytopathol 56(1): 405–426
5. LeBlanc Z, Waterhouse P, Bally J (2020) Plant-based vaccines: the way ahead. Viruses 13(1):5
6. Derevnina L, Kamoun S, Wu CH (2019) Dude, where is my mutant? *Nicotiana benthamiana* meets forward genetics. New Phytol 221(2):607–610
7. Li JF, Norville JE, Aach J, McCormack M, Zhang D, Bush J, J. (2013) Multiplex and homologous recombination–mediated genome editing in Arabidopsis and *Nicotiana benthamiana* using guide RNA and Cas9. Nat Biotechnol 31(8):688–691
8. Zhang Y, Cheng Y, Fang H, Roberts N, Zhang L, Vakulskas CA, Qi Y (2022) Highly efficient genome editing in plant protoplasts by ribonucleoprotein delivery of CRISPR-Cas12a nucleases. Front Genome Edit 4:780238
9. Molla KA, Sretenovic S, Bansal KC, Qi Y (2021) Precise plant genome editing using base editors and prime editors. Nat Plants 7(9):1166–1187
10. Xiong X, Liu K, Li Z, Xia FN, Ruan XM, He X, Li JF (2023) Split complementation of base editors to minimize off-target edits. Nat Plants 9(11):1832–1847
11. Perroud PF, Guyon-Debast A, Casacuberta JM, Paul W, Pichon JP, Comeau D, Nogué F (2023) Improved prime editing allows for routine predictable gene editing in Physcomitrium patens. J Exp Bot 74(19):6176–6187
12. Li Z, Zhang D, Xiong X, Yan B, Xie W, Sheen J, Li JF (2017) A potent Cas9-derived gene activator for plant and mammalian cells. Nat Plants 3(12):930–936
13. Pan C, Wu X, Markel K, Malzahn AA, Kundagrami N, Sretenovic S, Qi Y (2021) CRISPR–Act3. 0 for highly efficient multiplexed gene activation in plants. Nat Plants 7(7):942–953
14. Bombarely A, Rosli HG, Vrebalov J, Moffett P, Mueller LA, Martin GB (2012) A draft genome sequence of *Nicotiana benthamiana* to enhance molecular plant-microbe biology research. Mol Plant-Microbe Interact 25(12): 1523–1530
15. Yue JJ, Hong CY, Wei P, Tsai YC, Lin CS (2020) How to start your monocot CRISPR/Cas project: plasmid design, efficiency detection, and offspring analysis. Rice 13:1–13
16. Hsu CT, Cheng YJ, Yuan YH, Hung WF, Cheng QW, Wu FH et al (2019) Application of Cas12a and nCas9-activation-induced cytidine deaminase for genome editing and as a non-sexual strategy to generate homozygous/multiplex edited plants in the allotetraploid genome of tobacco. Plant Mol Biol 101:355–371
17. Li S, Xia L (2020) Precise gene replacement in plants through CRISPR/Cas genome editing technology: current status and future perspectives. Abiotech 1(1):58–73
18. Sawahel W, Fukui K (1995) Gene cloning in plants: innovative approaches. BioTechniques 19(1):106–110
19. Niazian M, Noori SAS, Galuszka P, Mortazavian SMM (2017) Tissue culture-based Agrobacterium-mediated and in planta transformation methods. Czech J Genet Plant Breed 53(4):133–143
20. Anami S, Njuguna E, Coussens G, Aesaert S, Van Lijsebettens M (2013) Higher plant transformation: principles and molecular tools. Int J Dev Biol 57(6–8):483–494
21. Demirer GS, Zhang H, Matos JL, Goh NS, Cunningham FJ, Sung Y et al (2019) High aspect ratio nanomaterials enable delivery of functional genetic material without DNA integration in mature plants. Nat Nanotechnol 14(5):456–464
22. Altpeter F, Baisakh N, Beachy R, Bock R, Capell T, Christou P, Visser R (2005) Particle bombardment and the genetic enhancement of crops: myths and realities. Mol Breed *15:305–327*
23. Lin CS, Hsu CT, Yang LH, Lee LY, Fu JY, Cheng QW et al (2018) Application of protoplast technology to CRISPR/Cas9 mutagenesis: from single-cell mutation detection to

mutant plant regeneration. Plant Biotechnol J 16(7):1295–1310

24. Woo JW, Kim J, Kwon SI, Corvalán C, Cho SW, Kim H et al (2015) DNA-free genome editing in plants with preassembled CRISPR-Cas9 ribonucleoproteins. Nat Biotechnol 33(11):1162–1164
25. Andersson M, Turesson H, Olsson N, Fält AS, Ohlsson P, Gonzalez MN, Hofvander P (2018) Genome editing in potato via CRISPR-Cas9 ribonucleoprotein delivery. Physiol Plant 164(4):378–384
26. Neelakandan AK, Wang K (2012) Recent progress in the understanding of tissue culture-induced genome level changes in plants and potential applications. Plant Cell Rep 31: 597–620
27. Qiu L, Wang Y, Qu H (2020) Loading calcium fluorescent probes into protoplasts to detect calcium in the flesh tissue cells of Malus domestica. Hortic Res 7
28. Kollárová K, Kusá Z, Vatehová-Vivodová Z, Lišková D (2019) The response of maize protoplasts to cadmium stress mitigated by silicon. Ecotoxicol Environ Safety 170:488–494
29. Yagi N, Yoshinari A, Iwatate RJ, Isoda R, Frommer WB, Nakamura M (2021) Advances in synthetic fluorescent probe labeling for live-cell imaging in plants. Plant Cell Physiol 62(8): 1259–1268
30. Islam MR, Youngblood M, Kim HI, González-Gamboa I, Monroy-Borrego AG, Caparco AA et al (2024) DNA delivery by virus-like nanocarriers in plant cells. Nano Lett
31. Nayak N, Mehrotra S, Mehrotra R (2024) Isolation and transfection of nicotiana benthamiana mesophyll protoplasts for fluorescent protein visualization
32. Dai Z, Wang A (2022) Isolation and transfection of plant mesophyll protoplasts for virology research. Plant Virol Methods Protocols 2400: 43–53
33. Wu FH, Hsu CT, Lin CS (2023) Targeted insertion in Nicotiana benthamiana genomes via protoplast regeneration. In: Plant genome engineering: methods and protocols. Springer, New York, pp 297–315
34. Yuan J, Li D, Liang Y, Meng Y, Li L, Yang L, Li J (2024) An optimum study on the laser scanning confocal microscopy techniques for BiFC assay using plant protoplast. Bot Stud *65*(1):2

Chapter 24

Regeneration of Transgenic *Nicotiana benthamiana* Raised from the Genome-Edited Protoplast

Shrinithi Kannan and Jebasingh Tennyson

Abstract

Plant genome editing is an emerging technique that has revolutionized plant genome engineering which helps to edit the plant genome precisely for the development of traits in many crops. Specifically, with clustered regularly interspaced short palindromic sequence (CRISPR)-CRISPR-associated protein (Cas) system, a progressive improvement in genome editing has been achieved with protoplast. Though protoplast isolation, transfection, and regeneration are available for many plants, regeneration of protoplast for many plants remains major challenge. In this methodology chapter, we outlined the construction of sgRNA for genome editing, transfection, and regeneration of transgenic *N. benthamiana* from the genome-edited protoplast and assay for gene targeting.

Key words Protoplast, Callus, Genome editing, *Nicotiana benthamiana*, PCR

1 Introduction

The genome editing in plants has evolved substantially during the past 20 years. ZFNs, TALENs, and the CRISPR-Cas system are the three generations of programmable nucleases that have been used to introduce selective DNA double-strand breaks (DSBs) in eukaryotic cells [1]. To investigate the function of any gene of interest, the cell's DNA repair system has been used to introduce insertions and deletions (indels) at the targeted DSBs. Previously, the insertions were "random" occurrences which are brought by "error-prone" DNA repair processes, but now with the genome editing techniques, the insertion is "nonrandom [2].

The CRISPR/Cas system is a genomic locus of tandem direct repeat sequences and protospacers, derived from invading elements [3]. It consists of CRISPR loci and a Cas9 protein, which forms a complex called single guide RNA (sgRNA) and determines the specificity of cleavage of the target sequence in the nucleic acid [4]. The Cas9 protein is an endonuclease associated with CRISPR

Nitish Kumar (ed.), *Transgenic Plants: Methods and Protocols*, Methods in Molecular Biology, vol. 2973,
https://doi.org/10.1007/978-1-0716-4803-2_24,

loci, responsible for double-stranded breaks (DSBs) at the site when targeted by a sgRNA [5]. This molecular mechanism opened up its potential applications in genome editing [6, 7]. *Nicotiana benthamiana* is a model dicot plant, which has vast applications in plant biology [8, 9]. Genome editing in *N. benthamiana*, involves *Agrobacterium*-mediated stable transformation to deliver DNA into target cells. However, mutant plants derived from *Agrobacterium*-mediated transformation could be considered genetically modified organisms (GMOs) [10]. Plasmids encoding Cas and sgRNA sequences or pre-assembled Cas-sgRNA ribonucleoprotein complexes can be delivered directly into protoplasts using transient transfection. Protoplasts are single plant cells obtained through cellulase and Macerozyme digestion. Polyethylene glycol (PEG)-mediated protoplast transfection and further regeneration from protoplast offers transgenic recombinant-DNA-free plants [11–15]. In this chapter, the materials and methods used for the successful transfection of Cas-sgRNA construct into protoplast and regeneration of transgenic *N.benthamiana* protoplast into plants and the evaluation of genome editing are discussed.

2 Materials

2.1 Equipment

1. 0.1-mm-deep hemacytometer.
2. Autoclave.
3. Laminar flow hood.
4. Microcentrifuge and swing-out rotor centrifuge.
5. Optical microscope.
6. Orbital shaker.
7. Thermal cycler.
8. Ultra-sonicator.
9. NanoDrop or spectrophotometer.
10. UV transilluminator.

2.2 Material

1. 0.22 μm syringe sterilization filter, with a polyethersulfone (PES) membrane or a mixed cellulose ester (MCE) membrane.
2. 15 mL centrifuge tube.
3. 50 mL centrifuge tube.
4. 40 μm nylon mesh.
5. Cell culture dishes (35 mm [diameter] × 10 mm [height]; 55 mm [diameter] × 15 mm [height]; 90 mm [diameter] × 15 mm [height]).
6. Glass Petri dish (60 mm [diameter] × 15 mm [height]).

7. Glass screw cap culture tubes (round bottom, 13 mm [diameter] × 100 mm [length] (*see* **Note 1**).
8. Test tubes (20 mm [diameter] × 150 mm [length]).
9. Wide-mouth glass bottle with cap (50 mm [diameter] × 80 mm [height]).
10. Cell spreader.
11. Mortar and pestle.
12. Genomic DNA plant purification kit.
13. Gel purification kit.

2.3 Chemicals

1. Tryptone.
2. Yeast extract.
3. Bacto Agar.
4. 1 M hydrochloric acid HCl (*see* **Note 2**).
5. Cas9 nuclease, *Streptococcus pyogenes* (*see* **Note 3**).
6. 5X polymerase chain reaction (PCR) mixture reagent.
7. Proofreading and high-fidelity DNA polymerase.
8. Ethanol (absolute for analysis).
9. 75% (v/v) ethanol.
10. Tween-20.
11. Agarose.
12. Chloroform.
13. Alcohol (24:1).
14. Isopropanol (chilled).
15. Ethanol (70%).
16. TE buffer (10 mM Tris-HCl, 1 mM EDTA, pH 8.0).
17. RNase A (10 mg/mL).
18. Liquid nitrogen.

2.4 Preparations of Reagents

Prepare all solutions using ultra-pure water (18 MΩ). Sterilize by autoclaving at 121 °C or filter with 0.22 μm syringe. Maintain the reagents at room temperature (22–26 °C) and all the prepared solutions, are operated in a sterile environment.

1. 1 M potassium hydroxide (KOH): Dissolve 0.56 g of KOH in ultra-pure water, and adjust to a final volume of 10 mL. Sterilize the solution by passing through a 0.22 μm syringe filter. The solution can be stored at room temperature for 6 months.
2. 0.1 M 2-(N-morpholino) ethane sulfonic acid (MES), pH 5.7: Dissolve 1.95 g of MES in ultra-pure water, and adjust pH to 5.7 with 1 M KOH and the final volume to 100 mL. Sterilize

the solution by passing through a 0.22 μm syringe filter. The solution can be stored at room temperature for 1 month.

3. 0.6 M potassium chloride (KCl): Dissolve 4.473 g of KCl in ultra-pure water, and adjust to a final volume of 100 mL. Sterilize the solution by passing through a 0.22 μm syringe sterilization filter. The solution can be stored at room temperature for 6 months.
4. 0.2 M potassium chloride (KCl): Dissolve 1.49 g of KCl in ultra-pure water, and adjust to a final volume of 100 mL. Sterilize the solution by passing through 0.22 μm syringe filter. The solution can be stored at room temperature for 6 months.
5. 10 mM magnesium chloride ($MgCl_2$): Dissolve 0.203 g of $MgCl_2$ in ultra-pure water, and adjust to a final volume of 100 mL. Sterilize the solution by passing through 0.22 μm syringe filter. The solution can be stored at room temperature for 6 months.
6. 10 mM Tris (2-carboxyethyl) phosphine hydrochloride (TCEP): Dissolve 0.287 g of TCEP in ultra-pure water, and adjust to a final volume of 100 mL. Sterilize the solution by passing through 0.22 μm syringe filter. The solution can be stored at room temperature for 6 months.
7. 0.5 M glucose: Dissolve 4.50 g of glucose in ultra-pure water, and adjust to a final volume of 50 mL. Sterilize the solution by passing through 0.22 μm syringe filter sterilization. The solution can be stored at 4 °C for 6 months.
8. 0.8 M D-mannitol: Dissolve 7.29 g of D-mannitol in ultra-pure water and adjust to a final volume of 50 mL. Sterilize the solution by passing through 0.22 μm syringe filter. The solution can be stored at room temperature for 2 months.
9. Half-strength Murashige and Skoog (MS) basal medium with vitamins: Dissolve 1.1 g of commercial MS stock powder and 15 g of sucrose in water; adjust pH to 5.7 with 1 M KOH, and adjust the final volume to 500 mL. Sterilize the solution by passing through 0.22 μm syringe filter or by autoclaving at 121 °C for 20 min. The solution can be stored at room temperature for 2 months.
10. 1 mg/mL kinetin: Dissolve 20 mg of kinetin in 2 mL of 1 M KOH, and adjust to a final volume of 20 mL with ultra-pure water. Sterilize the solution by passing through 0.22 μm syringe. The solution can be stored at 4 °C for 1 year.
11. 1 mg/mL IAA: Dissolve 20 mg of IAA in 2 mL of 1 M KOH, and adjust to a final volume of 20 mL with ultra-pure water. Sterilize the solution by passing through 0.22 μm syringe. The solution can be stored at 4 °C for 1 year.

12. 1 mg/mL naphthalene acetic acid (NAA): Dissolve 20 mg of NAA in 2 mL of 1 M KOH, and adjust to a final volume of 20 mL with ultra-pure water. Sterilize the solution by passing through 0.22 μm syringe filter. The solution can be stored at 4 °C for 1 year.
13. 1 mg/mL 6-benzylaminopurine (BA): Dissolve 50 mg of BA in 2 mL of 1 M KOH and adjust to a final volume of 50 mL with ultra-pure water. Sterilize the solution by passing through 0.22 μm syringe filter. The solution can be stored at 4 °C for 1 year.
14. 2 M calcium chloride ($CaCl_2$): Dissolve 29.40 g of $CaCl_2{\cdot}2H_2O$ in ultra-pure water, and adjust to a final volume of 100 mL with ultra-pure water. Sterilize the solution by passing through 0.22 μm syringe filter. The solution can be stored at room temperature for 6 months.
15. 20% sucrose: Dissolve 10 g of sucrose in ultra-pure water, and adjust to a final volume of 50 mL with ultra-pure water. Sterilize the solution by passing through 0.22 μm syringe filter. The solution can be stored at 4 °C for 1 month.
16. 3 M sodium chloride (NaCl): Dissolve 17.53 g of NaCl in ultra-pure water, and adjust to a final volume of 100 mL with ultra-pure water. Sterilize the solution by passing through 0.22 μm syringe sterilization filter. The solution can be stored at room temperature for 6 months.
17. 20 mg/mL 5-bromo-4-chloro-3-indolyl-β-D-galactopyranoside (X-gal): Dissolve 100 mg of X-gal in dimethylformamide (DMF), and adjust to a final volume of 5 mL with DMF. The solution can be stored in the dark at −20 °C for 1 year.
18. 50 mg/mL isopropyl β-D-1-thiogalactopyranoside (IPTG): Dissolve 250 mg of IPTG in ultra-pure water, and adjust to a final volume of 5 mL with ultra-pure water. The solution can be stored at −20 °C for 1 year.
19. 100 mM N-2-hydroxyethylpiperazine-N-2-ethanesulfonic acid (HEPES): Add 1 mL of commercial 1 M HEPES (pH 7.2–7.5) solution to 9 mL of ultra-pure water.
20. CTAB extraction buffer (2% CTAB): 2% CTAB, 100 mM Tris-HCl (pH 8.0), 20 mM EDTA, 1.4 M NaCl, and 2% (v/v) β-mercaptoethanol (added fresh before use).

2.5 Preparation of Solutions

1. Cas9 assembly buffer: Add 10 mL of 100 mM HEPES pH 5.7, 12.5 mL of 600 mM KCl, 5 mL of 10 mM $MgCl_2$, 5 mL of 10 mM TCEP, and 5 mL of glycerol, and adjust to a final volume of 50 mL with ultra-pure water. Sterilize the solution by passing through a 0.22 μm syringe filter and store at −80 °C.

2. W5 solution: Add 10.3 mL of 3 M NaCl, 12.5 mL of 2 M $CaCl_2$, 5 mL of 0.2 M of KCl, 4 mL of 0.1 M MES, pH 5.7, and 2 mL of 0.5 M glucose in autoclaved sterile water, and adjust to a final volume of 200 mL with water. The solution can be stored at room temperature for 2 months (*see* **Note 4**).
3. Transfection (PEG- Ca^{2+}) solution: Dissolve 4 g of PEG 4000, 5 mL of 0.8 M mannitol, 0.5 mL of 2 M $CaCl_2$, and 1 mL ultra-pure water to a final volume of 10 mL. The solution can be vortexed to accelerate the dissolution and sterilized by passing through 0.22 μm syringe filter with a PES membrane or MCE membrane. The PEG- Ca^{2+} solution must be freshly prepared before each protoplast transfection (*see* **Note 5**).

2.6 Preparations of Growth Medium

1. Callus induction medium: Half-strength MS solution supplemented with 3% (w/v) sucrose, 0.4 M mannitol, 1 mg/L NAA, and 0.3 mg/L kinetin, pH 5.7. Sterilize the solution by passing through 0.22 μm syringe filter.
2. Shoot-inducing medium: For liquid medium, to half-strength MS supplemented with 3% (w/v) sucrose, add 1 mL of 1 mg/mL BA, adjust the pH to 5.7 with 1 M KOH, and bring final volume to 500 mL. For solid medium, add 5 g Bacto Agar (final concentration: 1% [w/v]. Sterilize both the liquid and solid media by autoclaving. After sterilization, let the solid shoot-inducing medium cool down to 50 °C, and aliquot into cell culture dishes (90 mm [diameter] × 15 mm [height]) with approximately 25 mL of medium in each dish.
3. Root-inducing medium: MS supplemented with 3% (w/v) sucrose, add 0.5 μL of 1 mg/mL BAP, 2 μL of 1 mg/mL Kinetin, 1 μL of 1 mg/mL IAA, adjust pH to 5.2 with 1 N HCl, and bring final volume to 1 L. Add 10 g Bacto Agar (final conc. 1% [w/v] and 1 g/L activated charcoal, and sterilize by autoclaving. After sterilization, let the root medium cool down to 50 °C, and aliquot 10 mL into test tubes or 50 mL in wide-mouth glass bottles with caps.
4. 2xYT solid medium: Dissolve 16 g of tryptone, 10 g of yeast extract, and 5 g of NaCl in water, and adjust to a final volume of 1 L with ultra-pure water. Add 20 g Bacto Agar (final conc.2% [w/v]), and sterilize by autoclaving. After sterilization, let the medium cool down to 50 °C, add appropriate antibiotics, and aliquot 20 mL into cell culture dishes (90 mm [diameter] × 15 mm [height]).

3 Methods

All processes should be performed in aseptic conditions within laminar air flow chamber. Wash the young leaf (explants) of *Nicotiana benthamiana* with sterile water, isolate the protoplast from leaf explant by Cellulase R-10 and Macerozyme R10 enzyme solution, and check the density of the protoplast before transfection. (Refer chapter 23: Isolation and Transfection of Mesophyll Protoplasts from *Nicotiana benthamiana* leaves.)

3.1 Construction and Synthesis of CRISPR/Preassembled Cas-sgRNA and Donor DNA

Select the target sequence of 20-nucleotide sequence in the genome that is adjacent to a protospacer adjacent motif (PAM). For the Cas9 enzyme of *Streptococcus pyogenes*, the PAM sequence is "NGG" (where N can be any nucleotide). The target sequence must be upstream of the PAM in the target gene.

1. Select and screen the target sites using the CRISPR-Pv.2.0 website, which provides candidate sgRNAs by calculating off-target effects and ranking based on high on-target scores to minimize unintended genome edits (*see* **Note 6**) [16].
2. Use the RNAfold web server to predict the secondary structure of the single guide RNA (sgRNA) sequences to ensure they form the correct structure for effective CRISPR/Cas9 binding and activity [17, 18].
3. The single-guide RNA (sgRNA) used in the CRISPR/Cas9 system can be synthesized commercially, similar to primers. For in vitro synthesis of sgRNA, develop oligonucleotides that include the T7 promoter sequence, the protospacer sequence (target site), and the sgRNA scaffold sequence for RNA synthesis. T7 RNA polymerase requires a 5′ G for proper transcription initiation. This ensures that T7 RNA polymerase can initiate transcription properly, producing a functional sgRNA [19, 20] (Fig. 1a).
4. Alternatively, sgRNA can be produced from a vector that contains the sgRNA template. PCR is performed with a proofreading and high-fidelity DNA polymerase (such as Phusion DNA polymerase or Pfu DNA polymerase). A reaction volume of at least 50 μL is recommended to obtain a sufficient yield of amplified DNA.
5. Amplified product is run in 0.8–1% agarose gels visualize it on UV transilluminator to ensure that the PCR worked. Slice the desired DNA fragment from the gel with a sterile blade and purify PCR products with a commercial gel purification kit.
6. Use 1 μg of purified PCR product as template for sgRNA synthesis with an In Vitro Transcription Kit (IVT) kit using T7 RNA polymerase (*see* **Note 7**). Follow the instructions of

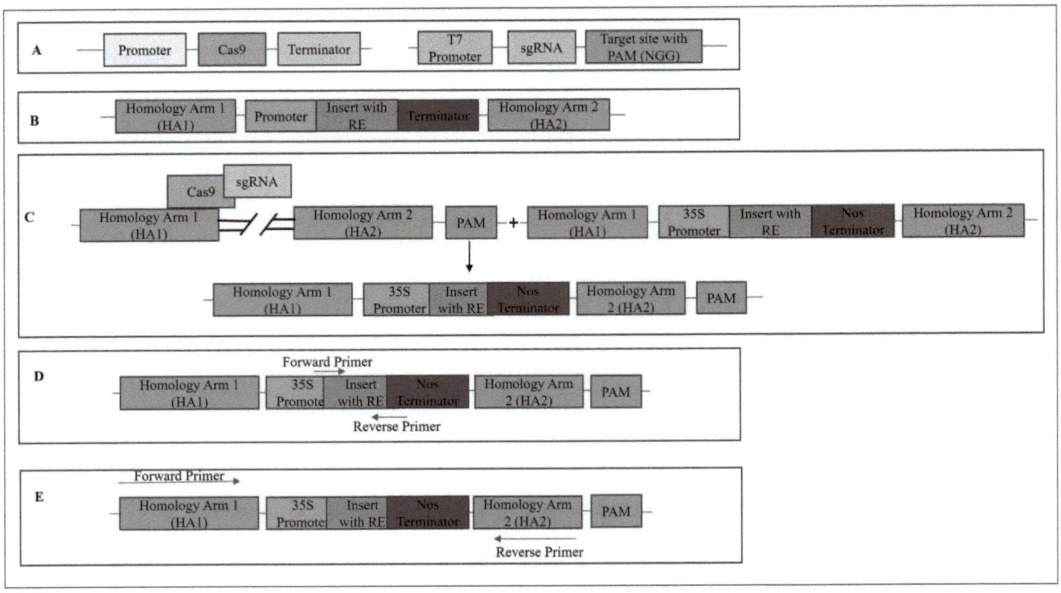

Fig. 1 CRISPR Cas assembly and transgene insertion in the genome of *Nicotiana benthamiana*. (**a**) Construction of CRISPR/preassembled Cas-sgRNA. (**b**) Construction of desired donor DNA containing novel Restriction Enzyme (RE); (**c**) CRISPR/Cas9 system induces a double-strand break at the target site of the genome by binding of sgRNA with PAM sequence. (**d**) To check whether the donor DNA is inserted at the target site, the PCR has to be performed with the transgene-specific primers (marked with arrow). (**e**) Insertion of donor DNA into the genome with homology flanking sequences and the insertion can be confirmed by the PCR with forward and reverse primers (marked with arrow) designed at the flanking homology sequences

the kit and incubate the synthesis reaction at 37 °C for overnight.

7. Remove the DNA template by adding 1 μL of RNase-free DNase I (1 U/μL) into the action mixture, and incubate at 37 °C for 15 min to 1 h.
8. Add 300 μL of absolute ethanol to the DNase I-treated sgRNA mixture, and adjust volume to 500 μL with DEPC-treated water.
9. Purify sgRNA with a small RNA purification kit according to the instructions. The sgRNA is dissolved in 20 μL of DEPC-treated water.
10. Run sgRNA mixture in 1.5–2% agarose gel, and measure the RNA concentration in a NanoDrop spectrophotometer. The 260/280 value should be approximately 2.0. The recommended sgRNA concentration is above 400 μg.
11. Construct the donor DNA for targeted genome integration by designing a sequence that includes the desired gene flanked by homology arms of 500–1000 bp. These homology arms are complementary to the sequences adjacent to the target site in the genome, allowing the cell's repair machinery to accurately

align the donor DNA with the cut site. In addition to the gene of interest, the donor DNA should contain a novel restriction enzyme for further screening and the construct should include regulatory elements (e.g., promoter, terminator) to ensure that the transgene is properly integrated and expressed (Fig. 1b).

12. After the CRISPR/Cas9 system induces a double-strand break at the target site in the genome, the cell initiates the homology-directed repair (HDR) pathway. The donor DNA with the flanking homology arms serves as a template for repair, enabling precise insertion of the desired gene into the target site. The homology arms ensure that the gene is integrated at the correct location in the genome, allowing for accurate transgene insertion and preventing off-target modifications (Fig. 1c).
13. Cas9/sgRNA RNP are assembled immediately prior to nucleofection by combining equal volumes (1.35 μL each) of 88.3 μM sgRNA and 40 μM Cas9 protein at a ratio of 1:2.2, and then incubate for 10 min at 37 °C.

3.2 Preparation of Protoplast from N. benthamiana Leaf Mesophyll

Refer to Chapter 23.

3.3 Transfection of Cas9/sgRNA and Donor DNA into Protoplast

1. After determining the protoplast density, centrifuge the protoplast to remove the W5 solution for 3 min at 200× *g*.
2. Drain out as much W5 solution from the protoplast pellet (*see* **Notes 8** and **9**).
3. Using a hemacytometer, adjust the cell density to 3×10^5/mL after resuspending the cells in the transfection solution.
4. Fill a glass screw cap culture tube with the Cas9/sgRNA of 1: 2.2 ratio and 50 μg donor DNA. Use Cas9 assembly buffer to adjust the volume to 20 μL.
5. Carefully add 200 μL of protoplasts (about 1.2×10^5 in cell density) to the tube, and stir for 20–30 sec.
6. Include the same amount of PEG-Ca^{2+}solution (220 μL).
7. Gently tap the tube wall thoroughly to coat the protoplasts with PEG-Ca^{2+} solution.
8. Allow the nucleofection mixture at 24–26 °C for 30 min.
9. Add 3 mL of the W5 solution. In order to stop the nucleofection reaction, thoroughly mix by gently rocking or inverting the tube for 20–30 sec.
10. Centrifuge for 3 min, at 200× *g*.

11. After removing the supernatant, gently wash the protoplasts with 3 mL of W5 solution, and resuspend them.
12. Centrifuge for 3 min, at 200× *g*.

3.4 Development of Callus from Transfected Protoplast

1. After centrifugation, remove the supernatant from transfected protoplasts.
2. Add 2 mL of callus induction medium to resuspend 500 μL of transfected protoplasts and transfer into a cell culture dish (35 mm [diameter] X 10 mm [height]).
3. Incubate in the dark for 3 weeks at 24–26 °C (*see* **Note 10**).
4. After 3-week incubation in callus induction medium, the protoplasts are regenerated into 2 mm calli.

3.5 Initial Evaluation of Callus for Insertion

1. Targeted insertion by the CRISPR/ Cas-sgRNA complex can be initially analyzed using callus raised from the transfected protoplast. The genomic DNA can be extracted from the callus using a DNA Extraction Kit (DNeasy Plant Mini Kit).
2. Use a NanoDrop to measure the concentration of genomic DNA, which should be between 25 and 100 ng/μL (*see* **Note 11**).
3. Utilize 1–2 μL of genomic DNA isolated from the callus, and blend with 0.8 μL (10 μM) forward primer, 0.8 μL (10 μM) reverse primer (Fig. 1e), 4 μL 5X PCR Taq mixture, and 13.4 μL H_2O to yield a final volume of 20 μL. Set the PCR program according to the amplification size and melting temperature of primers. Since *N. benthamiana* is a diploid, it possesses two sub-genomes. Specific primers for the various sub-genomes must be considered while designing PCR primers (*see* **Note 12**). Subject the PCR products for agarose gel electrophoresis to ensure that the expected amplicon size is met.
4. PCR products from the callus are purified by passing through the gel extraction kit (according to the manual supplied along with the kit).
5. For confirming transgene insertion, the donor DNA contains a novel restriction enzyme site(s); perform a restriction enzyme site gain assay for 20 μL reaction with 1 μg of PCR product, 1 μL of a restriction enzyme(s) that specifically recognizes the restriction enzyme site in the transgene, and 2 μL (1× final concentration) of 10X restriction enzyme buffer, and adjust the volume to 20 μL with nuclease-free water. Incubate at 37 °C for 1 h—overnight. Successful digestion of the PCR product confirms the insertion of the transgene at the target site.
6. To further verify transgene insertion, subject the PCR products for the capillary sequencing. The appearance of multiple peaks in the sequencing chromatogram approximately 3 bp upstream

of the PAM site, when compared to the non-transfected control, suggests successful genome editing by the CRISPR/Cas-sgRNA complex. This confirmed callus can then be used for regeneration into fully edited or transgenic plants.

3.6 Regeneration of Plant from Callus

1. Tilt the culture dish at 5–10° angle and remove the supernatant (*see* **Note 13**).
2. Add 3 mL of liquid shoot-inducing medium into the culture dish, and resuspend the calli with by pipetting. Adjust the shoot medium to the final volume of 10 mL.
3. Incubate the calli at 26 °C under a light density of 75 μmol m^{-2} s^{-1} with a 16 h light/8 h dark photoperiod.
4. After 4 weeks, green calli can be observed in liquid shoot-inducing medium and regenerated into 5 mm calli.
5. Pick up green calli using tweezers, and transfer them to solid shoot-inducing medium in conical flask.
6. Subculture green calli every 4 weeks on fresh solid shoot-inducing medium.
7. After subculturing one to two times, 100–500 shoot clusters of 1-cm size regenerative buds and leaves can be obtained.
8. Transplant the regenerated explants with leaves to root-inducing medium in conical flask. Adventitious roots will grow after 1 month.
9. Transfer the rooted regenerated plants to a pot containing peat moss, vermiculite, and per litter in a ratio of 1:1:1 (v/v/v). Grow the plants in the greenhouse or in a growth chamber at 26 °C under a light intensity of 70 μmol m^{-2} s^{-1} and 16 h light/8 h dark photoperiod.

3.7 Analysis of Regenerated Plants

1. Perform genomic DNA extraction from the 100 mg young and healthy transgenic *N. benthamiana* leaves, freeze the leaves in liquid nitrogen, and grind them to a fine powder using a mortar and pestle.
2. Transfer the ground tissue to a 1.5 mL microcentrifuge tube, and then add 600 μL of pre-warmed CTAB extraction buffer.
3. Incubate the mixture at 65 °C for 30 min, gently mixing every 10 min, and then add 600 μL of chloroform: isoamyl alcohol (24:1) to the tube.
4. Vortex the mixture briefly (10–15 s) to emulsify and centrifuge the tube at 12,000 rpm for 10 min at room temperature.
5. Carefully transfer the upper aqueous phase to a new 1.5 mL tube without disturbing the interphase.

6. Add an equal volume of chilled isopropanol to the aqueous phase. Invert the tube gently for several times to mix and incubate the tube at −20 °C for 30 min (or at 4 °C for 1 h).
7. Centrifuge at 12,000 rpm for 10 min at 4 °C, discard the supernatant, and wash the DNA pellet with 500 μL of 70% ethanol.
8. Centrifuge again at 12,000 rpm for 5 min and discard the ethanol. Allow the DNA pellet to air-dry for 10–15 min, and resuspend the dried DNA pellet in 50–100 μL of TE buffer.
9. Add 1 μL of RNase A (10 mg/mL) to the resuspended DNA, and incubate the tube at 37 °C for 30 min to remove any residual RNA. The extracted DNA can be stored in −20 °C.
10. Perform the PCR on the extracted DNA for the transgene according to Subheading 3.5, **step 3**.
11. To check whether the donor DNA is inserted at the targeted site, perform the PCR with the primers: (1) transgene-specific primer and (2) primer from the flanking homology sequence (Fig. 1d, e).
12. To determine the transgene presence in regenerated plants, compare wild-type (non-edited) and homozygous edited alleles (with identical edits or transgene insertions in all alleles) by conducting Sanger sequencing of the PCR products. Analyze multiple sequencing results to identify biallelic (both alleles edited or containing different transgene insertions/edits) and heterozygous allelic events (one allele wild-type, the other containing an insertion or edit). Use the Poly Peak Parser website [21] to distinguish overlapping peaks and assess the nature of the edits or insertions in biallelic and heterozygous plants.

4 Notes

1. For all experimental procedures involving protoplast isolation and transfection, use round-bottom centrifuge tubes.
2. Since hydrochloric acid is monoprotic, 1 M HCl is equal to 1 N HCl.
3. *E. coli* BL21 (DE3) produced the Cas9 recombinant protein using plasmid pMJ915 (Addgene #69090). Cas9 protein preparation was carried out according to 19, 20.
4. 500 mL of W5 solution are dispensed at a time and divided into five bottles. 0.22 μm Stericup vaccum filter can be used for sterilization in large volumes.
5. Catalog no.95904, BioUltra, Sigma-Aldrich, USA for PEG 4000. This PEG quality and purity are essential for obtaining

high transfection efficiency, with higher purity leading to higher transfection efficiency. Fresh PEG-Ca^{2+} solution should be made at least 1 h prior to transfection. When compared to an older solution, a freshly made PEG-Ca^{2+} solution increases the efficiency of protoplast transfection [22]. The PEG-Ca^{2+} solution should be filtered in a 0.22 μm syringe filter, PES membrane, or MCE membrane as autoclaving is not the optimum method for sterilizing it.

6. CRISPR-P 2.0. These customized settings could facilitate sgRNA design for various CRISPR-Cas9 systems in plant genome editing.
7. For T7 RNA polymerase solutions, do not vortex.
8. Transfection efficiency may be decreased by the W5 supernatant. With less W5 solution remaining, the transfection efficiency will rise.
9. Replace the W5 supernatant with transfection buffer prior to PEG-Ca^{2+}-mediated transfection.
10. The majority of protoplasts should stay intact during the isolation, transfection, culturing, and harvesting processes if they are derived from healthy leaf materials.
11. Measuring the absorbance ratio at 260 and 280 nm determines the purity of DNA (a good purity falls between 1.80 and 2.00).
12. The GC content of the primary pairs determines the annealing temperature.
13. Subculture at the right moment.

References

1. Gaj T, Gersbach CA, Barbas CF (2013) ZFN, TALEN, and CRISPR/Cas-based methods for genome engineering. Trends Biotechnol 31(7):397–405
2. Shamshirgaran Y, Liu J, Sumer H, Verma PJ, Taheri-Ghahfarokhi A (2022) Tools for efficient genome editing: ZFN, TALEN, and CRISPR. In: Applications of genome modulation and editing, pp 29–46
3. Kim H, Kim JS (2014) A guide to genome engineering with programmable nucleases. Nat Rev Genet 15(5):321–334
4. Barrangou R (2013) CRISPR-Cas systems and RNA-guided interference. Wiley Interdiscip Rev RNA 4(3):267–278
5. Cong L, Ran FA, Cox D, Lin S, Barretto R, Habib N et al (2013) Multiplex genome engineering using CRISPR/Cas systems. Science 339(6121):819–823
6. Doudna JA, Charpentier E (2014) The new frontier of genome engineering with CRISPR-Cas9. Science 346(6213):1258096
7. Sander JD, Joung JK (2014) CRISPR-Cas systems for editing, regulating and targeting genomes. Nat Biotechnol 32(4):347–355
8. Bombarely A, Rosli HG, Vrebalov J, Moffett P, Mueller LA, Martin GB (2012) A draft genome sequence of *Nicotiana benthamiana* to enhance molecular plant-microbe biology research. Mol Plant-Microbe Interact 25(12): 1523–1530
9. Goodin MM, Zaitlin D, Naidu RA, Lommel SA (2008) *Nicotiana benthamiana*: its history and future as a model for plant–pathogen interactions. Mol Plant-Microbe Interact 21(8): 1015–1026
10. Woo JW, Kim J, Kwon SI, Corvalán C, Cho SW, Kim H et al (2015) DNA-free genome editing in plants with preassembled

CRISPR-Cas9 ribonucleoproteins. Nat Biotechnol 33(11):1162–1164
11. Andersson M, Turesson H, Olsson N, Fält AS, Ohlsson P, Gonzalez MN et al (2018) Genome editing in potato via CRISPR-Cas9 ribonucleoprotein delivery. Physiol Plant 164(4): 378–384
12. Fossi M, Amundson K, Kuppu S, Britt A, Comai L (2019) Regeneration of Solanum tuberosum plants from protoplasts induces widespread genome instability. Plant Physiol 180(1):78–86
13. Shimatani Z, Kashojiya S, Takayama M, Terada R, Arazoe T, Ishii H et al (2017) Targeted base editing in rice and tomato using a CRISPR-Cas9 cytidine deaminase fusion. Nat Biotechnol 35(5):441–443
14. Hsu CT, Lee WC, Cheng YJ, Yuan YH, Wu FH, Lin CS (2021) Genome editing and protoplast regeneration to study plant–pathogen interactions in the model plant *Nicotiana benthamiana*. Front Genome Ed 2:627803
15. Wu FH, Hsu CT, Lin CS (2023) Targeted insertion in *Nicotiana benthamiana* genomes via protoplast regeneration. In: Plant genome engineering: methods and protocols. Springer US, New York, pp 297–315
16. Liu H, Ding Y, Zhou Y, Jin W, Xie K, Chen LL (2017) CRISPR-P 2.0: an improved CRISPR-Cas9 tool for genome editing in plants. Mol Plant 10(3):530–532
17. Gruber AR, Lorenz R, Bernhart SH, Neuböck R, Hofacker IL (2008) The Vienna RNA websuite. Nucleic Acids Res 36(suppl_2): W70–W74
18. Ma X, Zhu Q, Chen Y, Liu YG (2016) CRISPR/Cas9 platforms for genome editing in plants: developments and applications. Mol Plant 9(7):961–974
19. Lin S, Staahl BT, Alla RK, Doudna JA (2014) Enhanced homology-directed human genome engineering by controlled timing of CRISPR/Cas9 delivery. elife 3:e04766
20. Huang RS, Shih HA, Lai MC, Chang YJ, Lin S (2020) Enhanced NK-92 cytotoxicity by CRISPR genome engineering using Cas9 ribonucleoproteins. Front Immunol 11:1008
21. Hill JT, Demarest BL, Bisgrove BW, Su YC, Smith M, Yost HJ (2014) Poly peak parser: Method and software for identification of unknown indels using sanger sequencing of polymerase chain reaction products. Dev Dyn 243(12):1632–1636
22. Yoo SD, Cho YH, Sheen J (2007) Arabidopsis mesophyll protoplasts: a versatile cell system for transient gene expression analysis. Nat Protoc 2 (7):1565–1572

Chapter 25

Agrobacterium-Mediated Genetic Transformation in Crop Plants

Sankari Mohana, Reshma Anjum Mohammed, and Golla Narasimha

Abstract

Plant transformation has emerged as a key technique for investigations on plant gene function as well as crop development. Since the early 1900s, *Agrobacterium tumefaciens* and several related *Agrobacterium* species have been identified as plant diseases. Nevertheless, the ability of *Agrobacterium* to transfer DNA to plant cells has just been used for plant genetic engineering in the last 20 years. Initially, *Agrobacterium* was only used to transform dicot plants, but as the plant transformation protocol changed over the years, it was also used to transform monocots such as wheat, rice, and maize. The process of *Agrobacterium*-mediated plant transformation is quite complicated and developed, involving the genetic determinants of the host plant cell as well as the bacterium. The *Agrobacterium*-mediated gene transfer method acts as nature genetic engineer to understand basic plant metabolic processes as well as to exploit the system for biotechnological applications.

Key words Plant transformation, *Agrobacterium*-mediated gene transfer, Plant metabolic processes, Biotechnological applications

1 Introduction

The third-most populous primary food crop, after rice and maize, is common wheat (*Triticum aestivum* L.). Food safety for a growing world population can be dramatically improved by appropriate wheat cultivation [1]. Though wheat is essential, it stood behind the other major grain crops in terms of funding and genomic tool development for development [2]. An effective transformation technique is a vital resource for the functional investigation of genes in plants. The most popular biotechnological technique for crop improvement and plant gene function investigation is *Agrobacterium*-mediated genetic transformation, particularly for species for which genome sequencing data is not yet accessible [3]. The *agrobacterium* genus resides in various species such as *A. tumefaciens*, *Agrobacterium rhizogenes*, *Agrobacterium radiobacter*, and *Agrobacterium rubi* based on host range and symptoms

Nitish Kumar (ed.), *Transgenic Plants: Methods and Protocols*, Methods in Molecular Biology, vol. 2973,
https://doi.org/10.1007/978-1-0716-4803-2_25,

of disease. This bacterium is now commonly used to transform a wide range of essential horticultural and agricultural species, and the list of species that can be transformed by *Agrobacterium*-mediated transformation appears to be extending constantly [4].

The gram-negative, soil-borne plant pathogen *Agrobacterium tumefaciens* tends to deliver foreign DNA fragments into the plant's genome. A soil phytopathogen called *A. tumefaciens* carries transferred (T)-DNA from bacterial cells into host plant cells via a bacterial type IV secretion system (T4SS). This mechanism infects plant wound areas leading to crown gall disease [5]. Two genetic factors on the bacterial Ti-plasmid must be expressed for the plant genetic transformation process to succeed. The T-DNA, which is the first crucial component, is identified by conserved imperfect repeats of 25 base pairs at the extremities of the T-region, known as border sequences.

The genetic transformation of plant cells by *Agrobacterium* is based on the molecular transfer of a large tumor-inducing (Ti) or rhizogenic (Ri) plasmid residing in the bacterium, which is integrated into the plant nuclear genome (Fig. 1). The size of Ti plasmids ranges from 200 to 800 kbp [6, 7]. The T-DNA also termed as transferred DNA which is located on Ti plasmid and the size varies about 10–30 kbp [8]. The T-DNA region flanked by right and left borders which consist of three biosynthetic genes

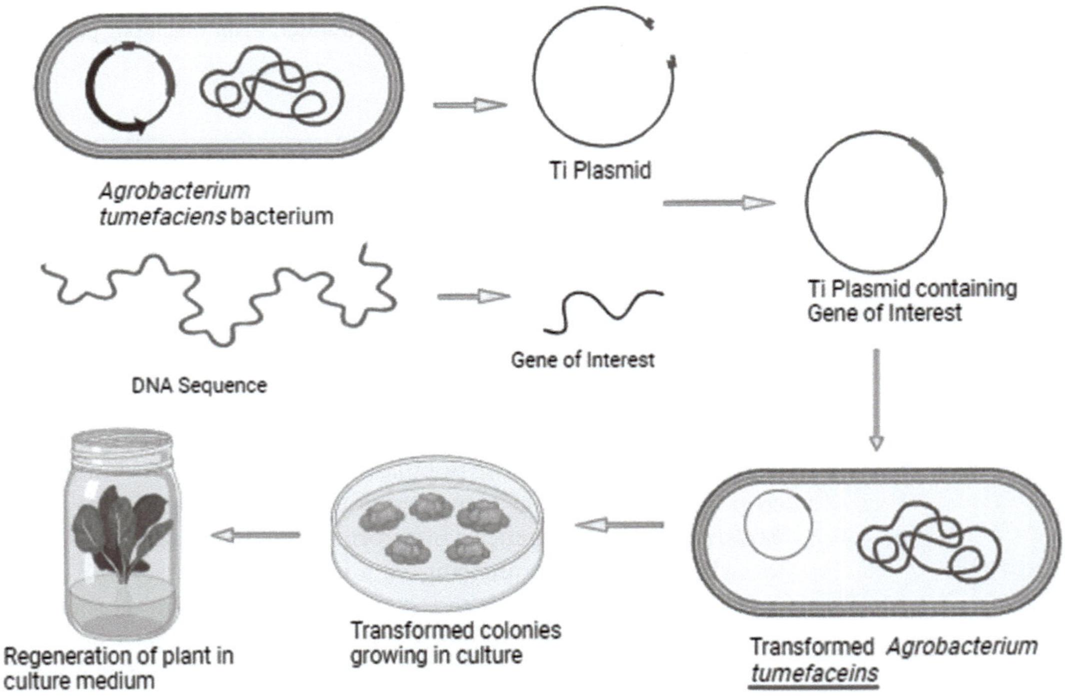

Fig. 1 *Agrobacterium*-mediated gene transformation method in plants

such as auxin (*aux*), cytokinin (*cyt*), and opine (*ocs*) is referred to as oncogenes which determine the tumor phenotypes [9, 10]. The second element is known as the virulence (vir) region, and it consists of at least seven main loci (virA, virB, virC, virD, virE, virF, and virG) that represent various components of the bacterial protein machinery to process and transfer T-DNA [11–13]. Vir genes consisting a number of proteins with various purposes such as vir A and vir G recognize plant-derived products and make expression of other vir genes. Vir C and vir D involved in T-DNA excision and processing, vir E plays major role in coating and protecting T-DNA through transfer, vir B is involved in establishment of type IV secretion system which is likely for transfer of T-DNA to plant cells, and vir E2, vir D4, and vir F take place in integration of T-DNA into nucleus of plant cell [5, 14, 15]. Steps involved in T-DNA transfer and integration are depicted in Fig. 1.

2 Materials

Prepare the solution using Milli-Q water (18.2 MΩ), and use all chemicals in analytical reagent grade. Prepare and store all reagents at 4 °C or room temperature based on condition requirements. Carefully follow all waste disposal methods and regulations while disposing of waste.

2.1 YEP Broth Preparation

5 g L^{-1} yeast extract, 5 g L^{-1} peptone, 5 g L^{-1} sodium chloride, 5 g L^{-1} sucrose; maintain the pH to 7.2 (*see* **Note 1**).

1. Weigh 5 g of yeast extract, 5 g of peptone, 5 g of sodium chloride, and 5 g of sucrose, and transfer into a beaker.
2. Add distilled water to make 900 mL. Mix and adjust the pH to 7.2. Finally, make it to 1000 mL by adding distilled water. Store at 4 °C in refrigerator.
3. The broth is supplemented with antibiotics.

2.2 YEP Agar Preparation

5 g L^{-1} yeast extract, 5 g L^{-1} peptone, 5 g L^{-1} sodium chloride, 5 g L^{-1} sucrose; maintain the pH to 7.2 (*see* **Note 1**).

1. Weigh 5 g of yeast extract, 5 g of peptone, 5 g of sodium chloride, and 5 g of sucrose, and transfer into a beaker.
2. Add distilled water to make 900 mL. Add 8 g of agar powder to make agar medium. Mix and adjust the pH to 7.2. Finally, make up to 1000 mL by adding distilled water. Store at 4 °C in refrigerator.
3. The broth supplemented with antibiotics [16, 17].

2.3 Agrobacterium Strain

Agrobacterium tumefaciens strain EHA105, *Agrobacterium tumefaciens* strain EHA105 harboring binary vector pCAMBIA 1301 plasmid required for transformation.

1. Acetosyringone (100 mM): Take 19.6 mg acetosyringone, dissolve in 500 μL of dimethyl sulfoxide (DMSO), and make it into 1000 μL. It should be stored in 20 °C in the refrigerator (*see* **Note 2**).
2. $CaCl_2$ (0.1 M): weigh 21.9 g of $CaCl_2$, and mix with distilled water. The reagent should be stored at 4 °C in the refrigerator.

2.4 Antibiotics Preparation

1. Kanamycin stock (50 mg/mL): Weigh 50 mg of kanamycin, and dissolve it in 500 μL of sterile distilled water. Finally, adjust the volume to 1000 microliter.
2. Rifampicin (50 mg/mL): Weigh 50 mg of rifampicin, dissolve it in 500 μL of methanol, and finally make it to 1000 μL of stock solution (*see* **Notes 2** and **3**).

3 Methods

3.1 Preparation of Agrobacterium (EHA 105 Strain)-Competent Cells

Carry out all the procedures in the laminar airflow chamber.

1. Take the YEP agar plate, streak *Agrobacterium* EHA 105 strain, and incubate at 28 °C for overnight (*see* **Notes 4** and **5**).
2. Take a single bacterial colony and inoculate it into 5 mL of YEP medium which contains 10 μg/mL of antibiotic rifampicin, and keep it in an incubator shaker overnight at 28 °C.
3. After this, take 1 mL of overnight culture, inoculate into 10 mL of YEP medium, and incubate 3–4 h once the OD reaches 0.2–0.6 at 600 nm [18] (*see* **Note 6**).
4. From this, take approximately 3 mL culture, centrifuge at 5000 rpm for 10 min at 4 °C, and discard the supernatant.
5. In the pellet, add 1 mL of 0.1 M $CaCl_2$ (ice cold), centrifuge at 5000 rpm for 10 min at 4 °C, and remove the supernatant.
6. Resuspend the pellet with 1 mL of 0.1 M $CaCl_2$, and distribute into different Eppendorf tubes (100 μL per Eppendorf tube) for further transformation experiment [19].
7. Further, take 100 μL *Agrobacterium* competent cells, and add plasmid construct (ex.pCAMBIA 1301) to carry out the transformation experiment by heat-shock method in a water bath at 42 °C for 45 s, and place in ice for 5 min (*see* **Note 7**).

8. Then add YEP Broth (900 μL) to the tube and incubate for 2–3 h at 37 °C. After centrifuging the culture at 4000 rpm for 1 min, add 100 μL YEP broth to the pellet, which was then spread out onto YEP agar plates with 100 mg/L of kanamycin.
9. The plates were then incubated at 37 °C overnight [17, 20].

3.2 Agrobacterium Gene Transformation by Co-cultivation Method

1. Take 50 mL of overnight culture (*Agrobacterium tumefaciens* strain EHA105 containing pCAMBIA 1301 vector) and centrifuge 5000 rpm at 4 °C for 5 min. Resuspend the pellet, with 5 mL of YEP medium which contains antibiotics (such as rifampicin (50 mg/mL) and kanamycin (50 mg/mL)). Discard the supernatant.
2. Further incubate the culture overnight at 150 rpm on an incubator shaker.
3. Take 50 μL overnight culture, inoculate into 100 mL of YEP broth which contains antibiotics, and incubate till the OD reaches 1 at 600 nm.
4. Centrifuge the culture (5000 rpm, 10 min), and resuspend the pellet 50 mL YEP broth containing 10 μM acetosyringone.
5. Take any explant and surface-sterilize the leaf. The leaf was rinsed with 70% ethanol, following a wash with sterile distilled water.
6. Again rinse with 2% mercury hydrochloride solution, and again wash with sterile distilled water. Eliminate the moisture content blotting with tissue paper.
7. Co-cultivate the explant into the above suspension culture through intermittent shaking at 12–14 (*see* **Note 8**).
8. After co-cultivation, inoculate the explant into a co-culture medium that contains acetosyringone (100 μM). Incubate the culture 2 days in dark condition [21–23] (*see* **Note 9**).
9. After incubation, transfer the explant into an MS medium. Every week subculture the explant on the same medium. Confirm the transformants via Histochemical GUS assay.

4 Notes

1. Sterilizing bacterial growth media is most commonly using autoclave-generated steam at 121 °C and pressure close to 15 lb/in2. 15 min to 1 h should be the recommended autoclave times. It should be mentioned that agar doesn't conduct heat well.
2. Antibiotic stock solutions should be filter-sterilized and stored at −20 °C. Antibiotics should be added to the medium when the temperature is less than 60 °C. Antibiotic-containing plate

media should be labeled with the preparation date and stored at 4 °C. For liquid media, add the antibiotic before inoculating the culture. Keep in mind that many antibiotics in plate media have a relatively short half-life.

3. The method of storing will define which tubes and labels are best. For accessibility, use small (2 mL) tubes; tubes can be glass or plastic with caps that fit tightly to limit air exchange and prevent contamination; it is convenient for ease of labeling with a pen; however, tape will work as well as a label depending on the method of storage; if strains are to be stored frozen, note that tape may fall off after a few years from use.
4. Inoculate the strain or culture in fresh liquid culture or medium, and preserve it for 2 weeks. Every 2 weeks once subculturing of the culture is a must.
5. The *Agrobacterium* strain will depend on the work and the characteristics of plasmid.

   The choice of *Agrobacterium* recipient strain will depend on the goal of your experiment and the characteristics of the plasmid to be transferred.
6. Usually cultures inoculate in the early hours because it takes hardly more than 4–8 h to reach the OD600 of 1.0. Most researchers will harvest the cell in the middle of the exponential growth phase or early stationary phase.
7. The washing step in the freeze/thaw method helps to remove the antibiotics and benefit to recovery of transformants.
8. Co-cultivation incubation should not exceed 48 h in the co-cultivation medium.
9. Acetosyringone concentration (20–200 μM) varies depending on the plant species and *Agrobacterium* strain. Adding acetosyringone to the *Agrobacterium* infection medium increases the GUS expression in infected explants.

References

1. Grote U, Fasse A, Nguyen TT, Erenstein O (2021) Food security and the dynamics of wheat and maize value chains in Africa and Asia. Front Sustain Food Syst 4:617009
2. Hayta S, Smedley MA, Clarke M, Forner M, Harwood WA (2021) An efficient Agrobacterium-mediated transformation protocol for hexaploid and tetraploid wheat. Curr Protoc 1(3):e58
3. Cordeiro D, Alves A, Ferraz R, Casimiro B, Canhoto J, Correia S (2023) An efficient agrobacterium-mediated genetic transformation method for solanum betaceum cav. Embryogenic callus. Plants 12(5):1202
4. Gelvin SB (2003) Agrobacterium-mediated plant transformation: the biology behind the "gene-jockeying" tool. Microbiol Mol Biol Rev 67(1):16–37
5. Hwang HH, Yu M, Lai EM (2017) Agrobacterium-mediated plant transformation: biology and applications. Arabidopsis Book 15:e0186
6. Fortin C, Marquis C, Nester EW, Dion P (1993) Dynamic structure of agrobacterium tumefaciens Ti plasmids. J Bacteriol 175(15): 4790–4799
7. Goodner B, Hinkle G, Gattung S, Miller N, Blanchard M, Qurollo B, Goldman BS, Cao Y,

Askenazi M, Halling C, Mullin L (2001) Genome sequence of the plant pathogen and biotechnology agent Agrobacterium tumefaciens C58. Science 294(5550):2323–2328

8. Barker RF, Idler KB, Thompson DV, Kemp JD (1983) Nucleotide sequence of the T-DNA region from the *Agrobacterium tumefaciens octopine Ti* plasmid pTi15955. Plant Mol Biol 2:335–350
9. Peralta EG, Ream LW (1985) T-DNA border sequences required for crown gall tumorigenesis. Proc Natl Acad Sci 82(15):5112–5116
10. Veluthambi K, Ream WALT, Gelvin SB (1988) Virulence genes, borders, and overdrive generate single-stranded T-DNA molecules from the A6 Ti plasmid of Agrobacterium tumefaciens. J Bacteriol 170(4):1523–1532
11. Wang K, Stachel SE, Timmerman B, Van Montagu M, Zambryski PC (1987) Site-specific nick in the T-DNA border sequence as a result of *Agrobacterium* vir gene expression. Science 235(4788):587–591
12. Rubin RA (1986) Genetic studies on the role of octopine T-DNA border regions in crown gall tumor formation. Mol Gen Genet MGG 202:312–320
13. Subramoni S, Nathoo N, Klimov E, Yuan ZC (2014) Agrobacterium tumefaciens responses to plant-derived signaling molecules. Front Plant Sci 5:322
14. Brencic A, Winans SC (2005) Detection of and response to signals involved in host-microbe interactions by plant-associated bacteria. Microbiol Mol Biol Rev 69(1):155–194
15. Kalogeraki VS, Zhu J, Stryker JL, Winans SC (2000) The right end of the vir region of an octopine-type Ti plasmid contains four new members of the vir regulon that are not essential for pathogenesis. J Bacteriol 182(6): 1774–1778
16. Aggarwal D, Kumar A, Sudhakara Reddy M (2011) Agrobacterium tumefaciens mediated genetic transformation of selected elite clone (s) of *Eucalyptus tereticornis*. Acta Physiol Plant 33:1603–1611
17. Sankari M, Hemachandran H, Anantharaman A, Babu S, Madrid RR, Fulzele DP, Siva R (2016) Identifying a carotenoid cleavage dioxygenase 4a gene and its efficient agrobacterium-mediated genetic transformation in Bixa orellana L. Appl Biochem Biotechnol 179:697–714
18. Konagaya KI, Nanasato Y, Taniguchi T (2020) A protocol for Agrobacterium-mediated transformation of Japanese cedar, Sugi (Cryptomeria japonica D. Don) using embryogenic tissue explants. Plant Biotechnol 37(2):147–156
19. Asif A, Mohsin H, Tanvir R, Rehman Y (2017) Revisiting the mechanisms involved in calcium chloride-induced bacterial transformation. Front Microbiol 8:296127
20. Amal TC, Karthika P, Dhandapani G, Selvakumar S, Vasanth K (2020) A simple and efficient Agrobacterium-mediated in planta transformation protocol for horse gram (Macrotylomauniflorum Lam. Verdc.). J Genet Eng Biotechnol 18(1):9
21. Parimalan R, Venugopalan A, Giridhar P, Ravishankar GA (2011) Somatic embryogenesis and Agrobacterium-mediated transformation in Bixa orellana L. Plant Cell Tissue Organ Cult 105:317–328
22. Konagaya K, Kurita M, Taniguchi T (2013) High-efficiency Agrobacterium-mediated transformation of Cryptomeria japonica D. Don by co-cultivation on filter paper wicks followed by meropenem treatment to eliminate Agrobacterium. Plant Biotechnol 30:523–528
23. Rani A, Panwar A, Sathe M, Kush A (2018) A modified in planta method of Agrobacterium-mediated transformation enhances the transformation efficiency in safflower (*Carthamus tinctorius L.*). J Plant Biochem Biotechnol 27: 272–283

Chapter 26

Characterization of Transformed Plants by GUS Assay

Sankari Mohana, Reshma Anjum Mohammed, and Golla Narasimha

Abstract

A reporter gene GUS *(uid* A) is a fusion protein used to study the expression pattern of transgenic plants. A gene fusion is a DNA structure (in vitro or in vivo) in which the regulatory sequences of one gene influence the transcription and/or translation of the coding sequences of another gene (a reporter). This histochemical assay is used to investigate the mechanism of plant transformation events, cellular expressions, function of the plant genome, and their level of expression in genes.

Key words β-Glucuronidase (GUS) assay, Histochemical tests, *Agrobacterium tumefaciens*-mediated transformation, Plant transformation

1 Introduction

The GUS (*uid*A) gene is a reporter gene was emerged from bacteria (*Escherichia coli)* that has been employed frequently in plants due to its stability, high sensitivity, and quick detection method [1, 2]. The enzyme glucuronidase breaks down glycosidic linkages to separate molecules of glucuronic acid from other molecules. The enzyme possesses the ability to cleave monosaccharides at either the middle or end of the chain, demonstrating both endo- and exo-glycosidase activity [3–7]. This technique has a higher variability and higher sensitivity than spectrophotometric methodologies [8–10].

The molecular weight of GUS is approximately 68KDa [11]. GUS A is a plant reporter gene that was first utilized and encodes β-glucuronidase (β-D-glucuronoside glucuronosohydrolase, E.C. 3.2.1.31). GUS A, along with GUS B, GUS C, and GUS R, is a member of an operon that was first identified from *E. coli* (6). Upstream of GUS A, GUS R is differentially transcribed and codes a particular regulator of the gus operon. GusB encodes a glucuronide-specific permease, while the product of GUS C is unclear in function [11, 12]. When a buffer comprising the substrate 5-bromo-4-chloro-3-indolyl ß-d-glucuronide (X-Gluc) is

Nitish Kumar (ed.), *Transgenic Plants: Methods and Protocols*, Methods in Molecular Biology, vol. 2973,
https://doi.org/10.1007/978-1-0716-4803-2_26,

introduced to plants that are expressing the transgenic *uid*A gene under the control of a cloned promoter sequence, GUS cleaves X-glucose into two indoxyl derivatives, which oxidatively dimerize to generate 5,5′-dibromo-4,4′-dichloro-indigo, the blue pigment emblematic of GUS staining [13]. It was made possible to establish the location of gene activity in the tissues and cells of transformed plants by histochemical analysis [14], and this assay evaluates the presence of enzyme activity in specific tissues.

2 Materials

2.1 Potassium Ferricyanide (100 mM, MW: 329.2 g/gmol)

Take 3.3 mg of potassium ferricyanide and dissolve in 100 mL of sodium phosphate buffer (*see* **Note 1**).

2.2 Potassium Ferrocyanide (100 mM, MW: 422.2 g/gmol)

Take 4.22 mg of potassium ferrocyanide and dissolve in 100 mL of sodium phosphate buffer.v (*see* **Notes 2** and **3**).

2.3 Sodium Phosphate Buffer (50 mM, pH 7)

Take 24 g of monobasic sodium phosphate, and dissolve in 100 mL of distilled water.

2.3.1 Stock A: Monobasic Sodium Phosphate (0.2 M, MW = 120)

2.3.2 Stock B: Disodium Hydrogen Phosphate (0.2 M, MW = 141.96)

Take 28.39 g of monobasic sodium phosphate, and dissolve in 100 mL of distilled water,

From this take 38 mL of Stock A and 62 mL of Stock B, and make the final volume using distilled water (*see* **Notes 4** and **5**).

2.4 Na EDTA (0.5 M, MW = 372.2)

Take 18.61 g of Na EDTA and mix with 100 mL of distilled water. pH—8 (*see* **Note 6**).

2.5 X-GLUC Stock Solution (2 mM, MW: 521 g/gmol)

Take 5.2 mg of X-GLUC, add 50 μL of DMSO, and make the final volume up to 5 mL.

2.6 X-GLUC Staining Solution

Take 5 mL of 2 mM X-GLUC Stock solution, add 3 mL of sodium phosphate buffer, and make it into 10 mL by adding 2 mL of methanol (*see* **Note** 7).

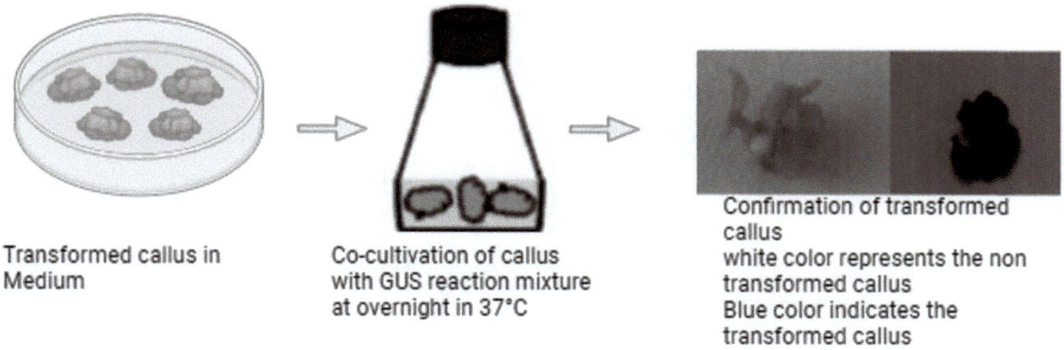

Fig. 1 Histochemical GUS assay for confirmation of transformed colonies

2.7 GUS Reaction Mixture

Take 40 mL of sodium phosphate buffer (50 mM, pH 7) and add 1 mL of potassium ferricyanide, 1 mL of potassium ferrocyanide, and 100 μL of NaEDTA. Make the final volume into 50 mL with distilled water. Finally, add 0.1% triton X-100.

3 Methods

1. Take the co-cultivated explant (*agrobacterium*-mediated transformation of explant) for the GUS experiment with some modification [11].
2. For histochemical confirmation, incubate the explant in the GUS reaction mixture overnight at 37 °C.
3. After overnight incubation, wash the explants with 70% ethanol.
4. Observe the blue color explants (Fig. 1).

4 Notes

1. Potassium ferricyanide and potassium ferrocyanide need to be prepared fresh and kept in room temperature in brown bottles.
2. Ferrocyanide and ferricyanide enrich the oxidative diminera-tion of two molecules of the indoxyl aglycone intermediate and are responsible for the formation of the indigo blue color which occurs in the reaction site of GUS. The concentration of the ferricyanide and ferrocyanide should be at least 1 mM for a good result.
3. Sterilizing the water is most commonly using autoclave-generated steam at 121 °C and pressure close to 15 lb/in2. 15 min to 1 h should be the recommended autoclave times.
4. Phosphate buffer gives a 20% higher result in GUS assay when compared to citrate buffer [15].

5. The pH of EDTA should be 8.0, adjust with NaoH.
6. GUS stain preparation: methanol gives 25% higher activity when compared to ethanol [16].
7. Overnight incubation at 37 °C gives good results.

References

1. Jefferson RA, Kavanagh TA, Bevan MW (1987) GUS fusions: beta-glucuronidase as a sensitive and versatile gene fusion marker in higher plants. EMBO J 6(13):3901–3907
2. Crow RM, Gartland JS, McHugh AT, Gartland KM (2006) Real-time GUS analysis using Q-PCR instrumentation. J Biotechnol 126(2):135–139
3. Ramadan AM, Eissa HF, El-Domyati FM, Saleh OM, Ibrahim NE, Salama M, Mahfouz MM, Bahieldin A (2011) Characterization of inhibitor (s) of β-glucuronidase enzyme activity in GUS-transgenic wheat. Plant Cell Tissue Organ Cult 107:373–381
4. Blázquez M (2007) Quantitative GUS activity assay in intact plant tissue. Cold Spring Harbor Protoc 2007:pdb.prot4688
5. Béziat C, Kleine-Vehn J, Feraru E (2017) Histochemical staining of β-glucuronidase and its spatial quantification. In: Plant hormones: methods and protocols. Springer New York, New York, pp 73–80
6. Kim KW, Franceschi VR, Davin LB, Lewis NG (2006) β-Glucuronidase as a reporter gene: advantages and limitations. In: Arabidopsis protocols. Humana Press, Totowa, pp 263–273
7. Fior S, Vianelli A, Gerola PD (2009) A novel method for fluorometric continuous measurement of β-glucuronidase (GUS) activity using 4-methyl-umbelliferyl-β-d-glucuronide (MUG) as substrate. Plant Sci 176(1):130–135
8. Borkhardt B, Skjøt M, Mikkelsen R, Jørgensen B, Ulvskov P (2005) Expression of a fungal endo-α-1, 5-L-arabinanase during stolon differentiation in potato inhibits tuber formation and results in accumulation of starch and tuber-specific transcripts in the stem. Plant Sci 169(5):872–881
9. Beaud D, Tailliez P, Anba-Mondoloni J (2005) Genetic characterization of the β-glucuronidase enzyme from a human intestinal *bacterium, Ruminococcusgnavus.* Microbiology 151(7):2323–2330
10. Cocking EC (2005) OBPC symposium: maize 2004 & beyond—intracellular colonization of cereals and other crop plants by nitrogen-fixing bacteria for reduced inputs of synthetic nitrogen fertilizers. In Vitro Cell Dev Biol-Plant 41: 369–373
11. Jefferson RA, Burgess SM, Hirsh D (1986) β-Glucuronidase from Escherichia coli as a gene-fusion marker. Proc Natl Acad Sci USA 83:8447–8451
12. Gallagher SR (ed) (1992) GUS protocols: using the GUS gene as a reporter of gene expression. Academic Press, San Diego
13. Dedow LK, Oren E, Braybrook SA (2022) Fake news blues: a GUS staining protocol to reduce false-negative data. Plant Direct 6(2): e367
14. Naleway JJ (1992) Histochemical, spectrophotometric, and fluorometric GUS substrates. In: Gallagher SR (ed) GUS protocols: using the GUS gene as a reporter of gene expression, Academic Press, San Diego, CA, pp 61–76
15. Sankari M, Hemachandran H, Anantharaman A, Babu S, Madrid RR, Fulzele DP, Siva R (2016) Identifying a carotenoid cleavage dioxygenase 4a gene and its efficient agrobacterium-mediated genetic transformation in *Bixaorellana* L. Appl Biochem Biotechnol 179:697–714
16. Vitha S, Beneš K, Michalová M, Ondřej M (1993) Quantitative ß-glucuronidase assay in transgenic plants. Biol Plant 35:151–155

INDEX

Nitish Kumar (ed.), *Transgenic Plants: Methods and Protocols*, Methods in Molecular Biology, vol. 2973, https://doi.org/10.1007/978-1-0716-4803-2,

W

X

FSC
www.fsc.org

MIX
Papier aus verantwortungsvollen Quellen
Paper from responsible sources
FSC® C105338

If you have any concerns about our products,
you can contact us on
ProductSafety@springernature.com

In case Publisher is established outside the EU,
the EU authorized representative is:
Springer Nature Customer Service Center GmbH
Europaplatz 3, 69115 Heidelberg, Germany

Printed by Libri Plureos GmbH
in Hamburg, Germany